**BLACK**

Scribed line negative

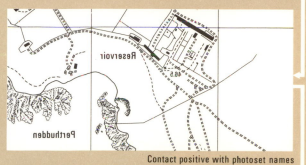

Contact positive with photoset names

Contact line negative

Final positive

*Based on cartographic materials produced for a four-colour topographic map of the Ythan Estuary, scale 1:7,500, Glasgow University, 1972.*

*The progressive printed impression is shown on the endpaper inside the back cover.*

Negative mask for 50% tint

Final positive

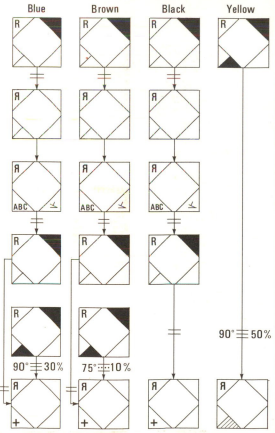

*For explanation of flow diagram, see Chapter 10.*

 **St. Louis Community College**

Forest Park
Florissant Valley
Meramec

Instructional Resources
St. Louis, Missouri

# CARTOGRAPHIC DESIGN AND PRODUCTION

SECOND EDITION

# CARTOGRAPHIC DESIGN AND PRODUCTION

## SECOND EDITION

## J. S. Keates

Longman
Scientific &
Technical

Copublished in the United States with
John Wiley & Sons, Inc., New York

**Longman Scientific & Technical,**
Longman Group UK Limited,
Longman House, Burnt Mill, Harlow,
Essex CM20 2JE, England
*and Associated Companies throughout the world.*

*Copublished in the United States with*
*John Wiley & Sons, Inc., 605 Third Avenue, New York, NY 10158*

*First published 1973*
Second edition 1989

**British Library Cataloguing in Publication Data**
Keates, J. S.
  Cartographic design and production. —
2nd ed.
  1. Map printing — Handbooks, manuals, etc.
  I. Title
686.2'83         GA150

ISBN 0-582-30133-5

**Library of Congress Cataloging-in-Publication Data**
Keates, J. S.
  Cartographic design and production / J. S. Keates. — 2nd ed.
     p.    cm.
  Bibliography: p.
  Includes index.
  ISBN 0-470-21071-0 (Wiley, USA only).
  1. Cartography.     I. Title.
GA105.3.K42   1988
526-dc19

Set in Linotron 202 9/11pt Times u/lc

Produced by Longman Singapore Publishers (Pte) Ltd.
Printed in Singapore

# CONTENTS

# PREFACE TO THE SECOND EDITION

Since the first edition was published many changes have taken place in cartography and map making, especially on the technical side. Research and literature in the English language has expanded considerably, from the detailed examination of particular methods to wide-ranging theories about cartography as a whole. In some respects this has been matched by an equivalent expansion in cartographic education, especially in Britain, at both graduate and technician levels. Although more attention is now being given to understanding the practice of cartography, it is to be hoped that the value of the study of cartography as a means of education will also be appreciated.

In order to provide a systematic and comprehensive arrangement, the basic principles of cartographic representation and design, cartographic technology and cartographic production are dealt with in the first three parts. This is now complemented by a new fourth part on applied cartography, covering the main types of map and chart.

Although in the First Edition it was possible to introduce briefly the fundamental processes of digital cartography, the rapid development of this field has meant that a complete account is now impossible within the confines of a few chapters. Consequently, although there are many references to the concepts and methods of digital cartography, no attempt has been made to cover the subject as a whole in this volume.

Despite the development and proliferation of new technical methods, the fundamental problems of cartographic representation, and in particular the complex relationship between information, representation and technology, are central to the study of cartography. It is hoped therefore that this edition will continue to provide students with a proper understanding of the theory and practice of cartographic design and production.

*J. S. Keates*
*September 1988*

# ACKNOWLEDGEMENTS

I am indebted to colleagues and former students for their contributions to the figures: in particular to Mr and Mrs M. Shand for the production of most of the diagrams and illustrations, and to Mr I. Gerrard for their photographic preparation: to Mr J. W. Shearer and the British Cartographic Society (Fig. 78); Miss E. Milwain (Figs 28, 32 and 34); and Miss S. Johnson (Fig. 203).

Acknowledgement is also made to the ByChrome Company, Inc. for Figs. 59 and 60; to Mecanorma Ltd. for the use of their materials in Figs. 109, 191 and 200; and to the Hydrographer of the Navy and the Controller of Her Majesty's Stationery Office for permission to use material from Chart 5011 *Symbols and Abbreviations used on Admiralty Charts*.

*J. S. K.*
*September 1988*

# PART ONE
# CARTOGRAPHIC REPRESENTATION

# 1 INFORMATION AND CARTOGRAPHIC REPRESENTATION

The informational content of a map is a function of several inter-connected factors. Information presented in a map must conform with the specific requirements of map structure, and this structure has both geometrical and symbolic characteristics. A map has a scaled, spatial arrangement based on an orthogonal projection of the Earth's surface on to a plane. The methods of representation by conventional signs or symbols are on the one hand conditioned by the characteristics of the phenomena, and on the other by the available information. These methods are expressed by symbols which have to be designed and modified by the need to adjust them individually so that together they provide a composition which reflects the map purpose. All of this is in turn affected by resources: resources for the collection of information, and resources for map design and production. This graphic representation is the primary function of cartography, and places cartography within map making as a whole.

The collection of information which can be represented in maps is usually a specialised activity, and has diverse origins. Some is produced by surveys specifically carried out to obtain information for maps. Some results from data acquired for other purposes. Although cartographers are closely concerned with the nature, characteristics and availability of these data, the collection of the primary information is not a cartographic activity as such. In this respect, cartography deals with a particular set of problems, and contains a body of theory and practice which is basically common to all kinds of map making.

## MAP STRUCTURE

The term 'map' refers to a two-dimensional graphic image which shows the location of things in space, that is, in relationship to the Earth's surface. It is distinguished from other kinds of representation in two dimensions – such as pictorial images and diagrams – in two principal ways. First, its perspective is an orthogonal projection of the Earth's three-dimensional surface on to a plane (Fig. 1A). And second, it does not describe or depict individual things, but represents them by signs which place them in classes or categories. In addition, because the Earth is nearly spherical, any map of more than a very small area must involve some distortion, and therefore employ some systematic means of representing the spheroidal or spherical surface on a plane. The projection is generally shown by selected parallels and meridians indicating the spherical coordinate system, the geographical graticule, or a local Cartesian grid based on the projection. Those maps which do not show any graticule or grid are still based on sources of information derived from maps on a coordinate system. For these the topographic 'outline' serves as a direct reference to the topographic surface.

## Maps, pictures and diagrams

Maps are not pictures because pictures, like photographs, have a central perspective, that is the entire scene is viewed or recorded as it appears from a single point (Fig. 1B). Pictures represent appearances, either real or imaginary, and therefore can only depict particular things, not the classes or categories to which they belong.

Maps are not diagrams, because the two-dimensional structure of a diagram can be used for many purposes. Although diagrams can also use coordinates, these can be varied at will, and are not restricted to a systematic representation of either the spherical or Cartesian coordinate systems used for maps, that is the geographical graticule or grid.

Given that the map has a specific structure which distinguishes it from pictures and diagrams, it is possible for the map to include both pictorial and diagrammatic elements. Particular classes of information may be represented by iconic symbols, sometimes referred to as pictorial symbols, in which some aspect of the appearance of a typical member of that class of objects is used as the basis of

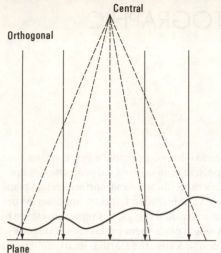

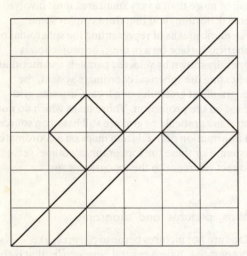

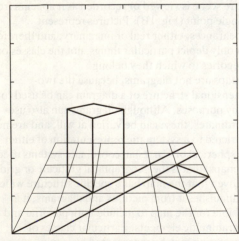

**1.** A. Orthogonal and central projection (vertical) on to a plane   B. Orthogonal perspective of map and central (horizontal) perspective of picture

symbolic representation. Diagrams can also be employed, and in some cases can be placed on the map, acting as symbols applied to either point or areal locations. Even so, the fundamental structure of the map is specific, and is not imitated by any other kind of two-dimensional representation.

It follows from this that the map is a purely artificial device constructed for particular purposes. Unlike a photograph, or a remotely sensed image, it does not depend upon what some physical medium will record. The map can be made to include whatever subject is of interest, and can be applied to past events, future expectations, or quite imaginary arrangements. It can show – and frequently needs to show – things that have no tangible existence, such as names and many boundary lines. In order to include some things, it can exaggerate them, and displace other and less important things to make room for them. It can operate over an enormous range of scale, and thereby serve many different purposes and objectives. To say that a photograph contains 'more information' than a map is to miss the point. A map is selective and arbitrary and can perfectly well represent things that cannot be photographed, or would not be resolved by a photograph at a particular scale.

## Map concepts

This range of capability makes the map a very powerful tool, and yet its use poses many problems. On the one hand the concept of a map is straightforward; the idea of indicating where things are on the Earth's surface by showing their relative positions, and therefore the distances and bearings between them, is basically simple. Not surprisingly it has been used by many different peoples through a long period of human history. On the other hand, orthogonal perspective and projection are quite sophisticated devices, which require understanding for proper map use. Indeed for a long period in Europe, pictorial representation and oblique views in central perspective (which imitate the way in which a landscape is perceived) were mixed up with maps, and the full evolution of the map in the modern sense is comparatively recent.

The two related characteristics of orthogonal or parallel perspective, and projection on to a plane, provide difficulties at different scales and with different types of map use. With large-scale topographic maps of relatively small areas, part of the real world can be viewed and compared visually with the map representation. The two perspectives are quite different: the constant scale of the map

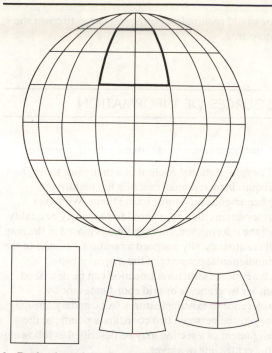

**2.** Projection of the three-dimensional coordinate system (geographical graticule) on to a two-dimensional plane

differs from the way that distance appears visually; and the map both generalises the detail and classifies the features by symbols. At such scales, it is the difference in perspective, and therefore point of view, which causes the greatest problems in interpretation.

At small scales, and especially where maps represent areas that cannot be perceived directly, the map acts more as a structural framework relating things to the Earth's surface. Here the main difficulty arises from the distortion of the sphere on to a plane (Fig. 2), and with maps of the whole Earth the artificial division of the continuous spherical surface. Although much is made of the inevitable distortion of area, distance or shape introduced by particular projections at small scales, the more insidious misrepresentation occurs through fitting the sphere on to a rectangle. It is clear, for example, that many Europeans 'see' Asia as a long way from North America, because of the division on most world maps through the Pacific ocean and at the poles, while remaining less aware of the proximity of the two continents across the Arctic Ocean.

In addition, because the map is two-dimensional, it can only suggest the third dimension, either of the Earth's surface or of phenomena related to it, in a limited way. Values can be given for a limited number of points, or along a limited number of lines. These indications can be used to interpret the relief of the surface, but it cannot be represented directly on a plane.

## Map and plan

The terms map and plan are rarely used systematically, although there should be a clear distinction between them. A plan only shows planimetry, the location of features as though they occurred on a plane. Therefore, there is no systematic description of the relief of the Earth's surface and its three-dimensional nature. A map does show this, even if only in a limited way, and therefore this three-dimensional representation forms the basis on which other things are located. Although plans are generally associated with large scales, where they are often specifically employed, it is perfectly possible to have large-scale maps and small-scale plans. Indeed many special-subject 'maps', including many statistical 'maps', show only a very limited topographic outline in plan, and indeed are plans, not maps. Plan is really a subset of map, and it would be an advantage if the terms were properly distinguished.

## Scale

A map or plan is dimensionally related to the real world through scale, which is the ratio between distance on the map and distance on the Earth's surface. It can be described in various ways: by numerical ratio, by which one unit of measurement on the map is given the equivalent number of units on the ground (e.g. 1 : 10 000); by a statement of equivalent distance on the map and on the ground, commonly using units of measurement suitable to their relations (such as 4 cm represents 1 kilometre); or by a scale bar, which graphically indicates distances on the ground at their map scale.

Although maps and plans are normally referred to by their basic scale ratio, the nominal scale has to be treated with caution. At comparatively large scales, and over short distances, both distances and directions on the map can be regarded as approximately true; that is they are close enough for most purposes. But for small-scale maps of large areas, the distortion introduced by projection inevitably means that the particular figure given for the scale ratio will only apply to particular points, lines or areas, as defined by the properties of that projection. This is most evident in single maps of the whole of the Earth's surface, which have considerable distortion, and for which the nominal scale ratio indicates the basis of construction rather than a means of comparative measurement.

### Scale and representation

Although the scale of the map indicates the ratio between the map surface and the area to which it relates, this does not apply to the symbols through which the informational content is expressed. Whatever the map scale, the symbols on the map must have minimum dimensions and contrast in order to be perceived and identified by the map user. There are minimum requirements for graphic legibility, regardless of map scale. Symbols cannot simply be made smaller and smaller as map scale diminishes. Consequently, although the overall nominal scale may be constant, the treatment of the map content is a function of not only the relative sizes of objects, but also the requirements of legibility and emphasis. On a medium-scale map, a main road will occupy far more space than its true ground dimensions would require, partly because a legible line must be used to represent it, and partly because it is deliberately emphasised as an important part of the map content. Similarly, many boundary lines are not actually marked on the ground, but located in relation to other topographic features or even coordinated points. Even so, if they are important dividing lines, and therefore high in the informational hierarchy of the map, they may be represented by relatively heavy lines to indicate their importance.

### Scale and map purpose

In a special-purpose map, or even in a multi-purpose map designed to serve many different user interests, the connection between scale, information and map purpose is a critical one. If a given level of information is necessary to satisfy the requirements of the map user, this has consequences in the scale used for the map. If this is insufficient, it may be impossible to devise a cartographic representation that will deal effectively with the information. In general, the most suitable scale is the smallest that will allow legible presentation. Extending the subject matter over a larger area inevitably entails more searching by the map user. There are many relatively simple small-scale maps of special subjects which would have benefited from a scale reduction, by which the relative positions of features – which are frequently the main object of study – would be more easily assimilated by fixations in central vision,

or would require less peripheral vision to cover the map area.

## SOURCES OF INFORMATION

### Characteristics of sources of information

The informational content of a map must satisfy two requirements: it must describe the features or phenomena and it must locate them. Whatever reservations may exist about the accuracy or quality of the information, once an item is placed in the map it is automatically assigned a position, because of the fundamental property of the map as a two-dimensional structure. Location can be described either by graticule or grid coordinates, or by reference to existing features. A boundary line may be plotted between two coordinate points, or the alignment of a section may be described as following the centre line of a river.

Sources of information vary widely in quality and type. There is a fundamental difference between information that has been obtained specifically for the purpose of a making a map, and information that has been obtained in the first place for another purpose. The two groups can be described as primary and secondary sources respectively.

### Primary sources

These provide information by carrying out a surveying operation specifically to obtain the data required for the map. This may involve direct observation and measurement on the Earth's surface, or interpretation and measurement on a suitable image. At the highest level this information should be complete and consistent, and correct within scale limits and human error. Points are measured in terms of their actual coordinate position, and linear features and outlines continuously. This type of survey applies not only to topographic and hydrographic surveying, but also to many specialised surveys, in which the boundaries of classified phenomena are located by observation and measurement on the ground, or interpreted and plotted through aerial photographs.

Both the classification and the locational information will be consistent with the scale and purpose of the intended map, or the intended map will be adjusted to the level of information that the surveying operation has found to be possible in the circumstances. A large-scale survey of an urban area will normally record all the lines at which vertical surfaces meet the ground surface, without any attempt to classify the objects so delimited. This gives the minimum descriptive information, for the plan is used as a basis for interpretation by the user on the ground. A survey of geomorphological features must identify and classify them in terms of their characteristics according to some geomorphological classification system. This may only be appropriate to a particular area or the judgement of an individual geomorphologist. Even so, the information should be consistent in itself.

## The third dimension

Although in theory the relatively static and visible land surface could be measured at all points (as in a digital elevation model), the information that can be shown on a map has to be selective, as the third dimension cannot be shown continuously on a two-dimensional surface. For topographic maps selected points or lines of equal value are chosen and measured. What can be interpreted by the map user depends on the scale and quality of this information, and its correspondence with the real surface. Three-dimensional phenomena which are invisible, intangible or subject to fluctuation or change are considerably more difficult to deal with.

In a primary survey, the objective is to obtain sufficient information, correctly located, to make a useful representation of surface variation in accordance with the user's needs. In practice this is not always possible, for either technical or economic reasons. A modern hydrographic survey will provide continuous lines of depth measurement in shallow water (profiles) along closely spaced parallel lines. From these survey data it is possible to select significant point values (depths) and also to interpolate bathymetric contours at selected intervals. The quality of data as a whole will be consistent. In former times, when depth measurements could only be taken as individual soundings, these could not be provided on a systematic basis for the coverage of large areas. They were unevenly distributed, and therefore contours could not be interpolated systematically.

When such surveys are carried out for map-making purposes, the positions of the points or lines of measurement are recorded at the time, so that the information can be properly located in the map or chart. This is quite different to a great deal of secondary data, in which the locational element is very limited.

## Secondary sources

In addition to the information collected for map-making purposes, a great deal of data is obtained about the human and physical environments, for government and local administration, planning and resource investigation, and scientific studies. Much of this information is spatial, but generally the information is not collected primarily to produce a map.

These sources of information cover a very wide field. Past events and conditions can be reconstructed through historical records and published evidence. For example, part of the frontier of the Roman Empire at a given time can be 'mapped' by a combination of surviving landscape features and literary evidence. For small-scale map making in particular, information about boundary changes, name changes and new developments may be obtained through a variety of reports.

The usefulness of such information for map making depends on both its descriptive and locational characteristics. This is most obviously the case with quantitative information. It may cover a range of quantities which are beyond the capacity of the map to show; its locational basis may be crude; and often it introduces problems that are a function of the nature of the data, not the means of cartographic representation.

The main categories to which such information refers are either discontinuous or continuous phenomena. Discontinuous phenomena include those that occupy discrete areas, as well as those that can be located at a point, or a small area that can be treated as a point at map scale. The output from a mine, or quantity of goods moved through a port, can be regarded as having point locations on small-scale maps. Some phenomena can occur only in specific locations. Very often the data consist of cumulative quantities which have been aggregated over a period of time, like gross annual output, or for which mean values have been determined, such as average annual output. Generally speaking the data can be properly located, and the cartographic problems are essentially those of representing quantities that may occur over a large numerical range.

Although the output from a mine can be recorded as a quantity, and therefore any figure is possible, there are other types of discrete phenomena that occur as individuals. Census statistics of population, whether human or animal, are aggregates produced by counting the number of individuals present in a given area at a given time. For enumeration purposes their location is fixed, even though the individuals are themselves mobile.

Whereas some kinds of continuous phenomena are relatively static, others may vary continuously in space, and in many cases also change over time. Information about these can only be obtained by measuring or recording continuously or intermittently over a time, at particular stations. Continuous records have to be aggregated or converted to mean values. Measurements of temperature and precipitation are obvious examples. The isolines which can be used to represent the continuous 'surface' are interpolated from a limited number of measurements at particular stations. The value of these isolines is a function of their density and distribution. Although such lines of equal value are nominally equivalent to the contours measured for a topographic map, they are likely to be radically different in reality. Ideally this difference should be expressed by the cartographic representation, but unfortunately this is rarely the case. Unless all the recording stations are also included, it is extremely difficult for the map user to make any assessment of the overall quality of the information.

## Intermediate sources

Although there is a relatively clear distinction between primary and secondary information, there are other sources that fall between these two extremes. The information obtained through remote sensing covers a wide range of products. Photogrammetric surveying can be regarded as a branch of remote sensing, as it produces an image of the topographic surface that is subsequently interpreted and measured, and in this respect is an indirect source of information. But a photogrammetric survey for map-making purposes is organised to obtain the desired information in three dimensions. If considerable ground detail is needed, then the aerial photography will be correspondingly large in scale in order to obtain the desired resolution.

With other types of remote sensing, the output can be a series of images of the Earth's surface, or of other physical phenomena which are distributed in space, but the resolution is fixed by the altitude and system. High-altitude systems will only record the dominant characteristics of relatively large unit areas, from a 'pixel' of about 30 m square in later Landsat imagery, to about 10 m in the more recent Spot imagery. The information so obtained has advantages of speed, uniformity and completeness,

but this has to be traded against a relatively low resolution of detail, and the fact that any system will only record physical signals within its sensitivity.

In a map, what is represented is not a function of plan area, but is based on a judgement of what is important from a human point of view. Whereas in primary sources of information, the surveying methods are organised to obtain the required information for the map, any 'map' from remotely sensed imagery is entirely dependent upon the physical characteristics of the system.

## Information sources and representation

These observations on information sources also serve to reinforce an important aspect of cartographic work, which often arises during the compilation stage. Although cartographers in general do not control data collection, and indeed data collection is a function of experts in particular fields, it is essential that they understand the cartographic requirements of maps in relation to data sources. This is particularly important in maps of specialised subjects. A cartographer cannot be an 'expert' in every field of human activity that may possibly involve maps. Therefore, a proper working relationship between the cartographer and those who provide or control the specialised information is vital. The deficiencies in some aspect of the data which become apparent at compilation stage can be explained to the specialist. It is the responsibility of the cartographer to make these issues clear. It is the responsibility of the specialist to make decisions about any changes or modifications to the information itself. Such a dialogue also assumes that the cartographer takes an intelligent interest in the information being represented, and above all does not assume a level of knowledge or familiarity which may be quite unfounded.

# REPRESENTATION AND MAP SYMBOLS

Methods of representation define the way in which the map symbols represent the phenomena. The contour line is a method of showing information about the elevation of a surface in relation to a datum. The 'dot map' uses point symbols to represent the distribution of aggregated quantities over an area. The choropleth map employs area symbols to indicate mean values over defined areas. So the line, point and area symbols provide the basis of representation, and in turn these symbols have to be given a specific form, dimension and colour in the map design.

## Symbols and design

A map represents features or phenomena by symbols, which are a particular category of signs. All human representation, expression and communication is carried out through the use of signs. With signs it is possible to refer to, to describe, and to organise concepts. The most fully developed and universally employed sign system is that of language which would seem to be fundamental to all forms of human expression and communication. Signs which are used graphically in an organised two-dimensional space operate in a different way to those used in verbal language. Although the term 'language' is often used to refer to any sign system, the differences between graphic and verbal description are more important than their similarities.

Such symbols, also referred to as conventional signs, operate by classification: that is, they group individual occurrences or features into groups according to some characteristics which they all share. The presence of a symbol for a canal on a map shows where any feature which can be classified as a canal occurs. It does not distinguish between them in terms of their minor variations, but places them in the same category. Such categories may be very general, and this applies to many topographic features, such as rivers, individual buildings, areas of woodland, etc. These descriptions are essentially qualitative, concentrating on distinguishing things by type or character. If the particular feature is important in the map content, even to the point of being the major map information, then the general category may be broken down into a number of sub-classes. These sub-classes may be distinguished either by a difference in quality or type, or by a difference in quantity or value. Some sub-classes will combine both. The group of symbols representing different types and classes of road may distinguish between all-weather and fair weather roads (a physical difference in construction and surface), by place in a road hierarchy (a value classification), by the number of carriageways (a numerical rating) or by road width (a physical dimension classification); or usually a combination

of all these factors. What is important is whether the factors used in the classification are appropriate to the information and function; not whether they are qualitative or quantitative.

## Symbols, concepts and conventions

General or multi-purpose maps usually employ symbols to represent classificatory concepts that are familiar to unspecialised map users, through the medium of normal education and experience, although whether this assumption is always justified is arguable. Specialised maps may introduce concepts that can only be understood given a familiarity with the concepts and terminology of that subject, or the activity which the map is intended to serve. Much of the information on a nautical chart is comparatively meaningless to a person unfamiliar with navigation, and a geological map will not offer much information to a map user who knows nothing about geology. In such a case the concepts involved, and the symbols that represent them, are the product of specialised study, and their meaning is only evident if the map user has, or is prepared to acquire, the necessary degree of understanding.

The symbols on a map are 'conventional' in the sense that the meanings attached to them are defined for the purpose of a particular map, and therefore they operate on the basis of a conventional agreement between the cartographer and the map user. Therefore, there is no standard 'language' or vocabulary, although the desire for this is often expressed. Map users may become familiar with certain maps, and therefore can use them easily because they can remember the symbols. Working with an unfamiliar set of symbols is more difficult. Here again, there are two opposing requirements. Ease of use would require familiarity, which would be helped by standardisation between maps of different provenance but the same type. But correct representation must take into account enormous variations in topography, degree of development of the built environment, and a huge variety of subject matter in specialised maps.

Some degree of standardisation is easiest to obtain in certain special-purpose maps where the user requirements are well defined, such as nautical charts. In other cases, the specialised classifications are themselves likely to change over time. The categories and symbols included in a 'land use' map are a matter of judgement according to a particular scheme of classification; they may not be valid if a different approach is used.

## CARTOGRAPHIC REPRESENTATION AND MAP USE

The information in the map can only be acted upon by a map user if the map structure and the methods of representation are understood, and the specified meanings of the map symbols known (see Board 1984). These provide minimum conditions within which the user has to operate. In many cases, using a map is a sophisticated and complex activity, in which information selected from the map is combined with the user's knowledge to increase understanding, devise a solution, or solve a problem. Many maps are complex objects because of the nature of the information they represent. Consequently, the user may have to carry out a prolonged search to extract the relevant information, and may repeat this several times until the desired objective is reached. As this objective is defined by the user, only the user can determine when the map-using task has been completed.

The essential function of the cartographer is to decide the correct map structure for the particular subject matter or purpose: to find means of representation appropriate to the characteristics of the subject phenomena and the available information, making clear any limitations or deficiencies where necessary; and to present this in the form of a graphic design which can be acted upon by the user. To realise these objectives, the cartographer must make some assumptions about the prospective user's attitude, map knowledge and skill in interpretation.

## MAP MAKING AND CARTOGRAPHY

In the English language the terms map making, mapping and cartography are often used as though they were interchangeable. For example United Nations reports and conferences on 'cartography' may include virtually anything to do with maps. On the other hand, most attemps to define 'cartography' and therefore describe what it is that 'cartographers' do, emphasise that cartography in this sense is primarily concerned with graphic representation and construction. Unfortunately the ambiguous use of 'cartography' confuses the relationship of data collection and cartographic representation within map making as a whole. All maps depend in the first place on the collection of basic information about the Earth's surface by means of topographic and hydrographic surveys. This information is the first step. Other maps may be derived at smaller scales, and specialised maps may be constructed using the topographic information as the means of reference or location. Whatever type of map is concerned, the nature, quality, completeness and up-to-dateness of the information is vitally important. From the user's point of view, a map may be bad either because the information has a poor correspondence with the phenomena, or is inadequate or out of date, or because the graphic representation is poor. The satisfaction of the map user is dependent on both. Therefore cartography cannot be considered in isolation from sources of information, and good maps cannot be produced cartographically from inadequate sources of information.

Figure 3 is an attempt to show the relationship between information sources and cartography within map making as a whole. The primary information is input from surveying operations, which include specialised surveys such as geological surveys, carried out for map-making purposes. Existing maps can be used directly as a source of information, and the re-working of this information may be regarded as primarily a cartographic task. Secondary sources of information, although supplied from outside the realm of map making, can also be used directly by a cartographer, although it is more usual for a specialist to deal with and process this information first.

It is clear that the balance between the tasks of supplying information and cartographic representation vary enormously between different map types. A large-scale engineering project may require a very detailed survey of a small area, yet the cartographic representation of this may be quite straightforward and graphically simple. The compilation of a new map of world vegetation at a small scale, making use of many different sources of information, is likely to be a very lengthy task, requiring a complex multi-colour design and technically sophisticated production. This simply reflects the great variety of things covered by the term 'map', and the complex relationships between information sources, map making and cartography.

### Map composition

Before the production of any map can begin, it must be composed; that is, its basic form and content have to be decided. This applies regardless of what

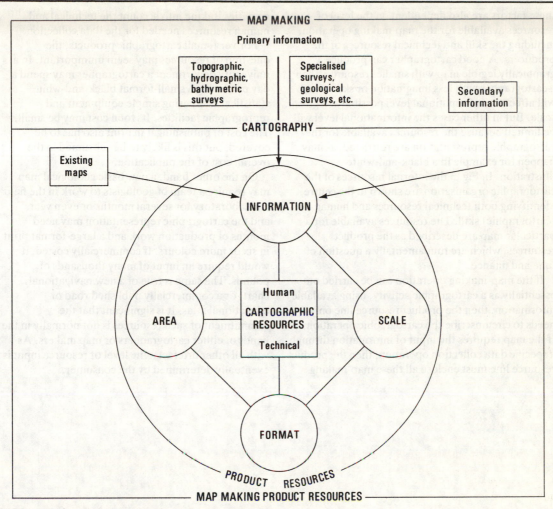

**3.** Map making and cartography: factors and variables

kind or level of technology is employed to produce it. This composition involves four basic factors: the geographical area, the level of information, the scale and the format (Fig. 3). The decision about geographical area comes first, as it is impossible to consider any map until this is known. The informational content is usually the link between cartographic representation and map making. If the information already exists, then the map is primarily a function of cartographic representation. But if specific information has to be collected, this must be input by other means. What is interesting, is that once the geographical area has been decided, a decision about any of the three elements of information, scale and format, will control the other two.

A map of the world (geographical area), which is to be included in an atlas, will have a fixed format, and so the maximum scale that can be obtained is restricted. Given this scale, the level of information that is graphically possible is also determined.

With a standard map series of a country at a given scale, the area and the scale are decided in the first place. The choice of scale will control the amount of detail that can be shown (informational level) and the format will be a function of the size of map (or series of sheets) needed to cover the area at that scale.

If a geological survey of a given area has produced a body of geological information, and a map of this is required, the scale will need to be large enough to present this level of information, and the resultant format will be a consequence of the chosen area and scale.

## Resources

In practice, of course, the relationships between

these factors are also dependent on the level of resources available for the map-making operation, including the skill and technical resources of the producers. A good cartographer can produce a graphically legible map with smaller resources than a cartographer with less imagination or skill, which will affect the informational level possible at a given scale. But in other cases the informational level will be limited because the resources available for the cartographic representation are restricted, as may happen for example in a black-and-white illustration. In Fig. 3 the internal resources of the cartographic organisation are shown in the centre, identifying both technical resources and human (cartographic) skill. The resources available for a particular map are described as the product resources, which are fundamentally a question of time and finance.

If the map-making operation can be carried out essentially as a cartographic activity, using available information, then the product resource line only needs to circumscribe the cartographic operations. If the map requires the input of information through a specific data collection operation, then the product resource line must enclose all these map-making activities, taking into account the technical and human resources needed for the data collection.

For very small cartographic products, the question of resources may seem unimportant. In a university department a cartographer may spend a day producing a small-format black-and-white statistical map, using simple equipment and reprographic facilities. Its total cost may be small. The cost of publishing it in print also has to be covered, but this is likely to be subsumed in the overall cost of the publication.

On the other hand, a new basic geological map may require a team of geologists to work in the field and laboratory for several months or even years, and the cartographic representation may need months of production work and a large-format print in six or more colours. If commercially costed, it would require an input of many thousands of pounds. The same is true of a new navigational chart, or a commercially published road or educational atlas. It is significant that the commitment of such resources is not normally in the hands of either cartographers or map makers. As with all other artefacts, the level of resource input is eventually determined by the consumer.

# 2 VISION AND PERCEPTION

In designing a map a cartographer uses visual perception, visual imagination and his own visual experience. The response of the map user is conditioned partly by the visual characteristics of the map, but also his own visual information processing abilities and aptitudes.

With a printed map, viewed by reflected light, radiation in the form of visible light reaches the observer's eyes, where it is focused by the lenses on to the receptive cells lying in the retinas. This pattern of stimulation is processed, first by the retinal cells, and subsequently at various stages in the brain. What is 'seen' is partly a function of this processed information, and partly a consequence of what the brain already knows, or expects. In this sense, visual impressions are not just 'seen' in some automatic way, but are constructed. This aspect of the sequence, the nature of the brain's response, is understood in principle but by no means fully explained in detail.

Vision depends therefore upon physical, physiological and psychological factors. Physical factors include the light which provides the source of energy and initiates the process, and any circumstances which affect it. Physiological factors include the eyes themselves, and the neural reactions of the nerve cells which respond to and transmit electrical impulses in the brain. And psychological factors include the learned responses of the brain, the attitude of the observer, and the effects of existing knowledge and experience. Whereas the physiological factors are common to most human observers, in the sense that they all have the same basic visual mechanisms, the psychological factors vary according to age, experience, education, and so on, and are individual.

## LIGHT

Light consists of radiant energy in the visible part of the electromagnetic spectrum, which lies between 380 and 770 nm (Fig. 4). The term is also often used to refer to radiation in adjacent wavelengths, for example 'ultra-violet light'. Normal daylight, which is regarded as 'white' light, contains radiation at all wavelengths in the visible spectrum. The wavelengths in different parts of the spectrum are perceived as different hues. Light reflected or transmitted by natural objects is rarely pure in hue, and the impression of a particular hue is normally due to the presence of a dominant wavelength. Some perceived colours do not exist in the spectrum, but are composed of two or more dominant wavelengths. Although the source of 'light' for the Earth is natural sunlight, light can also be produced by heating some materials to bring about the emission of visible light. Some artificial light sources do not emit over a continuous spectrum, but only at certain lines or bands. These are not normally used for viewing, but are important in some reprographic processes.

Colour temperature is defined as the temperature to which a theoretically perfect black body would have to be heated in order to emit light which would be the same as that produced by the light source.

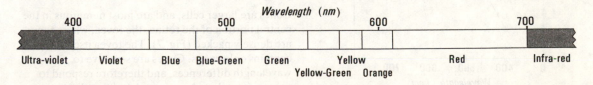

4. Visible spectrum

This is expressed in kelvins (K). At 2800 K, which is the colour temperature of an ordinary tungsten filament lamp, there is more emission towards the 'red' end of the spectrum, and the light is slightly yellowish.

As light reflected from a surface or object depends in the first place on the incident light, or illumination, it follows that it is impossible to make any measurements relating to the appearance of the image unless the light source characteristics are specified. In the analysis of colour, for example, it is usual to define a standard light source. The Commission International de l'Eclairage (CIE) has standardised these as follows: A represents incandescent lamps with a colour temperature of 2854 K; B represents noon sunlight with a colour temperature of 5000 K; and C represents average daylight, as would appear with an overcast sky, with a colour temperature of 6800 K (Fig. 5).

Light which falls upon a surface may be reflected, if the surface is opaque, or transmitted, if the surface is transparent or translucent. Reflection may be specular or diffuse. In a mirror-like surface, all the incident light may be reflected at one angle, and the angle of reflectance will equal the angle of incidence. Glossy surfaces, including some types of printing paper, reflect light at least partly in this

way. Matt surfaces, or those which are slightly uneven, give a more diffuse reflection, in which light is scattered in different directions.

## THE EYE

The primary components of the eye are a lens system with a circular aperture, the iris, which acts as a diaphragm, and the receptor system of the retina (Fig. 6). The retina contains an arrangement of two types of receptor cells, called rods and cones, which are themselves inter-connected to other types of cells which operate to process the signals detected by the rods and cones. These other cells lie in front of the receptor cells, and therefore the incoming light has to pass through them to reach the receptors. Rods are dominant towards the edge of the retina, and are sensitive to small amounts of light, but not to colour (wavelength differences).

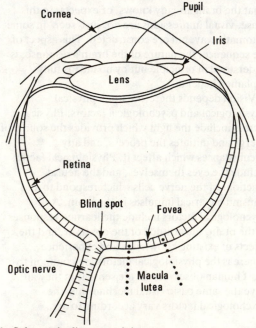

**6.** Schematic diagram of the eye

Cones are larger cells, and are most numerous in the central portion of the retina, the fovea, where they are densely packed (Fig. 7). The fovea itself contains only cones. Cones are sensitive to wavelength differences, and therefore respond to colour. It has long been postulated that different cones must be sensitive to different wavelengths, but

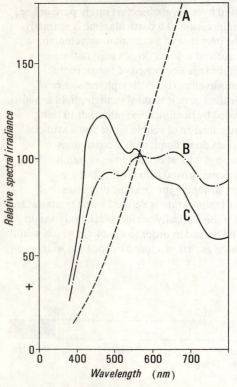

**5.** Spectrum of relative intensity

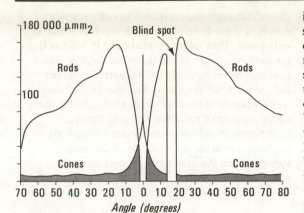

**7.** Retinal distribution of rods and cones

the exact arrangement of such cells in the retina has been difficult to determine. The blind spot is the point of exit of the optic nerve, and contains no receptor cells.

Maximum visual resolution and colour response is limited to a small area in the centre of the retina, and a large proportion of the other retinal cells and the processing in the brain is devoted to the information supplied through this area of central vision. The visual system can therefore respond to the entire visual scene, using the images from both eyes to give stereoscopic vision, or can concentrate on specific details in a smaller field of view by focusing on to the central retinal areas. Generally the eyes are moved from peripheral to central vision subconsciously, as need arises. Stereoscopic vision, through which distance and depth can be judged, is highly important for human activity, but is of no significance in the viewing of maps, which exist on a two-dimensional plane.

Conditions of lighting or illumination affect the visual system as a whole. As the cones do not respond to small amounts of radiation, sufficiently intense light is needed to make colour vision and acute vision possible. This is not normally a problem with map use, but conditions can exist where maps may have to be used with limited illumination. In such cases, both the resolution and colour response of the map user are affected.

## Visual acuity

For an object to be detected, it has to form an image of minimum dimensions on the retina, and if it is a small object, this must be in the central area. It also requires an adequate amount of light, because below a certain level of illumination, a sharp image is not formed at all. At very low levels of

illumination, an image can be formed, but with no sensation of colour. The minimum size which an object must have in order to be perceived is normally expressed in relation to the angle it subtends in the eye. It is generally accepted that under normal conditions of illumination this angle is one minute of arc for persons with normal vision. Visual acuity is usually tested using high-contrast targets (such as letters or sets of lines). On a map, however, the level of contrast is affected by background colour, and the colour of the image itself.

It would seem that visual acuity would depend upon the sizes of the cones in the fovea, as these are the minimum units that can detect a signal. However, it is known that signals from groups of receptor cells are transmitted to other retinal cells, and that the signals received and transmitted through the optic nerve are in fact taken mainly from areas, or groups of retinal cells, called receptive fields. This only emphasises that the reception and processing of signals in the eyes is a complex operation, and bears little resemblance to the uniform physical response of a sensitive film in a camera, although the analogy between the eye and a camera is often used.

## COLOUR

All the visible surface differences between images or objects viewed by transmitted or reflected light are aspects of colour. Although in ordinary parlance 'colour' is often used as though synonymous with hue, variations in lightness and darkness are just as much part of colour as variations in wavelength or hue.

Colour is tridimensional and trichromatic. In order to describe it, its three different attributes or properties have to be defined. The description of colour can take place in two ways. What is perceived as a colour is a sensation, and therefore has psychological elements as well as physical ones. Colour can also be measured quantitatively, using instruments, and this physical description of colour is obviously necessary where measurements have to be made and used, for example, in developing colour standards in industry. Although there are similarities between the two, there are also significant differences.

The attributes of colour as a sensation are hue, lightness and saturation. Hue is the easiest to describe, as it is the spectral characteristic of any colour, and the distinction between red, blue, green, etc., is quite clear. In a pure spectral hue, all radiation will lie within a narrow band in the visible spectrum, and therefore the hue will have a single or dominant wavelength. Colours made from a mixture of hues (such as brown) will have several dominant wavelengths.

Lightness (or value) refers to the perceived lightness or darkness of a fully saturated or partly saturated colour. It is clear that some colours are relatively light in appearance (such as yellow or orange), whereas others are dark (such as purple or brown).

Saturation (or chroma) refers to the purity of the hue. A colour which transmits or reflects radiation only at a single wavelength, or narrow wavelength band, is a fully saturated colour. In effect, it is not diminished by the presence of other wavelengths. Whereas red can be thought of as applying to a narrow band in the spectrum, pink is a red which has been decreased in 'redness'.

Achromatic sensations are limited to differences on a light–dark scale, in which there are no chromatic effects at all. In a 'black-and-white' image, black represents the absence of any reflected light, and white is the maximum value of reflected light; in between a series of intermediate grey tones are possible. These are all lightness or value differences. They can also be thought of as differences in intensity (or more properly luminous intensity), as the most intense light will be the white, and the least intense will be the black, which theoretically has zero intensity.

Apparent differences in colour lightness are due to the response of the visual system to radiation at different wavelengths, when the eye is adapted to normal daylight. The luminosity curve (Fig. 8)

shows that maximum response to radiant energy occurs in the middle of the spectrum, and is less at either end. Thus radiation at about 550 nm has the greatest visual effect, and therefore colours at or near this wavelength have the greatest apparent lightness. In this respect, differences in the black–white scale are differences in intensity, that is the actual amount of radiation reaching the eye. Differences in lightness for coloured images are connected with the wavelengths of the images viewed, when the luminance is constant at all wavelengths.

## Additive and subtractive primaries

It has been known for a long time that any colour can be matched by a combination of the correct proportions of red, green and blue light, which are therefore known as the additive primaries. This only applies to transmitted light. Wavelengths from these three primary hues combine additively, and their full combination yields white light. In a printed image viewed by reflected light, the pigments on the white paper act as filters, and subtract the wavelengths of the incident white light, apart from their own hue, which is reflected from the paper surface. These can also act in combination. Theoretically, suitable proportions of cyan, magenta and yellow can match any other colour, but they combine subtractively. In this respect, cyan = minus red; magenta = minus green; and yellow = minus blue. They can also be thought of as the intermediates between the additive primaries of red, green and blue (Fig. 9). As their effect is to filter out wavelengths, if cyan, magenta and yellow are superimposed, there should be no reflection at all, and black should result.

The total reflectance from the printed paper surface is also dependent on the paper itself.

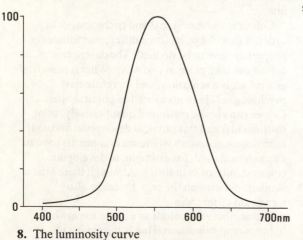

8. The luminosity curve

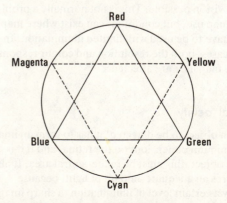

9. Additive and subtractive primaries

Makowsky (1967) shows that in normal map printing, the maximum reflectance from the white paper surface is about 82% of the incident light. One of the major difficulties with printed (or painted) images is that the pigments used are not spectrally pure, and therefore do not act as perfect filters. In practice, the combination of cyan, magenta and yellow does not produce a full black, and in colour reproduction an additional black 'printer' is usually added. Reproduction of a coloured image by using the subtractive primaries is referred to as the three-colour process; if black is added, then it is known as four-colour process.

## Colour description systems

The description of colour is important in cartography, as in the design and construction of a map the colour characteristics of symbols have to be specified. As multi-colour maps are produced in the colour-separated mode, in which the individual images constructed as map components are made achromatically, the colour appearance of the final map is a matter of judgement before it first appears as a proof or print. The printing colours in which the map is produced also have to be specified by one means or another. These printing colours must be known, as decisions about the treatment of map symbols cannot be made except in relation to specific colours.

There are two types of colour description systems: those based on visual judgement and those based on instrumental measurement. Both have a role in cartography. Visual systems depend on the judgement of colour differences by human observers, and although they are devised in terms of the properties of hue, saturation and lightness, the observer sees them as individual colours. The best-known examples are the Munsell and the Ostwald systems, named after their inventors.

### The Munsell system

The Munsell system is also three-dimensional, and employs the properties of hue, value (lightness) and chroma (saturation). According to the *Munsell Book of Color* (Macbeth 1976) 'The value (v) notation indicates the lightness or darkness of a color in relation to a neutral gray scale, which extends from absolute black to absolute white' and 'The chroma (c) notation indicates the degree of departure of a given hue from a neutral gray of the same value.' It is usually thought of or represented as a three-dimensional colour solid (Fig. 10). The

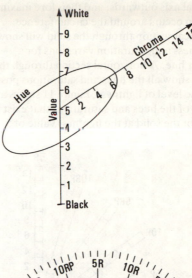

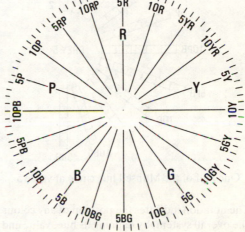

10. The Munsell system: arrangement of the hue circle

hues are arranged in a circle around the circumference. Values range on a vertical axis from 0 (black) at the base to 10 (white) at the apex: and chroma is arranged on a horizontal axis from the centre to the circumference.

The original scale of colours was constructed by hand in its complete form, and various copies and selections have been produced since. In the most common version, there are theoretically 100 equally spaced hues, 10 value or lightness intervals, and a varying number of chroma or saturation steps, depending on the hue, with a maximum of 16. The solid is not symmetrical, as hues have different positions on the lightness scale, and a varying number of possible saturations at any one lightness level. Achromatic values are represented by the central vertical axis, where there are no hue or saturation dimensions. Saturation increases from the central axis outwards, and therefore maximum saturation occurs around the circumference.

A vertical section through the solid will show all the lightness and saturation variations for a particular hue. A horizontal section through the solid will show all the hues and saturations possible at any one level of lightness. Figure 11 shows a selection of the hues and saturations that exist for one half of the solid at the lightness value of 5.

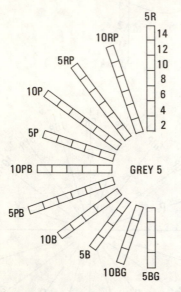

**11.** One half of the Munsell hue circle at value 5

The notation indicates the position of any colour in the overall system, by defining its hue, value and chroma. The hundred hues are divided into 10 groups, signified by *R* (red), *YR* (yellow-red), *Y* (yellow), *GY* (green-yellow), *G* (green), *BG* (blue-green), *B* (blue), *PB* (purple-blue), *P* (purple), and *RP* (red-purple). The hues within each group are numbered from 1 to 10. Therefore 5*B* 8/4 would indicate a pure blue (in the centre of the blue hues), with a high lightness value and a low saturation. More exact definitions make use of the full-scale divisions, for example 7.5*B* 8.5/8. The number of steps in saturation varies, but the strongest printing ink pigments have a maximum saturation of about 16.

The individual colours in the Munsell system are intended to represent all the colours that can be perceived as just noticeably different by a human observer, and therefore are theoretically at equal intervals. Despite the fact that there are many discrepancies in the original system, it has the advantage that it does approximate what the human observer perceives as colour.

## The Ostwald system

This system approaches the classification of colours in a different way. It describes a full range of colours between those with low values (theoretically equal to 0) and high values (theoretically equal to 100%). The notation consists of a description of hue and proportions of white and black, the white element being related to the concept of lightness, and the black element being related to the concept of saturation. The first number indicates the dominant wavelength of the hue, according to perceptually equal intervals decided by Ostwald, and is known as the Ostwald hue number. The first letter refers to the white content, *a* being equivalent to a white content of 89% (which is as light as is possible with pigmented surfaces), and the letters following in alphabetical sequence show decreasing white on a logarithmic scale. The second letter refers to the black content, *p* being equal to a black of 11%, and again a sequence of letters indicating diminishing quantities of black on a logarithmic scale. Therefore, the attributes of value and saturation are indirectly indicated by the proportions of white and black in the individual hues.

The Ostwald system has been used frequently for the formulation of coloured pigments, which are generally composed of available pigment hues which are then modified by adding quantities of white and black pigments. Some printing ink manufacturers use a similar basis for the description of the inks they have available. But as real pigments used for such purposes are spectrally impure, and some of them very impure, it does not provide a good basis for any complete or theoretical analysis of colour differences. Its possible application to a cartographic colour chart has been described by Brown (1982).

## Colour measurement systems

### The CIE standard colorimetric system

In this system the mixture of red, green and blue primaries is represented trichromatically, so that $X$ = red, $Y$ = green, and $Z$ = blue. These are arbitrary colours, arranged so that none of them is less than 0. The green primary, designated by $Y$, also defines the luminosity, both $X$ and $Z$ having no luminosity component. To describe the colours in a two-dimensional diagram (Fig. 12), two ratios are calculated to replace the tristimulus values of $X$, $Y$, $Z$. These are the chromaticity coordinates, in which $x = X/X + Y + Z$, and $y = Y/X + Y + Z$. When measuring coloured surfaces by reflection or

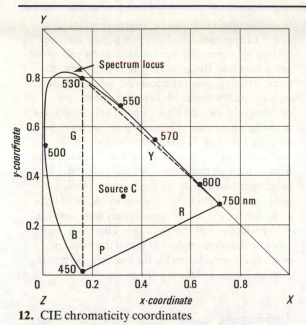

**12.** CIE chromaticity coordinates

transmission, the $Y$ value measured by reflectance is the luminous reflectance, and the $Y$ value measured by transmission is its luminous transmission. Therefore, these two coordinates can be used to describe the characteristics of any coloured light. Since the sum of $x$, $y$, $z$ is unity, any colour can be specified by $x$ and $y$ only. The chromaticity diagram shows that the theoretical primaries are represented by the corners of the triangle. Spectral colours are shown by the curve called the spectrum locus. This locus is nearly straight between 750 and 550 nm, but then curves round to the short-wave extreme. The straight line between the long and short wave extremes includes the mixtures of blue, red and purple. In the diagram, where $x = 1$, $y = z = 0$; and where $y = 1$, $x = z = 0$. The baseline, where $y = 0$, has no luminosity, and is known as the alychne or lightless line. The bluish-purplish colours associated with the lower straight section of the spectrum locus have little luminosity.

The areas associated with the different colours in the diagram vary in size and shape. The lighter colours (those with higher $y$ values) cover comparatively large areas compared with the darker colours which have low $y$ values. Therefore, two points a given distance apart in the upper part of the spectrum locus (in the green area) represent colours that are much closer than would appear on the diagram. This also fits in with the psychophysical response to colour, as small differences in lightness and saturation are difficult to distinguish with light colours, compared with dark.

The central point of the spectrum locus represents

the source of illumination, which has to be a defined white light source. Source $C$ has a colour temperature of 6800 K, and is one of the standard sources defined by the CIE for colour measurement. The degree of saturation of any colour is shown by its distance from the centre of the locus, the fully saturated hues lying on the locus itself.

Hues which lie along the base line of the locus do not have a 'wavelength', and can only be described in the system by drawing a line from the colour through the source and extending it until it meets the locus. This gives the complementary colour, and is described in the terminology by the addition of the letter $c$.

The CIE system provides a universal standard for colour measurement and description which can be applied both to transmitted and reflected light. After the Munsell system was measured according to the CIE standard it was reformulated and described as the Munsell renotation. The ability to convert from a Munsell description to the CIE coordinates is extremely useful as a means of giving a standard colour description.

### Other colour description systems

The display of coloured images, especially those which are used in visual display units such as televisions or television monitors, operate through the control of radiation to produce luminescence at particular wavelengths. Various methods have been developed to describe colours, based on red, blue and green additive primaries. In the YIQ system, the red, green and blue primaries are replaced by YIQ, and as in the CIE system the Y primary indicates the luminosity. This is, therefore, the component that controls the signal in black-and-white television.

Other systems include the HSV (hue, saturation and value) scheme, which makes use of a three-dimensional colour solid in the form of a hexcone; and the HLS (hue, lightness, saturation) scheme in the form of a double hexcone. Both these are similar to the Munsell colour solid, in that hues are arranged radially, and values range from black (0) at the base to white at the top.

### Colour terminology

A general problem in discussing colour is the lack of an agreed terminology. In colloquial speech, a great variety of terms such as 'shade', 'pastel', 'dull', etc., are used in a random and inconsistent fashion. In an attempt to reduce this confusion, the Inter-Society

Color Council of the United States and the National Bureau of Standards worked out a systematic terminology, known as the ISCC–NBS system. This uses terms for hues, and abbreviations, with modifiers or qualifying terms to distinguish variations in lightness and saturation. Hues are shown in the following table.

---

*ISCC–NBS abbreviations for hue names*

| | | | |
|---|---|---|---|
| Red | *R* | Purple | *P* |
| Reddish orange | *rO* | Reddish purple | *rP* |
| Orange | *O* | Purplish red | *pR* |
| Orange yellow | *OY* | Purplish pink | *pPk* |
| Yellow | *Y* | Pink | *Pk* |
| Greenish yellow | *gY* | Yellowish pink | *yPk* |
| Yellow green | *YG* | Brownish pink | *brPk* |
| Yellowish green | *yG* | Brownish orange | *brO* |
| Green | *G* | Reddish brown | *rBr* |
| Bluish green | *bG* | Brown | *Br* |
| Greenish blue | *gB* | Yellowish brown | *yBr* |
| Blue | *B* | Olive brown | *OlBr* |
| Purplish blue | *pB* | Olive | *Ol* |
| Violet | *V* | Olive green | *OlG* |

---

## Colour in printed reproduction

Although the terms hue, saturation and lightness define the properties of colour, the construction of images for reproduction in print has to take into account printing requirements. This affects the terminology used in colour description. A fully saturated line or solid colour image can be reproduced directly, but continuous variations in tone cannot. Therefore, if a lighter or less saturated area colour is needed, the continuous solid or tone is converted into a half-tone or tint. Half-tones and tints consist of arrays of very fine dots or lines, below visual resolution, which are integrated into an appearance of continuous colour. Because tints are formed by using particular densities of dot or line screens, they add a proportion of white to the colour, and result in an increase in lightness and a decrease in saturation. Technically this is quite easy to do, and tints for area colours in particular are widely used cartographically. The converse in colour terms is to produce a shade. This is a solid or tint of the colour, combined with grey (which can be a tint of the black). The effect is a reduction in lightness and a reduction in saturation.

This modification of colour appearance is useful both because it provides printable images and also because it makes possible the construction of light background colours from a range of darker hues.

Both solids and tints can be used in combination, and in cartographic production it is quite common to make a light green by a combination of a yellow tint and a blue tint. Because on virtually all maps, the lines, lettering, point symbols and solids require a high contrast with the background, the main printing colours used for a map are necessarily low in the lightness scale; that is they are dark and fully saturated colours. If other information is to be visible, colours applied to areas must be relatively light. Although it is possible to add additional light printing colours solely for this purpose (such as a light green or light brown), these would be of little or no use for the line images, as they lack contrast with a white or light background. Therefore, it is more economic to make full use of the primary printing colours chosen for the line images, making tints and shades of these as required.

## Trichromatic colour reproduction and map design

Trichromatic colour mixture and colour matching is employed directly in colour reproduction by the use of the three or four colour processes. Because the main requirement in the graphic arts in general is the reproduction in print of coloured tonal images, such as artists' drawings or photographs, it is sometimes assumed that all colour printing, including that for maps, should also be on that basis. But maps are not reproductions of something else, which already exists; and except in the special case of the photomap, they do not generally consist of images with a continuous tonal variation, in which any combination of colour and tone can exist at any point. As most of the information is carried by fine lines, small lettering and small point symbols, the principal requirement is for high-contrast hues, such as black, dark blue, dark brown and red. None of these is easily reproduced by the three colour process, which employs standard cyan, magenta and yellow. The cyan of the standard process colours is too light for fine lines; red can only be made by overprinting proportions of magenta and yellow; and dark brown would require the combination on a fine line of magenta, yellow, cyan and black. Apart from the technical problems and costs of converting a single fine line in one colour into a set of tints in two, three or four colours, the sharpness of the line images is reduced.

This is not to say that colour combinations cannot be used, or do not have a useful place in map construction. Dark blue, red, yellow and brown, if chosen as printing colours, can also combine solids and tints to make a variety of greens, oranges and purples. But the red, blue and brown can also be used directly for high-contrast line images. In cartographic practice, bichromatic combinations are more common than trichromatic

## VISUAL INFORMATION PROCESSING AND MAP USE

### The visual system

The radiant energy entering the eyes is first received and processed by the arrays of inter-connected cells in the retinas, before being transmitted along the optic nerves to the first stage in the brain, the lateral

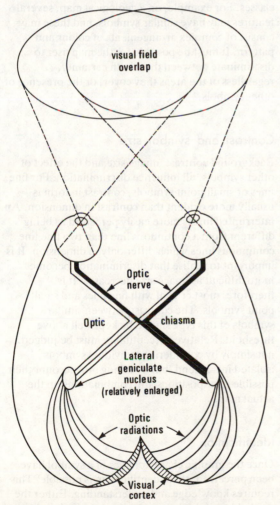

visual field overlap

Optic nerve

Optic          chiasma

Lateral geniculate nucleus (relatively enlarged)

Optic radiations

Visual cortex

**13.** Schematic diagram of the visual system

geniculate nuclei (Fig. 13). Here the two streams of information are combined, in order to achieve stereoscopic vision. From this stage they are transmitted into the visual cortex of the brain, which has further connections with other brain areas. Both the lateral geniculate nuclei and the visual cortex devote many more nerve cells to processing information supplied from the central part of each retina than to the areas of peripheral vision. They also appear to concentrate the processing on the detection of edges and contrast. Some of the cell groups in the cortex are specialised to respond to certain directions of movement, whereas others respond to certain shapes and angles. It seems clear that the incoming information is analysed and re-arranged, rather than just stored as a sort of replica of the complete visual scene. Because the information is analysed, it can also be recomposed. This makes possible the processes of transformation and orientation. Familiar shapes can be changed in scale, perspective and orientation, and still be 'recognised' by the brain, even though the perceived image does not exactly match that previously acquired.

### Eye movements

Various controls are involved in moving the eyes in a given direction, following a target, etc., but in looking at a stationary image, such as a map, they are usually directed for a moment at a certain area while the information is inspected, and then moved to another point or area. There can be a succession of movements between a peripheral view, which can see a large area in general, and a central vision view, when the gaze is concentrated on the detail of a much smaller area. For the brief period when the eyes rest on a given target they seem to be stationary. These short periods are known as fixations, and they can occur with a high frequency, every few milliseconds. As it is only during these short periods that the eyes are actually 'fixed' on the target, information provided to the brain is normally obtained through a succession of fixations. The complexity of this processing in map interpretation is described by Castner and Eastman (1985).

As one fixation rapidly succeeds another, and the stimulus of the previous one is lost, it is clear that the information must be stored, however briefly, in order for the brain to interpret it, and to decide whether to retain it or not. As many maps have a high density of information, and have to be examined carefully during use, the limitations imposed by the need to extract information through fixations are highly important. Although the visual field is perceived simultaneously, the detail in it can only be attended to and processed bit by bit. This requires the use of memory.

## Memory

Although the concept of memory usually suggests the long-term storage of information, it is clear that the normal process of vision must involve memory also, as it takes place over time. The storage of a fixated image long enough for it to be processed is referred to as brief visual memory, or iconic memory. Short-term memory can retain and operate on several items simultaneously for a short time. Some information can be stored in long-term memory, although this often requires a deliberate act of 'remembering', usually by repetition, or a very dramatic or significant experience. A great deal of learning is devoted to establishing concepts, information and ideas in long-term memory.

Looking at a map to obtain information must involve complex interactions between all three memory types. Information obtained briefly during fixations is processed and added to that obtained from previous and ensuing fixations. It can be acted on collectively by short-term memory, which can also recover and add other information already held in long-term storage. For example, a map symbol may be identified by matching it against a symbol shown in the legend and remembered through short-term memory, or by retrieving the symbol from long-term memory, based upon previous experience. The comparison of any two images which are not simultaneously present in the visual field must include a remembered image of one of them.

## Visual information processing

In order to extract information from the map, the map user must understand the map symbols and their relationships in space. To do this it is necessary to detect the individual symbols, discriminate between them, and correctly identify them, as a preliminary to any process of interpretation. Detection is apparently straightforward, but it is possible to specify the symbols in a multi-colour map so that detection may be difficult. Fine lines and small point symbols against coloured backgrounds may result in some symbols being 'missed', especially in peripheral vision, where there is no colour response, and acuity is lower. Obviously

detection, allowing for all the various arrangements and combinations on the map, is the minimum requirement. In general, map symbols are kept well above the threshold of detection, which is the absolute minimum at which detection is possible. Reduced illumination will also affect detection.

## Discrimination

In addition to detecting individual symbols, it is necessary to know whether they are the same or different. Discrimination requires a higher level of contrast than detection, as all the aspects of form, dimension and colour must be correctly perceived. Although many map symbols may have large and significant contrasts, there are often cases where two or more symbols are similar, but slightly different. This is always likely to be the case with maps which have many symbols arranged in classes and sub-classes. For example, on a geological map, several features may have similar symbols, and these may consist of complex arrangements of colour and pattern. It must be possible for the map user to discriminate between them with certainty, regardless of the areas they cover, or the presence of other symbols.

## Contrast and symbol size

Background contrast, image size and the effect of other symbols, all influence discrimination. For fine lines or small point symbols, contrast in form is usually more evident than contrast in dimension. An interrupted line is more easily perceived as being different from a continuous line than two fine line continuous lines which differ only in dimension. It is important to realise that discrimination becomes more difficult as image size decreases. It is, therefore, most critical with fine lines and small point symbols. The contrast between similar symbols of this type should be kept well above threshold. Relative perceptibility must be judged, not simply by considering individual symbols isolated in a legend, but by taking into account their possible juxtapositions and combinations on the actual map.

## Identification

Once the graphic characteristics of a symbol have been perceived, then identification is possible. This requires knowledge and understanding. Either the symbol is matched against the corresponding symbol in the legend, or against an impression of the symbol stored in memory from previous experience. In using an unfamiliar map, or a map dealing with an unfamiliar subject or area, it is often necessary for the map user to refer frequently to the legend at the initial stage, in order to identify symbols correctly. With practice this becomes a habit, and may be entered into long-term memory. Identification also requires that the meaning of the symbol is understood. An inexperienced map user may correctly perceive that the brown line on the map is defined in the legend as a contour line, but if he does not understand the meaning of the term 'contour' as used in cartography, he will not be able to proceed to any further interpretation.

Identification is made easier for the map user if the graphic variables are used systematically, so that similar things or features have a common graphic element in representation, and if the map content is sorted into distinct visual levels. This increases contrast, and makes symbol identification easier. Familiar conventions should be respected, especially in general-purpose maps, unless there is a strong reason for departing from them. Well-established conventions mean that the map user encounters what is expected, even if subconsciously, and there is a general tendency in visual perception for the brain to give the expected interpretation to the visual stimulus. Although any attempt at complete symbol standardisation encounters insuperable problems, a random use of the graphic variables in symbol design can provide the map user with too many unaccustomed demands in symbol identification, which tends to make the map 'difficult to read'.

## Identification and interpretation

Correct identification of symbols makes map use possible. Beyond this point the extraction of information from the map depends on the user's attitude, objective and existing knowledge. Different levels of map use exist, and not all types of map use involve the same degree of interpretation. In some cases the objective may be limited and simple; for example, finding a named place. In other cases, the user may need to examine the whole map, or an area of it, in order to reach an understanding of the topographic or other characteristics of the area, possibly matching this with existing knowledge, or using it in the solution of a problem. Interpretation presumes understanding, and this is not possible unless the map user is aware of the significance of scale, generalisation, map structure

and the symbolic nature of cartographic representation.

The fundamental objective of the cartographer is to make sure that correct identification is possible, and to assist the tasks of interpretation. This in turn is affected by legibility, the density of information on the map, and consistency of graphic organisation.

The overall design of the representation also may have aesthetic significance. The use of colour in particular can influence not only the ease with which the map can be used, but also the response of the map user either to the whole map or important features. Although the nature of this aesthetic response is highly subjective and individual, there is no doubt that map users are affected by the map design, which can lead them to strong preferences for some maps compared with others, even though the scale and informational content are equivalent. Robinson (1967) and Makowski (1967) attempt to describe some of the intangible but well-known aspects of the influence of colour on map perception.

## Identification and recognition

The realisation that something is familiar, or has been encountered previously, is often referred to as recognition. However, in cartography and map use there are two aspects of familiarity which need to be distinguished. The first is the familiarity of a symbol, described by the identification process. The second is that certain geographical shapes are unique and familiar, so people can 'recognise' the outline of the Mediterranean Sea, or North America, or other major physical components of the Earth's surface, or the topographic features of their own localities. Although both processes require the matching of what is perceived on the map against some representation held in long-term memory, they are sufficiently different to require distinction.

## Visual search

A map is an inanimate object. Information can only be extracted from it by the directed activity of the map user. In most maps, the information has to be sought, and the process of looking at a map involves also looking for the detail relevant to the enquiry. Visual search raises many problems. Obviously visual search is reduced if the map information is limited to what the user needs to know, or what the map author wants to say. This can be contrived in some cases if the map gives a simple locational description, for example in a news telecast, where the location of one place referred to in the news is shown by a simple map. The ultimate consequence of this of course would be to make a different map for every different enquiry or presentation. On the other hand, many maps give an account of phenomena over an area, and it is the locational relationships between different features which are the most important information. In such cases the user has to search the map, either to find the particular information that is pertinent to the task, or to extract the combination of elements that must be appreciated and understood.

Visual search involves both central and peripheral vision, but generally speaking this cannot be used as a basis for designing map symbols, as an item which is peripheral in one fixation may be central in another. And, of course, the information classified as irrelevant, and therefore ignored, by one map user, may be of critical importance to another.

Visual search is hindered if too much information is presented by very small symbols, close to the threshold of discrimination. This tends to slow down the process of correct identification. Conversely, it is also hindered if too many features or symbols are emphasised, as they compete visually for attention, and may cause an unnecessary distraction. The crowding of information in some areas will mean that search time is increased, and there is a danger that short-term memory will be overloaded. What is more difficult to appreciate is that if the map is at

too large a scale, time is wasted by searching through 'blank' spaces. For most maps, the most suitable scale is the smallest at which the total information can be legibly presented, as this enables more of the information to be perceived through central vision, with fewer fixations or head movements. In this respect the correct formulation of scale, area and information is critical to the map user, as well as the cartographer.

In the composition and design of the map the cartographer also 'searches' the information, using visual imagination, in order to anticipate the problems involved in representation. Experimental or trial versions of the map are devised in order to try out possible solutions. Although the cartographer cannot necessarily imagine all the tasks which a variety of users may undertake, he can scrutinise the map at the initial stage to ensure that the basic concept of scale is suitable, and that the map is legible under all conditions. Very often, even though adequate methods of representation have been chosen, the first colour proof of the map or prototype will demonstrate that some slight adjustments to the design are needed. This critical attitude is an essential part of cartographic activity. Although there is no such thing as a perfect or absolutely correct design, the cartographer has to use his own visual judgement in a critical manner, in order that the prospective map user can perform visual map-using tasks without difficulty or distraction.

# 3 MAP DESIGN AND CARTOGRAPHIC SYMBOLS

## CARTOGRAPHIC DESIGN

The term 'design' is used to describe a wide variety of activities, ranging from the design of a single object, like a jug, to the design of a complex arrangement in space, such as a new town or a transport network. Whatever the product, it is clear that something is created, and this creation has a purpose or function. In the sense that it is creative, it is distinguished from manufacture or production; but because it is overtly functional it is distinguished from the creativity characteristic of the fine arts.

It has been pointed out frequently that in the days of the manual production of complete objects, design and production appeared to be carried out simultaneously by craftsmen, although in fact such designs were usually the outcome of a long process of gradual evolution by experience (see Jones 1981). Mass production, and the introduction of new technologies, resulted in the separation of the design and production stages of most manufactured articles, and also in most cases the separation of the designer from the producer. Therefore, an architect devises plans and elevations for a proposed building, but the builder does the actual construction. It is obvious that the designer should understand the medium with which he works, and which provides one constraint on what he may do. In addition, the designer is normally working for a client, whether in or outside his own organisation, and this client may be the actual user, or the representative of the users. Therefore, the task that the designer is set is, at least in theory, conceived or stated by the client. But although the client may describe the objectives, or what he wants to have made, he relies on the designer to create the form.

Great innovators of course created things before any market for them existed, so that they could hardly operate through market research. Even so, what they devised or invented was conceived and designed with the intention of making something useful or marketable.

Whereas at one time design ability was regarded as a gift to be developed by practice and experience, in recent decades much attention has been given to examining the ways in which people learn and behave, and the use of 'scientific' models of human activity. There have been many attempts to apply such theories to the analysis of the design process, particularly in architecture, engineering and visual communication in general (Lawson 1980). The theories of cartographic communication are similar, because although they seek to analyse map making and map using on the basis of 'scientific' models, they mainly focus on the cartographic design stage as the central factor in determining whether a map will be satisfactory or not.

### Design and creativity

All such theories struggle with the most important characteristic of the design process, which is that the purpose of design is to create something which did not exist before, and which has to be realised before it can be examined or tested. A scientific approach can only be employed properly when the objective can be stated in advance, and described by measurement. In addition, the nature of a design problem is not necessarily self-evident, and often prolonged enquiry is needed before the factors relevant to the design are discovered.

To assume that the design function is simply to solve a set of problems posed by the client or consumer also understates the role of design. Generally speaking, a client can make a limited number of statements about desirable objectives or requirements, although such descriptions are rarely complete. As Petchenik (1983) points out, the fact that some of the objectives may be in conflict is not necessarily obvious to the client, although one responsibility of the designer is to make this clear. Most things made for human use compromise between conflicting requirements. Motor cars should be fast and economical; cameras should be sophisticated, robust and light in weight; shoes should be both comfortable and durable. Maps should be highly informative, yet 'easy to read'. In this respect, designers generally have to choose between alternatives in emphasis. Consequently there is no such thing as a perfect design, for each design represents only one possible solution.

Apart from the technical constraints of materials and processes, design is part of the overall production cost, and therefore consumes resources. Whereas production costs can be specified, there is no means of knowing how long it will take to conceive a 'good' design. It is generally true, and very often the case in cartography, that the resources devoted to the design stage are quite limited. A cartographer designing a new road atlas is unlikely to find a client willing to pay for five or ten experimental versions. In many cases the so-called conservatism of designers is a consequence of the limitation on experimental work. What has been found to be more or less satisfactory before is adopted as the basis, and only slight modifications made. Indeed design can be so restricted that changes are purely cosmetic.

On the other hand, lengthy investigation of design problems does not necessarily mean that a good design will result. In many cases, it is the ability to conceive the problem in a different way which leads to an improved solution. The ability to produce a large number of variations quickly with a computer does not necessarily have this quality, for original starting points still have to be generated by imagination. Scores of variations are of little use if the parameters are inadequate to begin with.

Although there have been many attempts to analyse design as a sequence of processes, the relationship between design and aesthetics remains intangible. Whatever may be said about 'functional' design in theory, the fact is that people judge artefacts and objects in terms of their appearance as well as their utility. This is true for motor cars, houses and clothes. It is clear that this subconscious aesthetic judgement is affected by a mass of cultural factors. It is often expressed through fashion, and may be uninformed or superficial, but it is there.

The ability to create something new, to devise a form or shape which had not been conceived before in that sense, also enlarges the designer's role beyond merely satisfying a consumer demand. Very often the demand does not exist until the object has been created. This ability to anticipate may also mean that the designer must take the risk of producing something in advance of public opinion, or make demands on the consumer which are novel. This is obviously the case with many graphic arts, as well as cartography, and in this sense designers do not simply follow taste but create it.

## DESIGN PROCEDURES

Although it may be impossible to explain how to design, it is instructive to examine the stages which seem common to many design procedures. These have been analysed and described in different ways by different writers, but they often refer to five stages, namely, first insight, preparation, incubation, illumination and verification, although a variety of terms is used.

### First insight

The first stage is to study the task in general, to discover if possible all the relevant factors, and to establish where the chief problems lie. If some of the requirements are irreconcilable, then it may be necessary to reformulate the product entirely. The cartographic equivalent of this can be regarded as the initial concept and planning of the map. The first requirement is to establish a proper relationship between the intended content, the geographical area of the map, the scale and the format (Fig. 3). Unless these are in harmony, nothing will be satisfactory subsequently. These relationships are in turn conditioned by the available technical resources (which will control what is possible graphically), and the nature of the desired end product . All these will be affected by production time and cost.

In many cases there is a lack of agreement between area, content and scale. The scale may be too small to provide an effective representation of the desired information. Either the scale (and the map format) must be enlarged accordingly, or the information split between two maps, or a decision taken to reduce the content. In most cases it is impossible to do this adequately unless there is some sort of compilation of the complete map information which can be studied. If the technical resources are limited – for example, if the map is to be produced in only two colours – then this will affect the range of graphic expression available, which in turn will affect the content and legibility of the map. On some occasions, if a map has to be produced to a deadline, its scope and content must be restricted to what is possible in the time available. The inter-play of these factors is of great importance when a new map is being considered. It may be even more difficult if the map is divided into a series of sheets to cover a large area, as then the range of topographic and geographic variations also has to be taken into account.

## Preparation

The second stage proceeds to develop and test hypotheses, to formulate possible solutions to parts of the overall problem, and even to imagine or visualise the product in a specific form. Avenues are explored and abandoned, as some previously unnoticed factor comes to attention. For a complex object, many different 'bits' may be designed, and then problems surface as the attempt is made to put these bits together. Often at this stage there appears to be a critical problem, to which there is no apparent solution.

The cartographic equivalent of this stage is to imagine, or even to experiment with, specifications which translate general ideas into specific symbols. Initially these may be thought about separately, but the design must be tested by trying to examine the various combinations in which the symbols will appear, that is against the actual spatial arrangement of the content. A line hue and gauge which seems suitable for contours when they are close together may be less effective in areas where they are widely separated. The choice of a particular area pattern must be modified by the fact that several other types of information are also present. At this stage, particular problems appear, which are themselves a consequence of a certain approach to the overall design. If the main roads are to have a red infill between black casings, and the built-up areas are to have a pink tint, what happens when the main roads become part of the built-up area, and intersect it? Is the red colour carried through the built-up area, or does it stop at the edge?

## Incubation

This leads to a further stage, which is now focused upon a particular design formula, but which seems to contain some difficulty that cannot be surmounted. If the general framework of the design is acceptable, then a solution may be found by slight changes in the specifications of other symbols. But on some occasions, it becomes evident that a different approach is needed. Conceiving this can be difficult. A new idea may be needed, but it cannot be forced by concentration, and if it does appear, often seems to come from nowhere. It would seem that the mind may operate on the problem even subconsciously, and indeed 'sleeping on it' often does seem to introduce the necessary span of mental relaxation. Leaving the problem aside for a while, and then coming back to it, also seems to help.

The generation of ideas is often considered to be the product of divergent thinking, the ability to go outward from a subject and to explore quite different possibilities. Although in studies of educational psychology a sharp distinction has been made between 'convergent' thinkers (rational, scientific, conformist), and 'divergent' thinkers (emotional, independent, unconventional), it is unrealistic to make such absolute distinctions. A good deal of design also requires an orderly and systematic approach. As Lawson (1980) puts it, 'The truly creative scientist needs something of the artist's divergent thought to see new possibilities, while for his part the artist needs to be able to apply the single-minded perseverance of the scientist to develop his ideas.' It is not easy to determine how the process of creative thought can be assisted, although a great variety of methods has been tried. It is clear that too close an acquaintance with existing similar products can have an inhibiting effect. Even too much experience with one type of product seems to reduce the possibility of new ideas. It is not uncommon for a newcomer, or an outsider, to make the most original suggestions.

Although this need to stimulate creativity is well known, and indeed is just as important to many 'scientists' as well as 'artists', it has rarely been studied in depth. Perhaps because it was so unusual, the Lund conference on creativity and context, in which a group of scholars and scientists attempted to analyse and describe their creative processes, makes fascinating reading (see Buttimer 1983).

It is clear that a familiarity with a wide range of design problems and solutions can assist in the generation of ideas. Being able to bring to mind a set of possible solutions to a recurrent problem at least provides some starting points. It is the generation or conception of these starting points which can be most difficult for the beginner. In this respect, the study of cartography is assisted by the close examination of a wide variety of different maps and map designs, considering them as solutions to problems.

## Illumination

Illumination is a term used in this context to refer to a sudden realisation of a different approach, which can lead to a new formulation of the whole design. A complete change in the specification of one symbol may lead to new possibilities for the composition as a whole. Once the idea exists, it is often accompanied by the feeling 'Why didn't I think of that before?' Good design solutions usually seem obvious after they have been devised, and indeed it is characteristic of a good design solution that the user remains unaware that there was any problem to solve.

### Verification

Verification normally refers to the full experimental or prototype stage, when at least a section of the map is produced to the full specification. For a small or relatively simple map, this may mean carrying out the whole production. For a larger or more complex map, or a sheet in a series, only a representative part may be completed in the first instance. Unless considerable resources are available, it is unlikely that this experiment will be discarded entirely if it turns out to be less than perfect. Design by iteration – by simply making a series of alternatives – is extravagant in resources. Cartographically, this stage is likely to result in a colour proof, in a form which at least approximates the map's intended appearance.

At this point some possible improvements may become apparent. A colour may be made slightly darker to improve contrast; a percentage tint may be made slightly lighter to give better contrast to the line image. These refinements often make the difference between an adequate design and a good one. In practice, it is rarely possible to pursue this refinement beyond a certain point, and indeed the desire to keep on changing and 'improving' may eventually destroy the original harmony.

## CARTOGRAPHIC SYMBOLS

### Graphic elements and graphic variables

The symbols on a map consist of points, lines and areas, and they can be varied in form, dimension and colour. Given a proper description of these terms, they can be said to encompass all possible variations in map symbols. Their locations on the map are controlled by their positions on the ground or in relation to it, and this aspect, although it may be modified, cannot be changed. Because the symbols are arranged in an ordered, two-dimensional space, they present information collectively, as well as individually, leading to the interpretation of relative position, distribution and structure. Consequently, the design of symbols, and the use of the graphic variables, has to take into account their possible juxtaposition and separation,

because this will influence their appearance in the map as a whole.

Although the design and specification of a map symbol may appear to constitute a single process, it is affected by two considerations, which form a sequence. The graphic variations which distinguish one symbol from another have to be systematically exploited; and these graphic variations have to be employed also to indicate the relationships between features, and to allow for the effect they have on each other when they appear in combination.

## TYPES OF SYMBOL

The descriptions point, line and area are not absolute, but are relative to scale and the characteristics of the feature being represented. On a large-scale plan a building may be shown by an outline defining its correct plan dimensions at ground level. This uses the line symbol as an outline to separate one surface from another. It could also be shown by an area symbol, without any outline, through a difference in colour (Fig. 14). At a much smaller scale, where the dimensions of the building, if reduced to map scale, would be less than the smallest symbol which would be legible, the building might be represented by a point symbol (Fig. 15). This would no longer show the dimensions of the building, but would state that a single feature

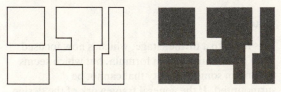

**14.** Outline and area symbols

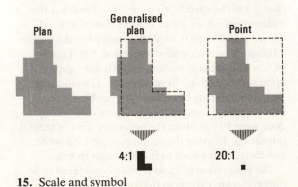

**15.** Scale and symbol

classified as a building existed at this point. Therefore, the use of the point or the outlined area is essentially a function of scale, relative significance and importance in the map content.

The same applies to linear features, such as large rivers, which are shown with correct areas at large scale, but by single line symbols at small scale.

## GRAPHIC VARIABLES

Symbols may vary in form, dimension and colour. The term form is preferred to that of shape, as in many cases topographic features do have characteristic 'shapes' which can be recognised, because they are familiar. It is desirable to distinguish between these two aspects. The form of a point symbol may vary from regular or geometric to iconic or pictorial. This will affect the size at which it is legible. The form of a line symbol may be continuous or discontinuous. The shape of the figure of an area symbol is determined by its actual extent. Form can only be indirectly applied in reference to a pattern of dots, lines, or other point symbols in a regular array, used to represent the area.

Colour applies to both chromatic and achromatic distinctions, therefore including those usually described as 'black-and-white'. The black-and-white map makes use of differences in lightness, even though there are no differences in hue. In a multi-colour map, perceptible differences for small point symbols are primarily contrasts in hue. Large point symbols may show distinctions by variation in lightness and saturation, but in such cases it is difficult graphically to distinguish them from area symbols. A single line symbol can be varied in hue, and slight variations in lightness and saturation can be introduced, but only by changing the form, for example from continuous to discontinuous. A 'dotted' line will appear lighter than a continuous line of the same hue and gauge.

All variations in colour can be applied to area symbols, which may differ in hue, saturation and lightness. In a black-and-white map the intermediate values are greys. The area of a symbol has a profound influence on its appearance, and the degree to which differences in hue, saturation and lightness are apparent. Generally speaking, the smaller the symbol, the greater must be the contrast in colour to be perceptible.

## POINT SYMBOLS

Form applies to all variations in the figure of the symbol. It may be regular (geometric) or irregular. For small point symbols, squares, circles (both outline and solid) and triangles are extensively used, as at very small sizes these simple geometric forms are most easily distinguished and discriminated by the visual system. Iconic forms imitate in a simplified fashion some aspect of the outward appearance of the feature, usually the profile, but they can also represent some other physical property, or even a concept associated with the nature of the feature (Fig. 16). A cross denoting a church is in one way symbolic for the Christian faith, but is also iconic due to the extensive use of the cross sign. This capacity of the map symbol to represent by association is extremely useful.

16. Graphic variables and point symbols

### Form and orientation

Point symbols which are not symmetrical in their proportions may also be used in different orientations. A rectangle may be horizontal or vertical; a tree symbol may be inclined to show windblown forest; a triangle may be placed on either its base or its apex.

### Complex forms

The form of a point symbol may be modified by addition or extension, or both. A simple geometrical figure, such as a square, may be extended by placing a cross on one side or edge; or by adding a circle or dot inside the outline. Such variations may be used to create a set of related symbols for a group of features.

## Point symbol dimensions

The dimensions of a point symbol may range from the minimum required for a perceptible symbol, that is the smallest size which can be detected and identified, to one which has been deliberately enlarged and made more complex. In many cases form and dimension are closely connected. Very small point symbols at the threshold of detection have no perceptible form, or at least the form may not be evident without close scrutiny. The general rule is that contrast in form requires a larger symbol than that which is just detectable. Large point symbols can be made more complex, but of course then intrude into adjacent map space. They can take on the graphic characteristics of area symbols, even though nominally they refer to the presence of something at a point, or occurring over a very small area at map scale. Point symbols which are deliberately enlarged are used to represent quantities in many statistical maps, for example by proportional circles, but variation in dimension is also a simple way of indicating relative importance. For features of the same type, larger is usually regarded as being more significant. On a small-scale general map, towns may be represented by a series of similar point symbols, the dimension being increased for both larger (in population) or more important towns. This relative importance, therefore, may be a reflection of an actual quantity, but also may be a judgement based on administrative or cultural significance.

## Point symbols and colour

For small point symbols, whether simple or complex, contrast needs to be high to maintain legibility, and therefore contrast in hue is the major colour distinction. Differences in saturation are only apparent if the symbol area is large enough. At small sizes, black, red or other high-contrast hues help to maintain legibility against light backgrounds, in terms of symbol dimensions and form. A small black square can be easily distinguished from a red square of the same size; it might be difficult to distinguish it from a dark blue one. In all cases, the variations in form, dimension and colour cannot be treated independently, as they affect one another. In order to maximise contrast and provide additional visual clues for the map user, very often two variables are used to contrast symbols which might be confused. For example, if there are two point symbols of about the same size, they may be contrasted by a difference in form (circle and square) combined with a difference in hue (black

and red). Complex point symbols often use two hues. For example the outline of a small circle may be black, and the circle filled in red.

# LINE SYMBOLS

The form of a line symbol is essentially a matter of continuity. There are a large number of possible variations in continuity, which range from a continuous line to a line of dots. The characteristics of any interrupted line are defined by the length of the segments and the spaces between them; both need to be specified to define the symbol. The simple interrupted line has a single segment and space structure. More complex interrupted lines may use different segmental lengths and/or forms in combination. The familiar 'dash and dot' line is a basic example of this more complex category (Fig. 17).

**17.** Graphic variables and line symbols

Line symbols can also be created in multiple, using two or more lines, and these may also have different forms and dimensions. Additions can be made to multiple lines. Extensions, such as cross lines, or short lines projecting to one side (illustrated by the railway track and canal symbols respectively) provide another set of possible variations on the basic line form.

## Line symbol colour

Fine lines, like small point symbols, need a high contrast with the background and with each other.

Consequently it is normal practice to use dark hues for the line image. The fact that many maps have much of the line information in black demonstrates this. The need for this contrast is so important that in most multi-colour maps the choice of printing colours is dominated by the needs of the line images. Multiple line symbols may also include colour combinations. A common example is the coloured infill between black or grey parallel lines, for example for classified roads.

## Types of line symbol

The two major types of line symbol are those representing linear features, and those which act as outlines to areas. The distinction between the two affects the characteristics of the symbol which may be appropriate both to the function and the graphic design. The line gauge for a linear feature may reflect its actual ground dimensions or its relative importance in the map content. A line which is held at secondary level (such as the contour line) is usually in a slightly lighter hue than the primary information, but in this case would need a slightly wider gauge to compensate for the lighter hue. The outline of an area represents a change from one surface to another. This 'edge' has no actual dimensions, and preferably should be as fine as possible. Again the line dimension will depend upon the chosen hue. Wherever possible, outlines of areas should be reduced in strength or omitted, and the contrast made by the use of a change in area colour or pattern.

The choice of line symbol is also affected by the characteristics of the linear feature which is represented. A topographic feature which is continuous, such as a permanent river, is appropriately represented by a continuous line: an interrupted or seasonal river should have an interrupted line. Similarly, where an area is delimited by a clear boundary or edge, a continuous fine line would be appropriate. But if there is a transitional zone along the edge, as happens with many types of natural vegetation, an interrupted line would suggest that this 'edge' did not follow any marked division in the landscape.

In some cases, lines represent measured values, not topographic features as such. The contour line is an obvious example. A contour line plotted photogrammetrically is a line of continuous measurement, and is correctly shown by a continuous line symbol. An interpolated contour derived from a set of point values is of a lower order, and therefore should be shown by an interrupted

line to indicate this distinction. Variations in line form are not used systematically in cartography to indicate the nature or quality of the source information, although this would be possible, and would be of benefit to the user. If all interpolated lines or lines derived from mean values were consistently represented by interrupted lines, the map user would be better informed about the nature and quality of the source information.

## Line symbol emphasis

As the larger part of the total map information is often carried by the line images, their treatment is very important in map design. A complex array of interweaving lines of much the same character is the most difficult image for the map user to interpret. Area figures have to be disentangled from linear features, and one line distinguished from another. The treatment of outlines, line gauge and line form can strongly affect the overall appearance and legibility of the map (Fig. 18). Due attention should be given to the exact specification of line gauge, form and colour, and full use made of the different lightness values of different hues.

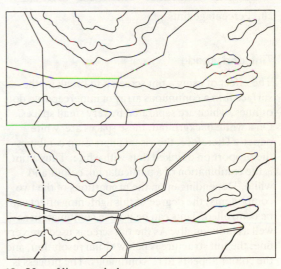

**18.** Use of line symbols

## Area Symbols

All variations in the surface appearance of an area can be described as differences in colour. On a printed map there are two main kinds of area symbol: a continuous surface colour, either solid or tint, or a pattern, which has a noticeable texture.

Although solid colour can be applied to an area, and is occasionally used in this way, on most multi-colour maps the primary colours of the printing inks are generally too dark to be suitable for colouring area. If other information is to be shown – as point or line symbols, or lettering – then the area colour must be light enough to retain contrast (Fig. 19). Therefore, it is common practice to reduce the saturation and increase the lightness of the area colours by making tints (Fig. 20).

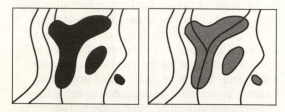

**19.**  Line symbols and contrast

**20.**  Percentage tints

## Tint and shade

The effect of using a tint screen is to break up the surface into a continuous array of imperceptible dots or lines, which are separated by very small spaces. On a white background, these spaces are 'white paper'. Therefore, the visual impression combines the proportion of inked dots with white. Thus a tint is the combination of a particular ink colour and white. Depending upon the proportions of the two components, the degree of this lightening effect can be controlled. Tints can be used in combination as well as individually. As the tint screens used are very fine, the tint structure is below visual resolution, and the colour appears to be continuous. The process is technically simple, and is widely used cartographically, primarily for the purpose of controlling area colour.

It is also possible to make a shade in a similar way. In this case the tint or solid area colour is combined with a tint of the black or a grey. This reduces both the lightness and the saturation.

In some maps, usually graphically simple specialised maps, the areas occupied by the area symbols may not include any other information at all. In such cases the requirement for a light

background does not apply, and the full range of solid colours and tints can be used to represent different values or categories.

If the only printing colour is black, it is highly unlikely that the solid black will be used to represent areas unless they are very small, and very important. But it is still possible to introduce tints of the black, which are greys, by the use of tint screens.

The number of perceptibly different tints which can be produced is affected by the lightness value of the particular hue. With a dark colour, such as a black, red or purple, it is usually possible to make four or five tints which can be distinguished even over small areas. On the other hand, with a light colour such as yellow or light green, it is dangerous to attempt to make more than two. Even if they can be distinguished in a legend, it is unlikely that they can be identified with certainty when seen in different locations on the map, especially if other highly contrasting colours are also present.

## Texture and patterns

With a tint or shade the surface colour appears to be continuous or 'flat'. It is also possible to indicate an area by introducing a pattern which is intended to give a visible texture to the surface. As with tints, such patterns may be either dark or light, depending on the structure of the pattern and the choice of colour or colours.

Most patterns are made by addition, by adding some colour to the white surface. They may consist of a regular array of point or line symbols, which themselves can vary in form and dimension. The total character of the pattern is a function of the form of the pattern unit (such as a round dot or an interrupted line); the dimensions of the pattern unit (the size of dot or the line thickness); the spacing or separation between adjacent units, and between lines of these units (which may not be the same); and the orientation of the rows of pattern units (Fig. 21). In place of a regular pattern unit, such as a dot or line, iconic symbols can be used, such as small tree symbols.

Combinations of different pattern elements can also be used, for example a row of fine lines alternating with a row of dots. In addition, different pattern arrays can be combined, for example by overlaying horizontal lines on vertical lines to produce a cross-line pattern. Given such a range of variables, the number of possible patterns is very great. If two or more colours are used in combination, an enormous range of graphic possibilities exists.

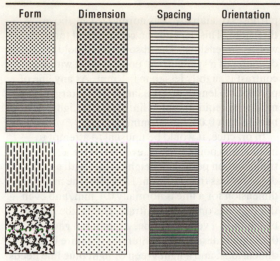

| Form | Dimension | Spacing | Orientation |
|------|-----------|---------|-------------|

**21.** Patterns

## Patterns by subtraction

Although it is common practice to compose patterns by addition, they can be very effective if used in negative form, that is by subtracting the pattern from the solid colour or tint, thus leaving the pattern in white or a different colour (Fig. 22). Technically this is quite straightforward, and it has the great advantage that it reduces any darkening effect, interfering less with other information. If two or more colours are used, then the subtraction of a pattern from one colour will leave the other colour exposed, producing a multi-colour pattern effect.

**22.** Patterns by subtraction

## Patterns and area symbols

Because patterns are composed of discrete elements, some of which can be placed at different orientations, they can also be visually disturbing. The aim should be to construct patterns which do not cause interference with the legibility of other information, and which can be visually integrated to give the impression of a continuous surface, even though the texture is perceptible. This can only be achieved if the pattern units and the spaces between them are small. In general, fine patterns are more effective than coarse ones. Although patterns are extremely useful, they have to be carefully controlled. Any pattern with a strong contrast can

have a disturbing effect on the overall map balance, especially if diagonal orientation is used. This is particularly the case in black-and-white maps.

Given these considerations, there are certain types of pattern which tend to be successful. These include a fine pattern of dots or lines, similar in appearance to a tint, but with a perceptible surface texture: patterns using iconic symbols in a relatively light printing colour, such as green or grey; these can be overprinted by darker line symbols; a fine pattern in combination with a light tint or a light solid colour, such as green or yellow: the combination of two patterns in different colours, usually placed at right angles, provided that the pattern weight and contrast is not strong; and patterns formed by subtraction, using combinations of colours which are about the same in lightness value. In maps where a large number of different area colours and patterns are needed, as in a geological map, the use of negative patterns can assist in solving the problem of achieving perceptible differences without introducing very strong or heavy patterns which can be visually disturbing.

# MAP DESIGN AND CARTOGRAPHIC OBJECTIVES

## Symbol design and map composition

In a multi-colour map, the design of individual symbols is not approached at random, but is related to the representational requirements of the different categories of information the map is to show, and their relative importance. In a general map, there is likely to be a basic distinction between the primary information, usually that describing the cultural features, emphasised as being in the 'foreground', and a secondary level, mainly the physical features, usually treated as 'background'. There may be slight differences in visual level within these two major groups. This is achieved by choosing appropriate symbols in terms of colour, dimension and form, in order to reach this balance. In special-subject maps, there is a distinction between the representation of the subject itself, which is the primary information, and the locational reference information, which is treated at a subordinate level. In both cases the strongest colours, with the highest contrast, are used for the principal information, and the lighter colours are used for the secondary information. Important items in the content will be emphasised by heavier lines and larger point symbols, and where necessary, more saturated area colours and stronger patterns. In general the order of emphasis is hue, dimension and saturation, and finally form, differences in form often being introduced to distinguish between features in the same general category.

The contrast between the primary hues is the most pronounced visual contrast, and therefore is used to separate the principal groups of content. On a topographic map, this is likely to mean employing black for planimetry and most of the point symbols and lettering; using dark blue for water features, brown for contours and green for vegetation. Additional emphasis may be obtained by adding a limited amount of red.

## Control of contrast

In conceiving this basic approach to the design, the starting point should be the minimum degree of contrast needed to make the symbols legible and perceptibly distinct. A suitable degree of emphasis can then be introduced for the most important subject matter. If the design begins with too much contrast, there will tend to be a progressive accumulation of heavy lines and dominating area colours. Eventually this may lead to a dull and unattractive design. The other reason for following this approach is that although the need for an increase in emphasis is usually apparent, it is often not so easy to appreciate that a problem can be solved by reducing the weight of another component, where the degree of contrast is excessive. If a line symbol does not show clearly against a coloured background, the problem may be solved by making the tint lighter, rather than making the line heavier. In addition, so far as a printed image is concerned, the accumulation of ink leads to a reduction in reflection from the surface, and consequently a dull appearance.

Both contrast and image quality can be affected by the type and whiteness of the printing paper used. Although theoretically most map design is conceived on the basis of contrast with 'white' paper, even the best offset paper reflects only slightly more than 80% of the incident light, and poorer papers considerably less. Any coloration of the paper will also affect the appearance of the printed inks.

## Legibility

In order to be legible, any symbol must be above the absolute threshold of detection, and for practical purposes of map use, it is normally well above threshold. This depends on contrast and form, as well as dimensions. The absolute threshold for the diameter of a symmetrical point symbol with high contrast is slightly less than 0.1 mm. As any line of perceptible length offers a larger visual target, the minimum requirement in line gauge is slightly less, being about 0.06 or 0.07 mm.

In practice, it would be unusual to attempt to include any point symbol, even a dot for a triangulation point or a height value, with a diameter less than 0.2 mm. Likewise, the minimum dimension for a line is normally 0.1 mm, and even this depends on a high contrast. Unless the line is black, or in a very dark hue, a minimum gauge of 0.15 mm is desirable. Lighter hues require larger dimensions, and a light green or light blue line needs a minimum gauge of 0.2 mm. Technically speaking, present reprographic and printing processes can effectively reproduce image elements to the limits of perception. Unfortunately, and possibly because of this, there is a tendency to make use of very fine lines regardless of their perceptual limitations. In practice, a line gauge of 0.1 mm should be regarded as exceptional, and reserved for particular purposes, such as the outlining of areas which are also differentiated by colour.

Where symbols are adjacent, or where a symbol is composed of two or more elements, it must also be possible to discriminate between them. This degree of separation is also affected by the dimensions, and therefore the contrast, of the components. The space between two parallel lines needs to be at least 0.15 mm to be discernible, and this should be increased if the lines are very fine. The same applies to line patterns. Dot and line screen tints are, therefore, deliberately constructed with the inter-line spaces less than this, so that the individual spaces are not resolved visually.

## Consistent categories

The subject matter of the map should be divided into major categories of related features, usually by treating a group of sub-classes as a series, and dealing with them as a whole. Generally the unifying graphic element is colour, or a pair of colours used together, where the different sub-classes are distinguished mainly by variations in dimension and form.

## Recognisable shapes

Very often an initial requirement of the map user is

to be able to 'locate' the area of the map in terms of its wider geographical position. An experienced map user will make use of all the clues provided by the configuration and arrangement of land/water masses, relief, named places, etc., adjusting his mental impression for scale. The inexperienced user may not be able to adjust to scale so easily, and may not make full use of all the visual clues present. This aspect of orientation depends largely on recognising the shapes of major topographic features. This is best assisted cartographically by making a clear distinction between land and water (Fig. 23), for this is a major differentiation in the landscape, and emphasises the outlines of recognisable major features. Recognition of shape is assisted by the separation of the surface into visually distinct areas by colour or texture. Often this does not need a strong contrast, as the visual system is extremely good at detecting such differences. The addition of a pale tint to water areas is normally sufficient to bring out the shape of the landscape.

**23.** Area symbols and separation of topographic features

### Reduction of the line image

The same process of reinforcing the contrast between major shapes can be helped by reducing the amount of information carried by line symbols. Linear features require line symbols, but outlines of areas can be avoided or reduced by making full use of area colour and texture, either with very fine

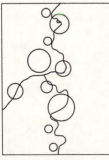 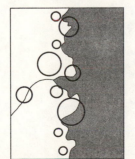

**24.** Outlines and areas

outlines, or without outlines at all (Fig. 24). The intermingling of linear features, point symbols and outlines leaves the map user the probem of clarifying area features and interpreting the 'figures' against the 'ground'. But if the areas are small and complex, then a fine outline will assist legibility.

### Use of white

Those maps which make extensive use of area colour can also treat the white or 'unprinted' surface as a specific part of the total composition. For example, in some topographic map designs which use surface colour extensively, the roads are emphasised by leaving them white against the adjacent coloured background. This is quite different to treating the white base as just so much empty space: it can be used positively. There is a basic distinction between a design hypothesis that begins with a line image, and then adds a limited amount of surface colour, and a design hypothesis that conceives the map as a complete colour composition, making deliberate use of white.

### Complementary symbolisation

If the foreground image consists mainly of point and line symbols, then contrast will be most easily obtained by making full use of area colour or fine patterns for the 'background' information (Fig. 25).

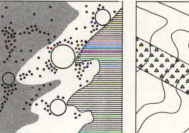

**25.** Complementary symbolisation

Conversely, if the dominant information consists of a range of area symbols, with variations in colour and pattern, then the other information can best be shown by line symbols. This can be very important in choosing the correct approach to a special-subject map, for the locational base information at the secondary level should not be in conflict with the principle information.

Confusion is bound to occur if both the foreground and background images are represented by area colours or patterns, or by lines and outlines.

For example, if there is a category of vegetation, such as forest, to be shown on a map, which also includes extensive area colour in the form of hypsometric or layer colours, then the forest should be in a pattern, so that the underlying hypsometric colour can still be perceived. Overlapping coloured areas will, of course, result in the combination of the two sets of colours.

## Degree of emphasis

The use of contrast in colour should be related to the visual dominance of important features. In most cases, the most prominent colours should be used sparingly. For example, if red is over-employed, its emphatic value is reduced. It is particularly important to ensure that elements that are small in extent – that is, which offer small visual targets to the eye – should appear in fully saturated hues with a high contrast. Red and purple are useful for emphasis, but will only be effective against relatively light or unsaturated backgrounds. The major contrasts visually should be reserved for differentiating major features or categories, not for minor differences between related features or sub-classes. A design that gives too many items a heavy emphasis defeats its own purpose. If everything is emphasised, then the map user is distracted, and the overall impression is one of confusion.

## Design testing

Throughout the evolution of the design of the map, there should be a careful consideration of perceptible differences, and the effect one symbol will have in association with others. Before the design is finally committed to a specification for production, it should be tested against all possible combinations and arrangements, that is in relation to the areas, configurations and juxtapositions which occur on the particular map. The inspection of a set of symbols viewed separately, as in a legend, is not an adequate way of testing how the symbols will appear on the map itself. Although it may be possible to perceive an area of light yellow in a legend 'box', it may be quite impossible to perceive it when reduced to a few square millimetres with lines of another colour superimposed on it. The imagination of the cartographer is most tested when trying to envisage the relationships of different symbol combinations and densities, which can only be judged effectively by relating the theoretical design to the actual arrangement of the information on the map.

# 4 GENERALISATION

As a map is always at a smaller scale than the phenomena it represents, the information it contains must be restricted by what can be presented graphically at map scale. This adjustment process is referred to as generalisation. It applies to all maps, and is fundamental to cartography. In addition to posing a major cartographic problem, it is also a powerful tool, because with generalisation it is possible to make small-scale maps of large areas, yet still selectively include the required information.

All maps are selective, and some highly selective, dealing with only one particular subject in detail. But even within the specified nominal content of a given map, some information will have to be omitted. In addition, complex irregular lines and outlines will have to be simplified. If many features of the same type are in close proximity they may be combined, which in turn can affect the basis of their classification. And because some map elements are judged to be more important than others, they may be visually emphasised, so that the symbols representing them are exaggerated in size, sometimes leading to the displacement of other symbols in the process. These processes – selective omission, simplification, combination, exaggeration and displacement – constitute the application of generalisation to the source information on which the map is based.

In many cases the generalisation processes operate in combination. In a small-scale plan of a built-up area, some individual buildings, usually small ones, may be omitted; the outlines of buildings are simplified; groups of adjacent buildings are combined; and the streets (spaces between buildings) which are retained may be exaggerated in width, leading to the displacement of the buildings fronting on to the street. As a consequence, instead of the concept 'built-up area' being interpreted by the map user from the array of tightly packed buildings, the generalised representation itself is now defined as 'built-up area', thus changing the basis of feature classification. Although details have been omitted, the basic character of the structure of the built-up area will be retained. This can only be done as a matter of judgement, because of the

understanding of what the concept 'built-up area' means.

In many cases, especially with small-scale maps, it is impossible to show areal features in accordance with their actual extent on the ground, because they are so small in relation to map scale that they could not be represented legibly. Abstract data, such as height figures, can be included in the map information, even though they have no tangible existence or real dimensions, and yet they must be represented by symbols which are large enough to be perceived. If the information is regarded as important, it is kept in the map by placing a symbol on the location concerned, thereby assigning it a position and describing it by classification. Therefore, its actual dimensions, if it is a real feature, are no longer the basis of either inclusion or representation. This is referred to by some writers as 'symbolisation', but the usage is incorrect, as of course all the information in a map is represented by symbols. The distinctions lie in what aspect of a feature serves as the basis of representation, not whether a mark on the map is or is not a symbol.

## FACTORS IN GENERALISATION

### Concepts of generalisation

Although generalisation is fundamental to map making, many different views about its scope and function have been expressed, probably because its ramifications are so complex in relation to both information and representation. There is no unanimity on the terms used to describe generalisation processes. A major difference of opinion centres on the question of whether or not generalisation also takes place at the data collection stage, thereby applying to all maps, or whether it is restricted to smaller scale derived maps. Some cartographers adopt the latter view, probably because the generalisation process is obvious with derived maps, and an inevitable cartographic task. The degree of generalisation which has already been applied to larger scale sources is not always evident. Correspondingly, many surveyors seem to believe that as long as linear features and outlines are plotted in their 'true' positions, no generalisation has taken place.

Although virtually all descriptions take into account selective omission and simplification, and recognise the need to maintain graphic legibility, two other factors are often referred to: character or characteristic, and relative importance. A definition given by Hettner (1910/1962) states that 'Cartographic generalisation is altogether something different to what a philosopher means by generalisation. Generalisation is first of all a question of restriction and selection of source material. This is achieved partly by the simplification of features on the map, partly by omitting small or less interesting objects.' This leaves unresolved the problem of deciding 'small' and 'less interesting'. Tobler (1964) indicates the need to '...capture the essential characteristics of a class of objects...' Both Lundqvist (1958) and Robinson (1952) also emphasise 'characteristic' as a fundamental concept in generalisation. The International Cartographic Association *Multilingual Dictionary* attempts a brief definition: 'Selection and simplified representation of detail appropriate to map scale and/or purpose', which again does not clarify 'appropriate' or 'representation'.

If these and other attempts to describe generalisation are analysed, the principal elements referred to seem to fall into two main groups: scale and graphic requirements (legibility), and characteristics and importance. The first pair are determining conditions; the difference in scale between the map and the real world phenomena, or the derived map and the source map, controls the available space, in which there are minimum requirements for graphic legibility. The effects of scale ratio, space and dimensions are in many respects measurable factors. The second pair are essentially judgements; they reflect the need to retain the essential characteristics (in terms of shape and configuration) of the phenomena represented, and also the fact that some things are judged to be more important than others, both within the same general class, and between classes. Therefore, some things will be retained, and exaggerated if necessary, and may be emphasised within the map design.

The exaggeration of symbol size is introduced partly to maintain legibility, and partly to accommodate relative importance. It also affects the

map design. A thick line representing a coastline inevitably leads to a greater degree of simplification than a very fine line. But if the coastline is represented in blue, it will need to be thicker than an equivalent black line in order to achieve the level of contrast commensurate with its importance in the total map content and design.

## Scale

It is normally true that the smaller the scale of the map, the greater will be the degree of generalisation. Certainly the consequences of generalisation are most evident at small scales. This tends to lead to rather simplistic views which confuse scale with derivation. There are large-scale maps (such as those at 1 : 10 000 scale) which are derived from even larger scale sources; and there are many basic maps at 1 : 50 000, 1 : 100 000 and even 1 : 250 000 scale. It is hardly reasonable to suppose that because a map at 1 : 25 000 scale has been produced from an 'accurate' survey it does not involve generalisation.

All surveys carried out either on the ground or by the interpretation and measurement of aerial photographs require rules for the omission of small features and areas, as well as rules for their classification. This applies to specialised maps, such as geological maps, as well as topographic maps. These rules generally apply to minor features, such as very small streams, areas of vegetation and land use, and small buildings. The detail recorded by the photogrammetric survey is controlled in the first place by the scale and, therefore, the resolution of the photography. The operator has to detect, identify, and then classify features through interpretation.

A comparative study of photogrammetric plotting by several different organisations, using the same photography and the same instructions, shows very clearly that differences in interpretation arise through different operators, and different degrees of generalisation result (Neumaier 1966). The instructions for this comparative exercise state that 'Only those features should be shown which are of importance for the 1 : 100 000 scale map. Features of minor importance should be omitted. In spite of such omissions, the characteristic features of the landscape have to be shown, especially within inhabited areas.' The priority of the relevant factors is summarised as importance, characteristics, and conditions of placing.

If displacement is necessary, the priority sequence is usually as follows: hydrographic lines, railways, main roads, minor roads, buildings, limits of vegetation. Therefore, if the edges of a road are displaced to accommodate a major road symbol, buildings fronting on to the road are displaced accordingly to keep their correct relative positions. Similarly, precepts are given for the treatment of isolated buildings and dispersed settlements.

Because photogrammetric operators can only view a limited area when actually plotting points and lines, it is difficult for them to monitor the relationships between elements over the whole area simultaneously. The possible alternatives are interpretation and generalisation during plotting; interpretation and generalisation before plotting; and generalisation after plotting. These alternatives highlight the fact that generalisation of features has to be adjusted both to their individual symbolisation and their relationships and balance over an area. In all cases, the information that results from the survey is a matter of judgement as well as measurement.

## Graphic requirements

It is clear that as scale is decreased, so is the total space available for the map symbols, but these cannot be reduced in proportion, as this would lead to illegibility. Legibility itself depends on symbol size, form and colour, which in turn affect contrast. It is also evident that the total amount of information concentrated within any area can affect the map user, especially if the representation approaches the threshold of discrimination. Cuenin (1972) gives a detailed analysis of the treatment of linear features on maps at a range of scales, comparing actual ground dimensions (width), the sizes of the corresponding symbols at a particular scale, the width which this would represent at ground level, and the corresponding coefficient of enlargement. What is interesting is that this proportional enlargement is not constant for a map at a given scale, but varies with the presumed importance and actual size of the feature. At a scale of 1 : 25 000 a main road is enlarged more than a canal, because the road requires a complex symbol to represent its characteristics and classification, and is also judged to be more important. But a footpath is enlarged by a greater ratio, because the minimum size of legible symbol required is large compared to the width of a footpath.

Apart from the fact that symbol dimensions are important graphic variables, and, therefore, are used to help to distinguish one symbol from another, the purpose of the map and the assumptions about the user's needs and interests also affect the graphic

exaggeration. Topographic maps, which place a great deal of emphasis on roads and their classification (in the belief that this information is valuable to the general user), are likely to exaggerate road symbols more than equivalent maps that are not designed under this assumption. Specialised road maps for route planning carry this symbol exaggeration even further. Exaggeration is particularly evident in many illustrative maps, in which the total content is first restricted and then major features emphasised. Therefore, graphic requirements are affected both by judgements of importance and the particular function and style of the map.

## Characteristics

The need to retain the characteristics of a feature is often referred to, yet the term is used in different ways. Because the primary information on a map is locational, the linear shape or ground plan of a feature is what the map shows most effectively, within the limits of scale. Roads may be straight or have sharp bends, railways have smooth curves. Coastlines and contours may be smooth or irregular. Some city areas are highly irregular in building shape and layout, others are symmetrical. But the 'shapes' of individual features in this sense are only one aspect of 'characteristic'. Some landscapes are characterised by the frequency of small features; some types of settlement are dispersed, others are nucleated. Thus the character of a geographical feature may reside in its distribution and frequency.

What is characterised can also have a wider connotation. A major road is described not only by its physical dimensions, but also by its hard surface, number of traffic lanes, and its position in a hierarchical classification of roads. Both its physical characteristics and some judgement of its importance will affect the symbol used in its representation, and therefore the degree of exaggeration and possible displacement.

The characteristics of a feature can also be regarded as important because they help to make possible the correct identification and recognition of different features, and reveal essential contrasts and distinctions in a landscape. Identification and recognition are processes of interpretation which require knowledge. So even at very small scales, where generalisation is at its maximum, the retention of characteristic elements demonstrates correctly the differences between features, and facilitates correct interpretation. The definition of what is 'characteristic' is a matter of human judgement or evaluation, helping the map user to place discrete and physically separate objects in the same class, and emphasising differences between them. Probably one of the most difficult cartographic tasks is to carry out selective omission when a region is characterised by numerous small features, either physical or cultural. Linear features can normally be retained by exaggeration if desired, and displacement if necessary. A highly irregular coastline should retain major indentations, and any angularity of form, at small scales, even though the minor irregularities have been removed by simplification.

## Importance

The degree of importance attached to a feature will affect both its inclusion and its graphic emphasis. In this respect, emphasis is a consequence of importance. Like characteristics, importance operates at two levels; both in the judgement of relative importance among features in the same class, and the judgement of relative importance between classes. Therefore, a single building may be unimportant in a built-up area, but very important in a sparsely populated one. A small village might be omitted in a densely populated region with many villages, but might be retained if it happened to be the terminus of a road in a rural area. Therefore, assessment of importance can affect selective omission and combination, and frequently leads to exaggeration and displacement.

As most maps are designed to have at least two visual levels, the degree of importance attached to different elements in the map content will lead to a greater graphic emphasis for some, often requiring larger symbols, and therefore more exaggeration. In addition, the basis for classification will change gradually with map scale. What is important on a large-scale map does not necessarily have the same degree of importance on a small-scale map. A large-scale plan of a built-up area does not represent roads by specific symbols, nor does it classify them. A medium-scale map covers a larger area, and is more concerned with routes between places, so a road classification is included, affecting the emphasis of road symbols. At a very small scale, only one class of road may be shown, and the emphasis in selection changes to a limited number of long distance or through routes. In this case, the basis for inclusion is not so much the existence of roads as features, but the function of connection between places shown on the map. In this way, scale, generalisation and map design are all inter-related.

## CARTOGRAPHIC SUBJECTIVITY

The practice of generalisation is often described as 'subjective' because individual cartographers use their knowledge and judgement to carry out processes of selective omission, simplification and so on. This is normally contrasted with the scientific principle of attempting to be 'objective', so that individual preferences or prejudices do not influence decisions or experimental results. The description of cartographic generalisation as being subjective is normally stated critically – the implication being that it is a defect that ought to be remedied.

A dictionary definition of subjective gives 'relating to self', and of objective as 'external to the mind'. Whether the external world can be known or understood except through the mind is a question that has perplexed philosophers for centuries. Even so, there is a fundamental contrast between the idea of subjective as being biased, uneven, irrational and incomplete, and therefore erroneous, and objective as systematic, factual, rational, 'scientific', and therefore correct. So far as maps are concerned, it is clear that the locational basis of the map, resulting from measurement and represented in a defined coordinate system, is objective, whereas the description and classification of features on the map is contrived for humanly conceived purposes, and is essentially subjective. Whereas the scale, and to some extent the graphic requirements of legibility are factors that can be treated objectively – that is, they can be described in finite terms – the factors of importance and characteristic are essentially judgements based on interpretation. Once a generalisation has taken place, it can only be described as either good or bad, not right or wrong, because the informational changes introduced have many possible alternatives, and there is no way of defining any absolute solution.

The photographic reduction of a line is objective, in the sense that it operates according to optical laws and is consistent; there is no human interference. But generalisation involves the deliberate changing of the representation to achieve certain ends. There is no absolute standard for generalisation, and indeed whether it is good or bad is also a human judgement. It is at least arguable that there is no such thing as a completely objective generalisation. In practice, the general intent of the objective argument is to make the generalisation process more consistent and systematic, thereby reducing the inconsistency that is inevitable with human operations. The discussion of objective generalisation has been promoted through the development of digital cartography, because for the information to be processed by a computer, specific instructions have to be given. Whereas programs for linear simplification have been devised, the operations involving selective omission, combination, exaggeration and displacement are more difficult to define, and therefore more intractable.

Proper generalisation depends on information and understanding. For topographic features – or at least for those which are a familiar part of the environment – it is generally accepted that their characteristics are sufficiently well known, and therefore do not require special study. Even so, a study of geology, physiography and geomorphology is often included in a cartographic curriculum. A knowledge of vegetation and land use, and of urban and rural geography, is also relevant. It is usually true that it is in the generalisation of less familiar environments that the process is most prone to error, essentially as a consequence of ignorance. To some extent this can be rectified by consulting larger scale topographic maps, but even this requires judgement in interpretation. In other cases, good larger scale source material is simply not available. The official topographic map series of a country, if available, are generally regarded as being the best source for topographic information, on the assumption that the survey organisation will be in the best position to make generalisation judgements about features within its own territory.

For specialised maps, the problems in some ways are more easily defined. Whereas most people are familiar with the features represented on topographic maps, at any scale, the classifications used on a geological map, or a nautical chart, are clearly within the province of specialised knowledge. Therefore, it is essential that any subsequent generalisation of such information should be directed by a specialist who understands both the meaning and significance of the information in detail.

## GENERALISATION PROCESSES

### Selective omission

The total number of features in a class, represented by a particular symbol, may or may not be represented in a map. In the case of a small-scale

road map, all the major roads will be shown, but many minor roads in built-up areas will have to be omitted. Small topographic features, or small areas in a special-subject map, may be omitted on the grounds that they are unimportant at map scale. The difficulty for the map user – and this is particularly true for medium-scale topographic maps – is that some features are shown while others are not, even though there is a symbol to represent them. An obvious example is the network of drainage channels present in a hilly region, in areas with an abundant rainfall and extensive surface drainage (Fig. 26). On medium-scale maps the inclusion of all of them would lead to a mass of short lines which would provide little useful information, and might interfere with the legibility of other detail. So some are omitted, but some are kept to indicate the general characteristics and distribution of the drainage.

**26.** Drainage at 1 : 50 000 scale reduced to 1 : 250 000

In normal practice the process operates by first selecting the major drainage lines and tributaries, so that the shape and extent of each drainage basin is correctly indicated (Fig. 27). Smaller streams are added in proportion to their frequency in different areas. At the initial stage, length of stream may be the first factor to consider, the larger ones taking

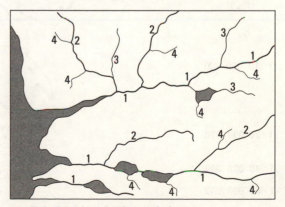

**27.** Priorities in selective omission

priority in the selection. But in the final adjustment, short streams which form part of the interconnections in a drainage system are also included, in order to complete the drainage network. Thus the judgement of relative importance is not simply a matter of size, but of local significance. The major problem with selective

omission is that it can be applied consistently over large areas, but varies locally. In the same way, isolated small buildings are retained as landmarks, even though buildings of the same type and size would be omitted in built-up areas.

The progressive effect of selective omission of settlements at small scales is illustrated by Fig. 28.

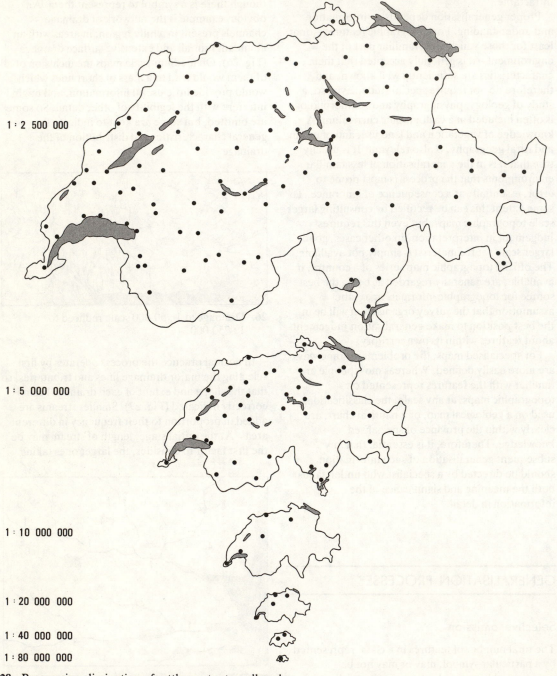

1 : 2 500 000

1 : 5 000 000

1 : 10 000 000

1 : 20 000 000

1 : 40 000 000

1 : 80 000 000

**28.** Progressive elimination of settlements at small scales

Not surprisingly, attempts have been made to make the process of selective omission more consistent, or more rational. These can be said to fall into two main groups: specific 'rules' that are said to govern the process by a quantitative decrease in information with scale; and methods that use extra information to judge relative significance.

## Rules for selective omission

It is clear that if a map is being compiled from a larger scale source map, then normally only a proportion of the features shown by individual symbols for one category will be retained on the derived map. The problem is to determine the ratio between the number of objects on the source map and the number retained on the derived map, and to take into account the factors which influence this. The relationship is clearly based on the scales of the two maps, and therefore their respective areas.

Cuenin (1972) points out that this process is considerably affected by scale. Using the terminology that

$N_i$ = number of objects on the source map, scale $1 : M_i$

$N_f$ = number of objects on the derived map, scale $1 : M_f$

then if there is a very small scale difference between the two, all the objects may be retained, so that $N_f = N_i$. This is more likely to occur at large scales, for example if a map at 1 : 5000 is derived from a source map at 1 : 2500.

If the same density of objects is to be retained on the derived map, then this is obviously a function of the relative areas of the two maps. This can be expressed by

$$N_f = N_i \left( \frac{M_i}{M_f} \right)^2$$

in which the number of objects retained is inversely proportional to the square of the scale relationship.

Cuenin (1972) then goes on to show that if the scales of the derived map and the source map are in the ratio of 1 : 2, then the proportion of objects retained on the derived map can be controlled by adjusting the equation.

If 70% are retained, then $N_f = N_i \sqrt{\dfrac{M_1}{M_f}}$

If 50% are retained, then

$$N_f = N_i \sqrt{\left( \frac{M_i}{M_f} \right)^2} = N_i \left( \frac{M_i}{M_f} \right)$$

And if 25% are retained, then

$$N_f = N_i \left( \frac{M_i}{M_f} \right)^2$$

With greater scale differences, then

$N_f = N_i \sqrt{\dfrac{M_i}{M_f}}$ is more common.

At small scales, usually less then 1 : 1 million, the relationship is complicated by the fact that the symbols used to represent the objects will already have been exaggerated, and that the symbols included in the derived map may also differ in size from those on the source map. To account for this, Töpfer and Pillewizer (translated with a critical commentary by Maling 1966), introduced two additional factors: a constant for symbolic exaggeration ($C_b$), and a constant of symbolic form ($C_z$).

These 'rules' are a rationalisation based on the examination of existing maps. Although they could be used to check a process of selective omission, to see if it was in accordance with what should be expected with a given map type and scale reduction, of course they do not explain which of the objects on the source map should be retained.

The operation of the rules has been demonstrated with both point and linear symbols. In both cases, poor selection may still lead to a poor generalisation, even if the quantitative rule has been followed.

## Factors in relative importance

The second approach is closer to the process of selective omission in the operational sense. Taking the example of the selection of named settlements on a small-scale map, it is clear that several factors are taken into account in deciding which to include and which to omit in a derived map. The selective omission is based on the judgement of relative importance. Major towns do not pose any problem, as they have a high priority in the total map content. But in a densely populated region there are usually many small towns and villages of much the same apparent size. Indeed, if a small-scale map is the source, they are all likely to be included in one classification, and represented by one symbol. Some of these may be selected by topographic factors that can be interpreted directly from the other map information: for example, position at a bridging point on a river; at the head of navigation; at the junction of several important routes; at the terminus of a route.

Other factors could be introduced on the basis of

external information, such as political, cultural or economic significance (Stenhouse 1979). This information may be available to the cartographer through his own knowledge and understanding, but this is unlikely to be consistent over large areas. But this type of information could be systematically compiled from other reference sources, and then used to provide a basis for selection. Given sufficient information, it could also be specifically directed in relation to the type of map. For example, in a tourist map, cultural, recreational and transport and communication information would have a greater effect on selection, and could be given a higher priority. If all the places were listed from a large-scale source in a small data base, and these factors assessed and then given a numerical rating, selections of places could be extracted according to defined factors, which in turn could be used as a basis for inclusion. It is the sort of task which could be dealt with relatively easily with a computer, as the total amount of data could be quite small, and the processing simple. For such a tourist map, all the towns which had any rating on the tourist scale of cultural interest would be automatically selected, and given priority over similar towns which did not have such a rating.

This type of procedure underlines one important fact in the cartographic treatment of information. In many cases the generalisation could be improved by additional input, by simply expending more time and effort on obtaining information, especially from non-map sources. The more rational treatment of some generalisation processes is fundamentally dependent on this.

## Simplification

Simplification applies both to linear features and the outlines of areas. The greater the sinuosity of the line, the greater is the effect of simplification. A straight line reduced in scale will still be a straight line, although shorter. A highly irregular line will suffer a progressive diminution in length as its minor irregularities are removed. Therefore, during this reduction it is very important that the characteristics of the feature concerned are retained. What has to be avoided is the replacement of irregular lines by smooth curves, and to avoid this exaggeration may have to be introduced in some cases.

If a section of irregular line is considered (Fig. 29), it can be described as having three main elements: its general direction, major forms and minor forms. The major forms are marked by a

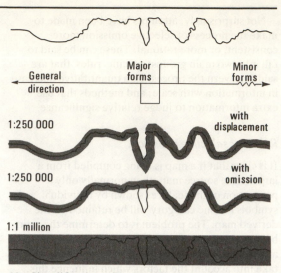

**29.** Simplification of an irregular line

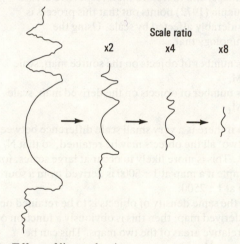

**30.** Effect of line reduction

sharp change in angle, and sufficient length of segment to make this noticeable. Simplification will lead eventually to the removal of the minor forms entirely, but the major forms should be retained as far as possible. Eventually, at a very small scale, only the major direction will be retained. Generalisation, therefore, progressively reduces irregular lines to regular ones, consisting of sections of straight lines or arcs (Fig. 30).

Where lines show complex features, such as coastlines, the retention of both major and minor forms also depends on the separation or spacing between adjacent line sections (Fig. 29). At some point a decision has to be made as to whether a particular space should be retained by exaggeration, or omitted. In order to maintain the major forms, it may be necessary to displace the lines slightly, so retaining the feature.

In practice, the omission of minor forms tends to happen automatically as the corresponding increase in line width with smaller scales reaches the point where the irregularities are smaller than the line dimension. If the major forms are angular, and marked by abrupt changes in direction, these should be kept, as otherwise the line character is lost.

### The analysis of simplification

During the simplification process, a cartographer can scan a section of line, and can change from concentrating on minor details to major forms at will, and can perceive a whole section of line simultaneously. Therefore, the operation of simplification can respond both to the detail of the line section and the overall line direction. In generalisation using digital data, the computer can only 'scan' the line data serially. It can be programmed to inspect a section of line several points ahead, and then ignore any changes in direction which are less than a certain angle and length. But it is difficult to make fixed rules about the extent of detail reduction. The cartographer can adjust the simplification to take into account the local characteristics, and consider the relationship of the line with other lines in the composition.

Line simplification using digital data and digital processing has been examined at length in many experimental studies, and several algorithms exist to perform line simplification (see Muller 1987). It is affected both by the nature of the original digitising and its subsequent processing. Small-scale line simplification is well illustrated by Pannekoek (1962), whereas McMaster (1986) and White (1985) deal specifically with data processing.

## Combination, exaggeration and displacement

Where many features of the same type are in close proximity, selective omission and simplification are often accompanied by combination, exaggeration and displacement. Whereas a large-scale map will show all areas of woodland above a certain minimum size, a small-scale derived map will omit small isolated areas, and group together adjacent areas into a continuous whole. So the small spaces are omitted and the general outline simplified (Fig. 31).

The effect is particularly noticeable in the

**31.** Combination

generalisation of built-up areas. Figure 32 shows how the building outlines on a large-scale plan would become quite illegible with a scale reduction of 25 times, and demonstrates the degree of generalisation needed to maintain a legible image. As the scale is reduced, the buildings are progressively grouped into blocks, major roads are exaggerated and minor ones omitted, until eventually all that is retained is a major road surrounded by continuous buildings.

Compared with line simplification, the generalisation of built-up areas is complex. This is revealed by the work of Christ (1976), in which the built-up areas on a topographic map at 1 : 50 000 scale were digitally generalised for a derived map at 1 : 200 000 scale.

The combination of features of the same type can only take place on the basis that their extension over the intervening 'spaces' is theoretically possible. Therefore, this type of combination cannot be applied to islands, which would convert water into land.

Linear features which are close together will still need to be represented by legible symbols at small scale, and a point can be reached when it is impossible to place the linear symbols side by side without displacing them to intrude into a larger area. If a river, canal, railway and road lie in close proximity in a narrow valley, then the symbols necessary to represent them legibly at a smaller scale may occupy more level space than actually exists on the derived map. In this case the contours which delimit the plain may also have to be adjusted, so that the linear features are not moved on to a slope.

At large and medium scale, where planimetric position of measured points (such as triangulation points) is important, linear features may have to be displaced in order to keep the measured points in their correct positions (see Christ 1978).

Displacement also occurs when small differences in position are important to the map user. A typical example is offset road junctions (Fig. 33). The relative positions of the minor roads may have to be exaggerated and displaced at small scale, in order to make this clear.

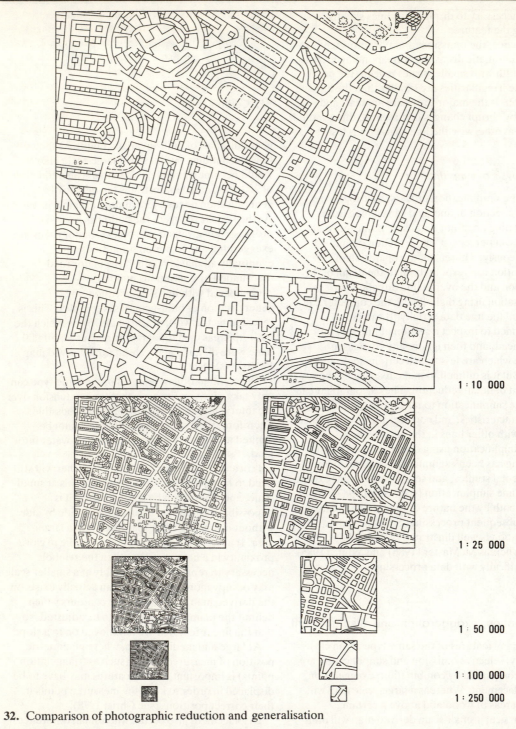

1 : 10 000

1 : 25 000

1 : 50 000

1 : 100 000

1 : 250 000

**32.** Comparison of photographic reduction and generalisation

**33.** Exaggeration and displacement of linear features

1:50 000

1:250 000
enlarged to
1:50 000

# THE PRACTICE OF GENERALISATION

The source material used for a derived map is normally at a larger scale, and sometimes a very much larger scale, so that the generalisation processes involved during compilation of the new map must anticipate the effect of reduction. The task can be approached in two ways. Generalisation can be carried out at the scale of the source map, and then reduced to the desired scale, or the source material can be reduced to the smaller scale and the generalisation processes carried out subsequently.

There are advantages and disadvantages with both methods. Working at the source scale gives a clearer picture of all the detail, and if the map is complex this can be important. If only a limited amount of information is required from a particular source map, then this can be extracted and the unwanted information omitted. The cartographer must anticipate the effect of scale reduction. Working at the reduced scale helps the cartographer to produce a generalisation that is appropriate to the derived map, but a photographic reduction of the source map may be difficult to interpret correctly, increasing the risk of introduced errors.

With the first method, it is essential that the line widths on the compilation for the different linear features and outlines are enlarged in proportion to the scale difference. If the source map is at 1 : 50 000 scale and the derived map at 1 : 250 000 scale (a scale difference of five times), then the lines used on the compilation should be five times as wide as the specified symbols. This will ensure that the minimum degree of simplification will be achieved, and will make clear where any displacement is needed to retain feature characteristics.

Because on a very small-scale map there is often little space to include important features in highly developed areas, it can be difficult to resist adding more information in areas where space permits. This will lead to an unbalanced generalisation, by applying different criteria in different parts of the map.

## Sequence of operations

As the generalisation of one element will affect the generalisation of others, it is necessary to follow the correct sequence in procedure. The basic topographic information is dealt with first. As the locations of many cultural features are relative to physical features, the generalisation of the physical features is the first task. The usual order is hydrographic features (shorelines and drainage), followed by contours and heights. Modifications to shorelines and drainage channels have to be followed where necessary by other terrain features. Locations of places on, or near to, water, will then be relatively correct on the derived map. Changes to the relief information will in turn affect those cultural features that are defined or influenced by the topography. For example, a boundary may be defined partly by following a river, and partly by following a watershed. Its position therefore moves in accordance with the generalisation of those features.

On small-scale maps, where many settlements may be indicated by point symbols, these must be placed in correct relationship to the terrain. Features that are connected to inhabited places, such as roads, tracks and paths, should be dealt with either after or in combination with the generalisation of settlements. Areas of vegetation and land use, which in developed regions are usually delimited by fences, roads, and other topographic features, are usually dealt with last, as their outlines depend on the positioning of both physical and cultural features. At large and medium scales, many minor internal boundaries are positioned in relation to surface features, including roads, streams and fences, and can only be compiled when the rest of the information is complete.

Features represented by point symbols, such as individual buildings, have to be placed in correct relationship to the topography. A water mill must remain adjacent to the stream, and therefore must be positioned in accordance with the course of the stream as it appears on the derived map. If a building lies between a river bank and a road, it must keep this position. Houses fronting on to a road must remain aligned on the road edge, displacing them if necessary to accommodate the larger road symbol.

## Consistency

Consistency in treatment is most difficult to achieve if a map is derived from several different source maps, especially if these themselves are at different scales. The generalisation of common features is likely to be different in detail on the different source maps, particularly with regard to selective omission. If a road taken from one map does not fit the topography as shown on another map, then the road alignment must be brought into agreement with the topography. Highly detailed boundary lines should not be added to a map where the treatment of the relief is very simplified.

With specialised maps, a major difficulty can be the different degree of generalisation between base map information and special information. Simple choropleth maps, which make very generalised statements, should not be shown with detailed topographic bases. A detailed topographic base, including outlines of major terrain and boundary features, may suggest a degree of specific detail quite out of keeping with the nature of the subject matter. Thus consistency requires above all an understanding of the nature and quality of the source material, and the characteristics of the features or phenomena being represented.

Successive generalisations can lead to a gradual reduction in the overall quality of the information, and the validity of its correspondence with the features and phenomena represented. Very small-scale maps are composed by the generalisation of other small-scale maps, and this re-generalisation process tends to repeat or even exaggerate errors once they have been introduced.

Figure 34 shows two examples of different generalisations of the same area, taken from four different maps. The variations are almost certainly due to a combination of different source material and successive generalisations by different cartographers. It is also clear that the cartographers concerned had different views about the level of detail proper to a given scale of representation. They demonstrate that, whenever possible, an effort should be made to return to much larger scale source material, especially the small-scale topographic series produced by national surveys, so that the fundamental topographic configuration – coastlines, major rivers, islands – is rectified where necessary.

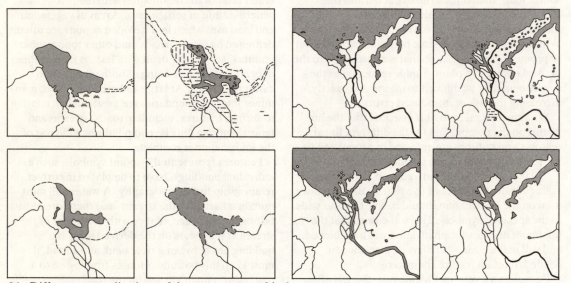

**34.** Different generalisations of the same geographical area

# 5 NAMES AND LETTERING

The giving of names is an important aspect of every human culture, shown by the rituals attached to naming, and the resistance to name changes. Names of places, or of other phenomena related to places on the Earth's surface, are an important component of many maps, and indeed the finding of a named place is a very common form of simple map use. Place names, that is those which refer to some area or feature of the Earth's surface, are the major group within geographical names, which include everything to which a geographical location can be given, such as the extent of a former empire or the territory inhabited by a tribe. Such names act as a geographical reference system, but are entirely different to map symbols. Whereas symbols in general describe only classes of features, names identify particular or individual features and places. Although coordinate systems provide a systematic method of defining location, names are individual, inconsistent, repetitious, unevenly distributed, and appear in many different languages and forms of script. Although they are abstract, they take up space on the map, and as a reference system they are highly inefficient. Despite all these deficiencies, they are constantly used to identify and refer to places and geographical locations, and indeed they are an essential part of the language and culture of any people.

## Types of place name

Toponyms are those names that exist in a particular language and refer to places and features within a country, or the linguistic area of a people. Exonyms are those names that are used in any language to refer to other places outside this territory in other parts of the world (see Breu 1986). In both cases, the name may consist of a single proper noun, or of a generic term accompanied by an adjective, noun, or phrase. Many proper nouns (such as Chippenham or Dublin) currently used in modern English are themselves derived from descriptive names in other languages, but have become modified over time. Names with generic terms, such as Sound of Mull,

incorporate a proper noun, but many others consist of a term together with an adjective or an adjectival phrase, such as Long Mountain, The Black Hills, The North Sea and Running Deer Creek.

## Sources of name confusion

The giving of names has generally been unplanned and unsystematic. Local names for small topographic features and settlements are introduced over time by local inhabitants or migrant peoples. Gradually names are added and others go out of use. Much name giving has been descriptive, and has led to endless repetition of common descriptive terms, such as Sierra Nevada, Burnside and Newtown. This has been extended by the transplanting of names by migrants, naming new places with those familiar from their former territory or place of origin, such as Dallas, Boston and Perth.

Explorers and travellers who were unfamiliar with the languages of the territories they entered often attempted to produce versions of what they thought to be local names in the vernacular. In addition, they introduced commemorative names, honouring their patrons or contemporary rulers. Consequently in the English language there are numerous names in different continents all including 'Victoria' in some way. Established names in other languages have often been corrupted in the process of using them incorrectly, so that either the pronunciation or the spelling has been changed. In the English language, some European names have been changed in pronunciation but not spelling (Paris); changed in spelling but not pronunciation (Rhine); or replaced by similar, but different, words (such as Warsaw, Lisbon, Rome, etc.).

In many parts of the world, the movement of peoples and their languages means that more than one name exists for the same place or feature. The combination of German and Polish names in central Europe, or of English and French names in parts of Canada, shows that different names either have existed, or do exist for the same place. In some countries, names are changed officially for political reasons. Therefore, establishing the 'correct' name for a feature or area is often far from simple, especially as names, once introduced, are likely to continue in use and in written records for a long time.

## Names and geographical areas

The difference between names that apply to a certain part of the Earth's surface, and those that refer to a territory with a given political system and administration, is also a source of confusion. The name 'Russia' can refer to the lands occupied or controlled by the Russian people, but the USSR includes not only 'Russia' but many other peoples as well. Soviet Union, like United Kingdom, when used by itself, describes a political–historical system, not a physical area. Great Britain refers correctly to one of the two main islands of the British Isles. Names of states or provinces used by other countries to refer to larger units introduce a further complication. In this respect, England is not a synonym for Great Britain, nor Holland for the Netherlands, although such expressions are commonly used. If the constitutional names of sovereign states are to be shown on a map, they should be treated consistently.

## Exonyms and conventional names

Most languages have names for other places and features in other parts of the world (see Ormeling 1980). The names of continents, oceans and seas, major physical features, and even regions, may be different in different languages, but such names are a proper part of the vocabulary of the language concerned. Thus English-language maps refer to the Baltic Sea, but the Swedes call it *östersjön* (Eastern Sea). In addition, many languages, and especially English, use names for places and features in other countries which differ from the correct local names. Sometimes this is due to the difficulty of pronouncing or transcribing the language; sometimes because of the survival of former names; and in many cases because the names are corruptions of place names taken from other languages. Opinions differ as to whether such names should be used on maps, or whether they should be firmly discarded in favour of 'correct' local names. Some argue that there is no reason why any place should not be referred to by an English-language name, if that is more convenient, and of course many names of this kind are perpetuated in literature, historical records and other documents.

Because of the wide spread of English-speaking peoples, some names that are conventional names in English have been widely adopted. Shanghai and Singapore may not be recognisable to Chinese or Malay peoples, but they are extensively used internationally. However, all maps in the English language should attempt to distinguish between proper names and English conventional names, and not leave the user under the impression that the conventional English name is the correct one.

## Foreign place names

Maps that include areas lying beyond the limits of the country of origin of the map must deal with foreign place names. If the other languages concerned use the same alphabet or form of writing, they can be written in a familiar script, even if they could not be pronounced without a knowledge of the language concerned. Names in languages using other scripts must be converted in some way to a form that can be identified and referred to by the map user. Various types of conversion can be used for different purposes. Translation is the process of giving the equivalent meaning. As many place names are meaningless, and as the change would replace the name by something quite different, translation is not a useful process in map making. Transcription is often used to describe the phonetic rendering of a foreign name. In some cases, for example for place names that have never been recorded before in writing, a transcription process may be necessary. Transliteration has a narrower focus, and can only be used correctly to refer to the conversion of the characters in one script (either letters or signs) to an equivalent set of letters or characters in another script; strictly this is only possible with languages that use an alphabetic or syllabic script.

## Systems of speech and writing

Linguistic sounds and forms of writing cover an enormous range, and any one language encompasses only a limited number of them. Whereas in a language like English, sounds are made mainly by expelling breath, using variations in duration and emphasis, other languages include implosive sounds, tone, pitch and stress. Most people are limited to the languages they know, and find it difficult to make sounds that may be common in other languages. In addition, methods of recording speech have evolved and developed in quite different ways. The main groups that can be distinguished are alphabetic scripts, syllabic scripts and ideographic scripts.

## Alphabetic scripts

In theory each sound should be represented by a different letter. In fact many sounds are duplicated by different letter combinations, and many letters change sound according to position and combination. In Arabic, for example, the forms of letters depend on whether they stand alone, at the beginning, within, or at the end of a word. In English use is made of both large (upper case or capital) letters and small (lower case) letters, and both are used in print. Russian Cyrillic has both, but only uses capitals in printed text, although lower case letters are sometimes used on maps.

Alphabetic languages can be divided into a number of groups. In the Roman alphabet all the letters are derived from classical Roman forms. Although the letters are constant, the sounds represented by them are not. Most Roman alphabets apart from English make use of diacritical marks and accents. From a cartographic point of view, these have to be carefully maintained in English-language maps for correct usage, and many cartographers are surprised by the large number of diacritical marks that exist. The principal non-Roman alphabets are Greek, Cyrillic, Arabic, Mongolian and Thai. Of these Cyrillic and Arabic are of major importance with regard to place names. Some languages which are basically Roman retain non-standard letters derived from earlier scripts, and Coptic and Gothic are mixed alphabets derived mainly from Greek.

Even with scripts with apparent similarities the alphabetic order is not the same. The difference is obvious with Greek and Cyrillic. Other scripts have a different reading direction, and Arabic, Hebrew and Urdu read from right to left.

## Syllabic scripts

True syllabic scripts, in which syllables are represented by signs, are rare (Amharic being the only major example), but combined syllabic and ideographic scripts exist. Japanese is a mixture of Chinese characters with the native syllabic language called kana. The combination is kanamajiri, and the characters are modified from Chinese. Other scripts combine letters and syllables, including Sanskrit, the Pali of Sri Lanka and Tamil in southern India. Hindi and Korean are the most important scripts of this type.

## Ideographic scripts

In these languages entire words or elements are given different 'characters', usually developed from a stylised pictorial or symbolic representation. Chinese is the main example. The total range of sounds is small, and therefore changes in tone and stress are very important. This is clearly demonstrated by the fact that although the written language is intelligible to all Chinese, regional speech forms are often unintelligible in other parts of China. The characters are written in columns downwards, from right to left.

## Transcription and transliteration

With even this brief review of some of the main forms of script that exist in the world's languages, it is clear that the representation of names on maps faces major problems. Although maps produced in other countries may be the basic source of cartographic information, some procedure must be adopted for converting the names into equivalent forms that can be understood and referred to by other map users.

It must be established immediately that names on such maps cannot operate as a guide to correct pronunciation, but are constructed solely to provide recognisable entities that can be used in the language of the map. Although there is no perfect system that can be univerally applied, consistent policies of name treatment are necessary on any map that includes foreign place names. As few alphabets have even the same number of letters, or incorporate the same range of sounds, it is not even possible to make an exact literal equivalence. For example, Russian Cyrillic has 32 letters compared with the 26 in English. Some of these can be properly represented by English letters, but others can only be approximated. The name of the city XAPbKOB is usually transliterated from Russian Cyrillic into English as Kharkov (or more correctly Khar'kov), whereas in German it is Charkow. Although the exonym Moscow is well established in the English language, MOCKBA would be correctly transliterated as Moskva.

The problem is complicated by the fact that many other organisations, national and international, also need to deal with place names in other countries, and have evolved their own methods for this purpose. Some of these are concerned with the spoken form, and therefore phonetic transcription; others need to work with many different languages simultaneously; and some have even developed their own versions of their own script for foreign-language purposes. For example, organisations such as the International Phonetic Association and the International Federation of Orientalists are clearly concerned primarily with spoken languages. The International Postal Union has to deal with virtually all languages simultaneously. The Russians and the Chinese have developed their own English-language versions of Russian Cyrillic and Chinese respectively, for international communication purposes. Modern Turkish was deliberately 'romanised', but uses several unusual diacritical marks, such as a capital I with a dot above it. Unfortunately these developments often do not coincide with the requirements that would be stated by cartographic linguistic experts who attempt to deal with names on maps. One consequence has been the replacement of 'familiar' English-language versions of Chinese place names (themselves in many cases conventional names based on transcription), by new versions that are supposed to approach more clearly the spoken Chinese forms. Yet the inability of foreigners who are not expert linguists to pronounce Chinese clearly makes this most unlikely, however helpful it may be to the Chinese.

The impetus for a consistent policy for the treatment of foreign place names has come mainly from government departments concerned with maps and charts, and international organisations with similar interests. The International Geographical Union, the International Geodetic and Geophysical Union, the International Hydrographic Bureau, and the World Meteorological Organisation, are all concerned with place names. Problems in dealing with place names, and foreign names, have been discussed in a series of international conferences organised through the United Nations.

## Standardisation of geographical names

Most agencies concerned with maps and charts of international extent encounter the problems of dealing with foreign names on maps, in addition to any problems that exist with regard to establishing correct place names in their own countries. This question has been discussed at several United Nations conferences since 1953, in which cartographic and linguistic experts attempted to put forward positive recommendations for the treatment of names on maps (Breu 1982). The first conference initially identified two main aspects of the problem: standardisation in individual countries, and transliterating or transcribing names from other languages.

Among the first recommendations made, the principal one was that countries should be encouraged to develop policies to deal with internal name problems, as without uniformity or consistency in source material it would be impossible to make progress with any proposals for international conformity or standardisation. Although complete standardisation could hardly be achieved bearing in mind the diversity of different languages worldwide, it was accepted that some progress could be made towards consistent and uniform treatment of names on maps (see Blok 1986). As linguistic experts are few in number, one task would be to develop guiding principles for domestic name standardisation and general policies for the treatment of names from other countries.

## Domestic standardisation

Correct names must be determined. This may involve deciding priorities between local usage and historical or written evidence, avoiding duplication, and deliberately replacing ambiguous or repetitive names. If necessary an authority should be created with the power to make decisions about all names used on official documents, including maps. Where more than one language is in use, a policy of priority, either nationally or locally, must be agreed.

With names of topographic features, the extent of the feature referred to by the name must be decided. Different sections of a river may have different names; some features are localities within larger features; some large features do not have agreed specific names in local usage. There must be a policy regarding the introduction of new names. Slightly different versions of a name may be due either to grammatical variations, or two different origins. Whether to use a systematic spelling of an existing name rather than that in use must be determined.

## International standardisation

Although it is clearly impossible for all countries to use exactly the same linguistic representations of place names, greater uniformity and consistency could be assisted if at least those countries with a common script used the same transliteration methods on their maps, and possibly accepted another widely used language at least for reference and explanatory information. For example, many European countries and the United States have developed transliteration systems for rendering place names from Russian Cyrillic into the Roman alphabet. A reduction in the great number of transliterated forms would help to reduce confusion. Where possible, therefore, international agreement should support policies of coordination and greater uniformity that have reasonable objectives.

The idea of an international language has often been discussed for secondary use on maps, but unfortunately such languages, even when devised, are not living tongues in the real sense, and do not contain any solutions to the problems of linguistic and topographic diversity. In practice, because of the widespread use of the English language for many purposes (for example for international air travel, marine navigation, etc.) many international series now accompany map sheets with English sub-titles and legends, and even some national map series have additional legends to cater for the needs of neighbouring countries.

Some countries have indirectly helped in this quest for greater uniformity by providing their own transliteration into another major language. For example, both the USSR and China have official Roman alphabet versions of their languages. Although in some ways these have created problems for map makers (because of the large number of changes they have introduced), in the long run this should lead to more uniformity in Roman alphabet versions of these languages.

## Policies for place-name treatment

In many countries the official mapping and charting agencies deal with areas of international extent, and therefore need some officially agreed policy for uniform place-name treatment. Such requirements have led to the formulation of rules or principles for the treatment of names on maps, especially where foreign place names are involved. In Britain the Permanent Committee on Geographical Names for British Official Use (PCGN) has representatives from all major government departments which have place-name interests, and its principal function is to advise those departments on geographical nomenclature. However, its recommendations and principles are frequently followed by other British map-making organisations, especially in dealing with foreign place names. It should be noted that the PCGN, like the United States Board of Geographic Names, only makes recommendations; it is not an authority like the Canadian Board of Geographic Names which actually decides and controls the treatment of names on official Canadian publications, including the national map series.

Most map publishers using the English language now accept the general principles advocated by the PCGN (Principles of Geographical Nomenclature), which include the following points:

1. English conventional names should be used for geographical features and regions, including water areas, of international extent, such as the Mediterranean, the Andes, and the names of independent states.
2. The names of administrative divisions of independent states, or any places lying within them, should be those used officially by the country concerned, for example *Roma, Schwarzwald*.
   If a conventional English name exists for the place or feature, it can be added in parentheses following the correct name, e.g. Livorno (Leghorn).
3. If a name within a foreign country contains a descriptive geographical term, this should not be translated into English, as this would cause a mixture of two languages. So *Cap Finisterre* is prefered to Cape *Finisterre*.
4. If a country uses any variety of the Roman alphabet, then place names within the country should be rendered exactly as they occur in that script, including all accented letters and diacritical marks.
5. In countries which do not use a Roman alphabet, an official national transliteration into Roman should be used if it exists, but if not a system recommended by the Committee should be used.

The acceptance of such principles by map-making organisations in general can reduce irrational and inconsistent treatment of place names, avoiding the introduction of alternative and erroneous versions, and gradually acquainting the map-using public with recognisable and standardised forms.

# LETTERING

As names are an important element on many maps, the design and use of lettering is equally important in map design. Lettering applied to a map involves two operations: first the design specification, which controls the appearance of each name, and second the selection and arrangement of the names on the map, which is part of the compilation process.

The appearance of the lettering is closely connected with the way in which it is constructed, and in this respect there have been several distinct periods in cartography, dominated by particular methods. During the long period when engraving was the principal means of map construction, the lettering was engraved by hand along with the rest of the map detail. Very fine lines can be produced by engraving, but the building up of larger and heavier letters was a relatively slow procedure. Styles tended to be limited to those that could be most rapidly produced by the engraving method.

With the advent of photography and lithography, the map image could be transferred to the printing plate, and in this phase drawing in ink gradually replaced engraving as the principal means of image construction. Very small lettering and fine lines were much more difficult, and as drawing on opaque base materials for subsequent photography on to glass or film became the standard procedure, the practice of drawing the original images at a larger scale also became established. This facilitated the drawing of lettering also, giving the cartographer more space and larger forms. Even so, hand lettering, in which each individual letter had to be separately drawn, was always a difficult and slow process, requiring great skill to reach standards of consistency and uniformity. The range of lettering styles tended to be limited because of the difficulty of controlling minute shapes in pen-and-ink drawing.

## Printed type

The forms of the letters which make up the alphabet have gradually become standardised through printing, and the printed image became the most important medium in communication. It was inevitable that at some stage hand lettering on maps would be replaced by the use of printed type. This has great advantages in consistency and uniformity of letter forms, arrangement and spacing, and can be produced rapidly from master images. It is the printed letter form which children learn to read, and these forms become imprinted at an early age. Consequently, it is not possible to depart from them significantly without a reduction in legibility, or even a complete loss of information. The best use of type therefore has become an element in cartographic design, and in modern practice typographic forms of lettering are universally used on printed maps.

Each alphabet, with its complement of figures and punctuation marks, has to be designed initially, and such a design is called a type face. These particular designs are essentially variations in form, and they can be classified into major groups, depending on how they make use of particular design features. In addition, to be used in print, each type face is made available in a range of dimensions, so that its size, weight and width can be controlled. Finally, lettering can be printed in different colours, according to the requirements of the map design. In this sense the graphic variables of form, dimension and colour also apply to lettering.

## Characteristics of type

In the Roman alphabet, two distinct letter forms are used, known respectively as capital (or upper-case) letters, and lower-case letters (Fig. 35). For the normal printing of text, these letters are arranged on a base line (Fig. 36). The shapes and proportions of the letters are controlled by the descender line, the mean line, the ascender line, and the capital line.

# Roman – Capitals and lower case
# ROMAN – CAPITALS
# *Italic – Capitals and lower case*
# *ITALIC – CAPITALS*
**35.** Capital and lower-case letters

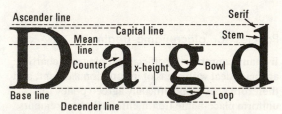

**36.** Characteristics of typographic letter forms

The distance between the base line and the mean line is known as the x-height, and is significant in the proportions of the letters. The positions of the ascender, descender and mean lines relative to the base line vary with different type faces. Figures are usually ranged on the base line, but in some type designs the figures also vary, with descenders and ascenders. The shapes of bowls, loops and counters are slightly different in different styles, but the basic elements of the letter forms are standard.

In addition to the distinction between capital and lower-case letters, the orientation of the lettering provides two possibilities: the upright letter form is known as 'Roman', and the slanting form is called 'italic' (Fig. 35). In some European languages using the Roman alphabet, a backward-slanting or reverse italic also exists, and this is sometimes used on maps, especially for the names of water features.

## Classification of type styles

The main classes of type style provide several recognisable groups in which the basic design elements are similar, even though the letter forms are distinct. These are mainly concerned with the way in which the strokes making up the letters are formed and terminated, and their relative proportions. The first major contrast is between serif and sans-serif type faces (Fig. 37). Serifs are the small extensions that terminate strokes. Serif forms are always accompanied by variation in the thickness and proportions of the strokes of which the letters are composed. In the group known as 'old style' the stroke thickness changes gradually from one part of the letter to another, and in symmetrical letters the thin–thick variation is oblique. In 'modern' faces there is a much sharper transition in stroke thickness, the contrast between the two is greater, and the stress is vertical. In sans-serif faces, not only are there no extended terminations to the strokes, but there is little or no variation in stroke thickness. This gives a much more simple form, which contrasts well with the serif type styles. Apart from these major classes, there are several intermediate ones. The most important is the slab-serif or 'Egyptian', which has the even stroke thickness of the normal sans-serif face, but single stroke serifs as well. In addition, there are a large number of 'display' faces, often produced only in capital letter forms, and at large sizes, which are widely used for decorative headings or eye-catching effects. Some of these depart radically from the conventional letter forms, but they are not used for continuous reading.

Individual type faces are named, sometimes after their originator, and although there are scores of them in use, printed text is dominated by a limited number. This is partly because some designs are much more successful than others; partly because

type, like other graphic elements, is affected by fashion; and partly because the manufacturers of type do not always produce a full range of all type variables and dimensions, so limiting the usefulness of some designs.

## Variations in typographic dimensions

Apart from variations in form, the dimensions of type also have to be varied for different purposes. The dimensional factors fall into two groups: those concerned with the dimensions of individual letters, and those concerned with spacing, which also has to be measured and arranged.

## Type size

The dimensional variations of type include height, width and weight. Height is the principal dimension, and is generally used to describe the overall size of type. Different systems of type measurement have been, and still are being, used, and a number of special terms are still retained to describe dimensional variations. These are of interest both to typographers (who plan and arrange type for printing), and compositors, who actually carry out typesetting operations. As in modern practice the type used for cartographic purposes is predominantly produced graphically for either printing or display, the measurement of type height is increasingly a function of standard measurement systems and units, rather than specialised typographic ones. Although in the past the typographic point measurement system (approximately 72 points to an inch) was also used for type production for maps, many modern photolettering machines for cartographic purposes describe the height of the type in millimetres, measuring from base line to capital line. Digital typesetting systems may use either.

During the long period when type was produced for printing by composing the characters in metal, the proportions and arrangement of characters on metal affected type measurement (Fig. 38). For example, the description 'twelve point' referred nominally to a size of 12 points, that is approximately one-sixth of an inch. In fact it described the face of the metal piece on which the character was based, and indeed different type faces had different letter sizes, even though they were all described as twelve point. Some types of photolettering or typesetting machines, which can be used to produce lettering for maps, still use series

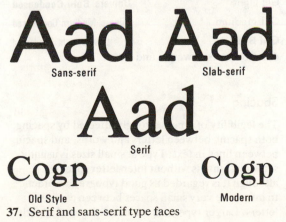

**Aad Aad**
Sans-serif          Slab-serif

**Aad**
Serif

Cogp          Cogp
Old Style          Modern

**37.** Serif and sans-serif type faces

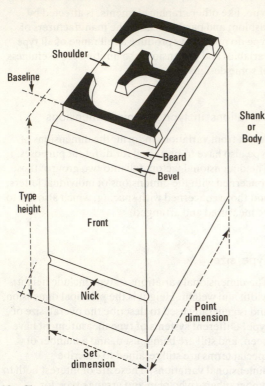

**38.** Characteristics of metal type

of type founts sized in points, and if this is the case, then due allowance has to be made for the difference between the nominal type size and the actual size of the lettering.

## Photolettering

For type produced in metal, every different type size and dimensional variation has to be produced separately. For a full range of sizes, widths and weights, this amounts to a large number of individual characters. When characters are produced photographically from negative masters, as happens with photolettering machines, size can be changed through a lens system. Reduction or enlargement applies to the whole character, not just its height, so that of course larger type is also thicker than smaller type. A typical type size for text in a book is 10 point. The actual size of a representative

**39.** Normal, condensed and extended versions the same type face

type face measured at this point size is about 2 millimetres. Lettering smaller than 10 point becomes increasingly difficult to read, and if smaller than 1.5 mm it needs a high-contrast image and close scrutiny.

## Type width and weight

Type can also be varied in width and weight. Most of the type set for reading has a 'normal' width, although this differs slightly with different type-face designs. The same type face can be produced in a narrower form, called condensed, and in a wider form, called extended (Fig. 39). Although extended type faces are generally used only for special or decorative effects, the condensed variation is often very useful cartographically, as it reduces the total space occupied by a name. Not all type faces have a condensed version at all sizes, which limits the usefulness of this factor.

The weight of a type face is partly conditioned by the type face design – some designs are normally heavier or lighter than others – but there are also specific variations on the weight of the standard version of the type face. It is possible to have a variation with finer strokes, usually referred to as 'light', and conversely a version with heavier or thicker strokes, called 'bold'. Extra bold is also possible, though again this is normally reserved for display purposes. The combination of condensed, medium and extended, and light, normal and bold, together with a full range of type sizes, gives a great number of graphic variables (Fig. 40). In typographic description, the normal weight and medium width are usually taken for granted and not specially described, but if the condensed, extended, light or bold versions are required they have to be part of the typographic description.

| | |
|---|---|
| Gill Light | **Univers Bold Condensed** |
| Gill Medium | Univers Medium Condensed |
| **Gill Bold** | Univers Light Condensed |

**40.** Contrasts in weight and set

## Spacing

The legibility of type is strongly affected by spacing, both spacing between letters and words, and spacing between lines in text. Type at small sizes is usually set 'solid', that is without inter-letter spacing, although it is regarded as good typographic practice to introduce very small spaces between capital letters. Larger type sizes are usually spaced slightly.

The space between words is usually one 'en', that is the width of the letter 'n' at that size, as this is the letter of average width. If parallel lines of type are set too close together, ease of reading is affected, and in practice a small amount of additional space is usually added between lines. Of course in reading text, people do not read letters or even words individually, but react very quickly to blocks or arrangements of shapes and spaces, scanning the printed text in sections, and jumping from one line to the next. Because reading is so assiduously practised, and therefore becomes a very sophisticated skill, the arrangement of letters and spaces is very important. Unusual type face designs, 'difficult' letters, letters crowded together or very small sizes, reduce the familiarity of the target, and rapidly affect the speed with which text can be read.

The lettering on a map produces quite different problems. Names are arranged individually, not in lines, and may be presented at different orientations. Contrast is affected by background colour and the presence of many other graphic elements. Names taken from other languages are unfamiliar, and therefore individual letters may need to be scrutinised, which departs from the practice of normal text reading. The circumstances of the legibility of lettering on a map are therefore quite different to those of legibility in printed text, and the selection of appropriate type faces depends on different factors.

## TYPE FACES AND CARTOGRAPHIC DESIGN

Lettering on a map is composed as part of the overall map design, and is used not only to represent names and figures, but to reinforce the classification of features and distinctions in importance between them. Although some maps may have few names, it is always likely that these may include a variety of styles and sizes, in order to make the lettering appropriate to the information on the map.

### Type styles

The general approach to the design of map lettering is to choose one or two principal type faces which contrast well with one another, and to use these for the names of features within the major classifications. For example, on a general map the names of physical features may be in a serif type face, and the names of all cultural features in one, or possibly two, sans-serif type faces. Within these two major styles, the typographical variables are used to distinguish between different feature classes and orders of importance. Where there is a series or hierarchy of sub-classes, as often happens with the names of towns, gradations of size are used in relation to the relative importance of the classes. Names extending over larger areas, and which are important either because the features they identify are large, or important in the administrative hierarchy, are usually shown in capital letters. In normal typographic practice the rule is to use capitals sparingly, so that they have a significant effect.

Differences in orientation are also employed to distinguish different feature categories. For example, if all physical feature names are shown in a serif type face, then the land feature names may be in Roman and the water feature names in italic. On multi-colour maps, colour distinctions are also used to relate the names to the features, and it is often useful to print the names of water features in blue, if blue is used for hydrographic features on the map.

### Choice of type faces

Names take up space on a map, and large names exercise a strong effect on visual search. Even so, on many maps, a large number of names of small features or less important places will need to be included at small sizes. Size has a considerable influence on the choice of appropriate type faces. At small sizes the relative proportions of ascenders and descenders to the x-height of the lower-case letters is significantly different in different type designs. Legibility is increased if the type face has a relatively large x-height, which keeps bowls and counters open. Short ascenders and descenders make the task of name arrangement easier, as names in crowded areas can be placed closer together. Type designs that have very fine strokes at small sizes, such as most of the 'modern' serif faces, may risk a decrease in legibility, especially if the reproduction is less than perfect. The overall 'weight' or average thickness of the letters is different in different type faces, and therefore those that are relatively fine and light are less successful if very small sizes are needed on the map.

For these reasons, some type designs are extensively used in cartography. For serif faces, Times New Roman is large on the body (that is, it has a large x-height) with short ascenders and descenders, and is a relatively heavy face at small sizes. Of the many sans-serif faces available, *Univers* is popular, partly because it is a very successful design, but also because it has been made available in a large range of weights and widths. The medium condensed version is useful for categories that need a degree of emphasis at small sizes, such as the names of towns and settlements, which usually have a high priority in the total map information. On maps with many classes of named features, a third type face is sometimes used for names of internal administrative divisions. The requirement is mainly for medium-sized capital letters in a face that will not be too obtrusive. Slab serif faces are useful for this purpose, as they have a distinctive character.

## Emphasis

Where additional emphasis is needed, a change in weight is often preferable to a change in size. If one name in a group needs to be distinguished, this can often be done by using a bold or semi-bold version at the same size. As with capitals, bold type should be used sparingly for effect, as its over-use will simply add to the overall weight of the lettering. Unfortunately, the bold version of many type faces is too heavy for cartographic purposes. Although some type designs have a semi-bold version, often this is not available at all the sizes needed. A change from medium to bold in the lettering is equivalent to using a heavier line gauge for line symbols. Large capital letters are usually heavy enough, without introducing the bold version as well, but bold can be appropriate for map titles. Selective use of bold can also be very effective in specialised maps and map illustrations.

## Lettering on maps

The lettering used on maps must have a high contrast with the background in all circumstances. For this reason, names are generally confined to the dominant line image colours, mainly black, dark blue, dark brown and red. They are never interrupted, as they always take precedence over other information. Relatively small differences in lettering size, indicated by height, are perceptible, and where a range of sizes is needed, they should be distinguished by a minimum difference of 0.2 mm. For maps with many names, the overall effect of type size and weight is considerable (Fig. 41). Small changes to type size can have a cumulative effect in the map design. Poorly designed lettering is not only confusing to the map user, but can interfere with the perceptibility of other map information.

| Name Name Name Name Name | **Name Name Name Name N** |
|---|---|
| e Name Name Name Name Nam | **e Name Name Name Name** |
| Name Name Name Name Nam | **Name Name Name Name N** |
| e Name Name Name Name Nam | **e Name Name Name Name** |
| Name Name Name Name Name | **Name Name Name Name N** |
| e Name Name Name Name Nam | **e Name Name Name Name** |

**41.** Cumulative effect of small differences in type size and weight

Despite the fact that printed type is so familiar, most people are only vaguely aware of its graphic characteristics. In the study of map design, it is essential that cartographers learn to scrutinise typographic forms and become familiar with them in detail, so that they can properly control and exploit the typographic variables within the map design as a whole.

**PART TWO**
# CARTOGRAPHIC TECHNOLOGY

# 6 IMAGE GENERATION AND REPRODUCTION

The term 'image' has several connotations. In one sense it refers to a likeness, in either two or three dimensions, of some natural scene, object or event, such as exists in a conventional photograph. In another sense it can refer to any representation deliberately made of objects or other phenomena. In this respect a map is also an image, two-dimensional in form, because it is a symbolised representation. Although there are important differences between iconic images and symbols, in a technical sense a graphic image exists as a physical structure, which allows the differentiation of different forms in a two-dimensional surface.

Many different processes, materials and instruments are employed to produce visible images, and therefore an analysis of how they can be formed is important in cartography.

Images may be generated as originals, or may be reproductions of something else that exists already. As a map has to be created, it is necessary to employ processes capable of generating the initial images required. Between this stage and the final production of the map, many other stages of reproduction may be involved, usually introduced to bring about some modification of the preceding image, to combine it with others, or to transfer it to another surface. Therefore, both generation and reproduction processes are necessary in map production.

## Visible images

There are two ways in which a graphic image can be perceived. In the conventional print formed by deposition on an opaque substrate (ink on paper), light is reflected from the surface to the viewer, and the image is formed by subtractive colour. In the standard television screen, rays of light are transmitted from the screen to the viewer, and the image is formed by additive colour. In the case of a photographic transparency projected on to a viewing screen, the image is formed by transmission through the transparency, but is perceived by reflection from the screen.

## IMAGE CHARACTERISTICS

### Type

There are two main types of graphic image: those that exist as continuous tonal variations, and those that consist of discrete elements (Fig. 42). They may be either chromatic or achromatic. Continuous-tone images have a continuous structure, and forms are represented by contrast in tone. With discrete images there is a discontinuous structure, and the image elements are separated by spaces. In a photograph of a scene, there is a tonal value at every point, lying somewhere between black and white in an achromatic image, or very light or very dark in a chromatic one. It follows that in a discontinuous image, there are separate image and non-image elements, and in a map these image elements may be composed of any combination of point, line and area symbols.

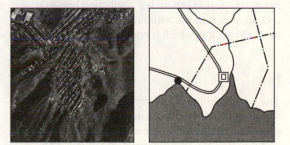

**42.** Continuous and discrete images

### Colour

#### Achromatic images

An achromatic image is one that transmits or reflects light without breaking it up into its component hues. The intermediate degrees are seen as greys, which can also be defined as variations in lightness or value. Theoretically, black reflects no light at all. Grey transmits or reflects a proportion of the incident light, all wavelengths being equally affected.

These intermediate values are also called tones, and a tonal image is one in which any tonal value can occur, and in which tones may grade imperceptibly into one another. An achromatic 'line' image is one in which only maximum and minimum colour (black and white in an opaque image), and points, lines or areas of maximum density are separated by 'background' or non-image areas. Areas of maximum density or opacity are known as solids. These discrete achromatic images are important cartographically, as most map production is carried out using images of this type.

### Chromatic images

Chromatic images also act by subtracting from the incident light, but in this case they operate as filters, transmitting or reflecting only the wavelengths that give the sensation of their particular hue. Therefore, some light has to be transmitted or reflected, and non of these materials (except black) is completely opaque to all wavelengths. For example, a red pigment, if it could be made spectrally pure, would subtract all other wavelengths except those of 'red'. Whereas the dyes used for some photographic materials are close to being spectrally pure, the pigments used in printing inks are mostly inefficient in this respect, and reflect not only their own wavelengths but some other wavelengths as well.

A monochromatic image is one in which all parts of the image have the same hue. The term is often used to refer to 'black-and-white' images, but correctly speaking there is a distinction between achromatic, which is without hue, and monochromatic, which is composed of a single hue.

### Form

The most familiar graphic images are in positive form, that is the parts that indicate form (the image elements) are dark against a light background, or opaque against a transparent background. The converse is a negative, in which the image elements are light against a dark background, or transparent against an opaque background (Fig. 43). Negative images are extensively used in cartographic

production, and therefore whether the form of an image is positive or negative has to be specified.

### Position

All graphic images that physically exist are made in or on some sort of supporting surface, which may or may not be transparent. A right-reading image is one that appears in its correct orientation on the side of the supporting material facing the viewer. If it 'reads' from right to left (in reverse) when on the side of the supporting material facing the viewer, it is a wrong-reading or reverse image (Fig. 43). This factor is important when dealing with images on transparent or translucent material, which therefore can be seen from either side. A reverse image can be read in the correct orientation, but only by viewing it through the thickness of the base material. As in contact copying the image should be in complete contact with the new material, the side that actually bears the image should be placed in contact. Therefore, in organising the production of a map through several stages of processing, the image position needs to be considered as a specific factor to be decided.

### Permanence

As graphic images will eventually disintegrate, or fade to the point where they are no longer usable, the degree of permanence has to be considered. There is a fundamental distinction between printed images composed of pigments on paper, which may last for many years, and displayed images, which are sustained for a limited duration. Some types of reprographic material provide images of limited duration, or which may fade when exposed to sunlight for long periods, and therefore this may have to be taken into account both in deciding the nature of the final image required, and the durability of the components from which it has been made.

## IMAGING SYSTEMS

So far as cartography is concerned, there are several ways in which imaging systems can be classified, according to their characteristics, mode of formation, application, and so on. At one time it was relatively easy to distinguish between

**43.** Positive and negative forms: right reading and wrong reading (reverse)

reprographic processes, employed primarily to reproduce an existing image on some other surface or in some other form, and machine printing processes, used to generate a number of facsimile copies. Today the whole field is complicated by the large number of devices that can be used to produce visible images, and the fact that a term such as 'print' has a variety of meanings. A studio photographer refers to a print as a single photographic copy, usually on paper, which is the final outcome of the photographic process. A computer or word processor may have attached a 'printer' that produces a single 'hard copy' of the information stored. A device that directs jets of coloured ink at a sheet of paper may be referred to as an ink-jet printer, or even an ink-jet plotter. But a 'drum plotter' also generates a permanent image at high speed.

There is still a basic distinction between the generation of an image, by which it is constructed and appears for the first time, and a reproduction, either as a copy of something else, or a copy of another existing image. The video camera records and reproduces scenes or pictures. The cartographer can construct a map, or map components, from a variety of sources of information which are not themselves images.

For cartographic purposes it is possible to distinguish between printing (the production of multiple copies, usually at high speed) and plotting, used to construct a single image. A plotting device, if capable of operating at high speed, may also be used to produce multiple copies, but on each occasion it has to go through the complete cycle of image construction. A plotting device can be used to generate an image, as it constructs the image step by step, in a manner similar to that of manual image generation, though it may operate at a far greater speed. So although a map may be drawn by hand (which is a means of image generation), a plotting device can also be used, depending on the nature of the source information.

A duplicate of an image can be produced by simply repeating the image generation process, but normally a duplicate is the product of a reprographic process. In this sense, an image can be copied by tracing, which is a form of manual duplication, but this tends to be erroneous and inefficient. But because many images may be constructed and then modified during map production, reprographic processes are of great significance. Even in 'automatic', or computer-based cartography, the image displayed is formed by directing radiant energy at some sensitive material, and is therefore the product of a reprographic process.

## Classification of imaging processes

Given the great variety of materials and processes used for cartographic purposes, it is necessary to examine and classify them in detail. Any such classification has to take into account the method of image formation, that is the physical means of creating it; the mode of image formation, distinguishing between processes that are sequential and those that are simultaneous; and finally the control of image formation, which may be by manual, reprographic or digital means. Both manual and reprographic image formation are analogue, in the sense that they operate by constructing or reproducing an image analogous to the one required; that is, an image is formed and then used to produce other images. With digital control, the information that is used to construct the image is held in digital form, and can only be processed through a computer. What is stored in computer-accessible form is not an analogue of the image, but a digital record.

# TYPE OF IMAGE FORMATION

## Formation by deposition

In this method a material such as an ink or pigment is deposited on a supporting base material (Fig. 44).

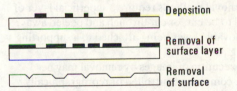

Deposition

Removal of surface layer

Removal of surface

**44.** Image formation

Virtually all forms of drawing and painting use this means, and frequently the image is formed on an opaque medium such as paper. The pigment may lie on the surface or may penetrate the surface of the substrate to some degree. Under manual control, the process is very flexible, and can provide a great variety of continuous and discrete images, achromatic or coloured. It is most valuable in creating images, especially those that are the product of original thought or expression, and least valuable for the uniform copying of images, in which it is both slow and inefficient.

The method can also function through computer-controlled operation, using either a plotting device driven through *x* and *y* coordinates, or in the raster mode.

In most varieties of this method there is actual physical contact between the tool or instrument containing the pigment or vehicle and the base or supporting material on which it is deposited. Control of this contact is vital in both manual and computer-controlled methods. In conventional printing there is a transfer of printing ink from the printing plate – directly or indirectly – to the paper or other material, which also involves contact between the two surfaces. In electrostatic systems the pigment, consisting of fine particles, may be attracted to the imaged surface by opposite polarity, without direct contact; and in ink-jet systems, very fine particles of ink are discharged on to the paper surface.

When image formation takes place through deposition using tools or plotting instruments, the action is sequential; that is, the image is constructed piece by piece, even though with some plotting machines this may take place very rapidly. When a printing surface is used, the action is more likely to be near simultaneous, as the image is actually formed along a line of contact or a rolling band.

## Formation by removal

Images formed by deposition are most frequently positive images on either opaque or translucent bases. Those formed by removal are usually negative, and are rarely produced on opaque bases. Most removal methods require a superficial layer of material to be cut, or cut through and peeled away, this removable layer being attached to a supporting base (Fig. 44).

The execution of images by removal may be manual, computer-controlled or reprographically controlled. In basic manual methods tools with sharp edges can be used to cut through the thin layer. Similar tools and methods can be used by the more sophisticated computer-driven plotting machines. Chemical etching first requires the superficial layer to be protected by the exposure and hardening of another superficial coating; this then allows only the required image lines or areas to be dissolved by the etching solution, usually an alcohol.

This type of image formation by removal is also characteristic of some printing surfaces. The engraved plate exemplified the operation of removing metal by manual means, though it was often assisted by chemical etching with acid. The manufacture of relief printing surfaces, as in letterpress images, also historically used the same

process of first protecting the surface and then etching away the non-image areas to leave the image in relief. Manual formation by removal is a sequential operation, like drawing, but the controlled etching methods are simultaneous, as they operate on all parts of the image at the same time.

## Image formation by photoreaction

The technology of photoreactive imaging systems covers an enormous field of materials, processes and instruments, many of which combine mechanical, optical, electronic and photoreactive components. Those which are of particular interest cartographically are the large number of photoreactive methods used both for the final map product and also for the intermediate processing stages.

Processes using these materials are also referred to as photosensitive systems or even light-sensitive systems, but the supply of energy to bring about chemical or physical change in the material is not limited to the visible spectrum. Basically, the absorption of radiant energy by a molecule results in excitation and the release of electrons. The electrons eventually bring about changes that are, or can be made, visible, as a difference in colour, opacity, resistance to solvents, and so on. Changes in conductivity are used in electrostatic systems (also called electrophotographic systems), and the effect of electron ejection is employed in forming the type of image made visible by the phosphors of the cathode ray tube. Both these are photo-electric effects.

Photoreactive materials differ considerably in speed and efficiency, and in the types of image that they can produce or reproduce. In a photochemical reaction, the ratio between quanta absorbed and molecular reaction governs the speed of image formation. In some systems, most notably that of the silver halides of conventional photography, only the initial phase of image formation depends on radiant energy, and the process is then amplified by chemical development. Conversely, with materials such as diazo, all the energy needed to produce the image must be supplied by the radiant energy. The measurement of speed of reaction between different materials has to take into account not only initial formation, but the requirement of the overall process. Silver-halide materials have the great advantage that they require a very short initial exposure, and the image formation is completed subsequently. The speed of reaction of such materials, usually given a sensitivity rating on an ISO/ASA scale, refers only to the initial exposure.

Spectral sensitivity describes that part of the electromagnetic spectrum to which the material reacts. Ordinary silver-halide emulsions have a spectral sensitivity from about 200 to 500 nm, but this can be extended to about 1300 nm by the addition of dyes. Diazo materials are sensitive to relatively short wavelengths of between 300 and 450 nm, whereas the phosphors of the vidicon tube are sensitive to between 400 and 800 nm.

## Radiant energy

All photoreactive systems require a radiation source to supply the necessary energy. The radiation source may be natural or artificial, that is either natural light or heat, or some artificially supplied radiation through a lamp or an electron beam. This may be emitted by the phenomenon being recorded, reflected, or transmitted. Although natural illumination is important for aerial photography, in cartographic production sources of radiation are normally lamps, which can be controlled in spectral characteristics, intensity and duration.

The production or initiation of an image requires an exposure, the effect of which depends upon intensity and duration. If only a short exposure is possible, as in many types of natural photography, or the recording of moving objects, then a highly sensitive material must be used which can react to small amounts of energy. Under studio conditions, stationary images can be exposed with intense lighting and for relatively long periods, so that exposure using artificial lamps does not operate under the same limitations as daylight photography. Some useful photoreactive materials require a great deal of energy to be supplied to produce the required material change, and this can be achieved with suitable sources of illumination.

## SOURCES OF ILLUMINATION

Electromagnetic radiation includes the wavelengths of visible light, and both shorter (ultra-violet and X-rays) and longer (infra-red) wavelengths. This radiation can be described either as a stream of photons (a photon being a single quantum of energy), or as a continuous wave. In practice both descriptions are used. The fundamental

characteristics of any radiation are its wavelength and intensity, and in all processes using photoreaction it is necessary to use sources of radiation that emit sufficient energy at the wavelengths to which the material is sensitive. Most radiation sources 'spread' the emitted radiation, but coherent light sources are those that emit an almost parallel beam over a very narrow band of wavelengths.

All electromagnetic radiation results from either incandescence (the heating of a substance) or luminescence, in which photons are produced by excitation caused by a stream of electrons. Certain phosphors emit light when excited by short wavelength radiation, and this is the phenomenon of photoluminescence. If the light emission is only sustained during excitation it is termed fluorescence, but if it persists it is called phosphorescence.

## Luminosity

Lamps that provide point sources of radiation do not emit equally in all directions. Theoretically, a radiation source should emit spherically. This luminous intensity is measured by the candela.

In practice, lamps used in reprographic work emit a beam of radiation in one direction, and this luminous flux is measured by the lumen. Centre beam is the direction of the radiation at maximum flux, and the edge of the beam is regarded as the angle where the intensity is half as great as at centre beam. This is important in judging the 'coverage' of an artificial light source used to expose an image.

Illuminance, usually referred to as illumination, is the measure of the amount of light reaching any surface, whereas luminance is the amount of light or radiation emitted from a surface, whether by emission, reflection or transmission.

Brightness is a psycho-physical response related to illuminance, but the human visual system does not respond in a linear fashion to measurable changes in intensity.

In map production, where very often large images are exposed to point light sources, reasonably equal illuminance of the whole area is important. The most even illumination can be attained by placing the source of illumination as far as possible from the surface, but the inverse square law means that intensity decreases as the square of the distance. Therefore, a compromise has to be reached between lamp distance, which controls intensity of illumination, and an even distribution.

The luminous efficiency of an artificial lamp is the relationship between the power input and the luminous flux produced, and is described as lumens per watt.

## Spectral distribution

Sources of illumination may emit energy over a continuous spectrum, which may be very wide (such as 'white' light) or a relatively narrow band; at particular points in the spectrum (usually referred to as lines); or at lines or groups of lines within a band (Fig. 45).

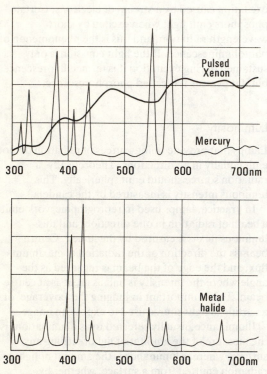

**45.** Relative wavelength intensity of pulsed xenon, mercury vapour and metal-halide lamps

## Artificial sources of radiation

The main types of radiation source used in reprographic work are tungsten (including tungsten-halogen), carbon arc and gas discharge lamps of several types.

Tungsten lamps have a continuous spectrum, but the ordinary tungsten lamp has a low colour temperature (2900 K), and is deficient in short wavelengths. It is low in luminous flux, generates considerable heat and is relatively inefficient. The photoflood lamp is a tungsten lamp operated at a higher temperature, which increases its luminous flux and efficiency, but reduces the life of the lamp.

In the tungsten-halogen lamp, the envelope surrounding the filament is made of quartz to withstand higher temperatures, and the iodine in the envelope vaporises and combines with the tungsten vapour from the filament. These lamps have a higher intensity and longer life.

Carbon arc lamps provide both a high intensity and strong radiation in the violet and ultra-violet wavelengths. The light is produced by passing an electric current through two carbon rods that are held slightly apart. When the rod tips are heated to the vaporisation point of the carbons, intense light is produced across the gap. The spectrum of emission is almost continuous, but with a much greater intensity in the violet and ultra-violet region. These lamps were widely used for a long period for graphic arts reproduction, partly because they approached the spectral distribution of 'daylight' more closely than the other lamps previously available, and partly because many photoreactive materials are most sensitive, or only sensitive, to radiation at short wavelengths. However, the disadvantages of variable output, heat and fumes have meant that they have generally been supplanted by more advanced types of lamp.

There are two main types of gas discharge lamps, pulsed xenon and mercury vapour. The pulsed xenon is a development of the electronic flash lamp. The tube is filled with xenon gas at low pressure, with an electrode at each end. A flash of high intensity but short duration is produced when the gas is ionised, using the energy stored in a capacitor. With a pulsed xenon lamp the flash is repeated at a rate of 100 or 120 times per second. The effect is of a continuous light, with a continuous spectrum. Although high-powered lamps require cooling, the pulsed xenon emits over a spectrum similar to daylight, and is cleaner in use than the carbon arc.

There are many types of mercury vapour lamp in use, the most simple being a glass tube containing mercury and argon. Electric current is passed through the tube, which has an electrode at each end, and vaporisation of the mercury produces an intense light with a discontinuous spectrum. Emission is limited to 'lines' in the ultra-violet, blue and green areas of the spectrum. In high-pressure lamps, which have quartz envelopes, the spectral range can be improved, for example by adding cadmium or zinc to increase longer wavelength radiation, but is still discontinuous.

Mercury-halide lamps include metallic halides such as lithium iodide, thereby increasing the range and number of 'lines' or emission within the visible spectrum. Spectral coverage is good, and such lamps are widely used in reprographic work.

The fluorescent lamp is a modification of the low-pressure mercury vapour lamp. It is usually made in the form of a long tube, and the inside of this is coated with a powder that fluoresces in response to irradiation by ultra-violet wavelengths, which causes it to emit radiation at longer wavelengths. The spectrum is a combination of the mercury type and a strong background radiation at all wavelengths of visible light. Different spectral characteristics can be produced by different tube coatings. These lamps can be assigned an apparent colour temperature, whereas the mercury vapour lamps cannot.

### Lasers

Lasers emit near-coherent light beams over a very narrow spectral band. The 'ruby' laser is high powered, whereas those most frequently found in reprographic work are gas lasers, such as helium-neon and argon, which emit at 633 nm and 458–514 nm respectively (the actual wavelength being affected by temperature). The emitted beam is close to being parallel, and is a continuous wave.

### Exposure and control of illumination

When an image is given an overall exposure, several factors have to be taken into account, including the characteristics of the sensitive material, the intensity and duration of illumination, and the angular coverage of the image area. As the lamp-object distance, and lamp intensity are generally fixed, the major control lies in the duration of the exposure. Exposure must be intense enough to provide practicable exposure times; it must be approximately uniform over the whole image area; and the angle of illumination must be arranged to avoid specular reflection from the copy or the glass cover of the copyholder. Duration is best controlled by using an integrating light meter, which measures the total amount of radiant energy received, and can vary the duration of exposure to compensate for any variations in intensity. The most advanced integrated light meters automatically shut off the lamp when the predetermined exposure has been received.

It is also possible to expose photographic material by the use of a light beam, either by employing an illumination source, such as a laser, which emits a beam and can be controlled in direction, or by directing a lamp through an aperture or slit. This method is important in automatically controlled plotting machines, which can produce an image by moving a light beam over the surface of a sheet of photographic film. It has the great advantage that there is no direct contact between the photo-exposing head and the film surface, and the line exposure can be varied by adjusting the beam width, or dividing it into multiple lines through a beam splitter.

The use of a laser or light beam is also important when plotting or scanning in the raster mode, as the beam can be controlled to plot or not to plot at every point in the raster array.

## MODE OF IMAGE FORMATION

An image can be produced either sequentially, that is, step by step, or simultaneously, all parts of the image being generated at the same time. All drawing operations are sequential, whereas most reprographic materials are processed through exposure simultaneously. Sequential processes may operate in vector, by which each element of the image is constructed separately, or in raster, by which the entire image surface is traversed in a regular array of points or cells. For an image consisting of discrete elements, in which much of the area may be non-image or 'background', vector mode is efficient and avoids traversing empty spaces. For a continuous image, where there is a value at every point, raster is efficient, as it can be operated at high speed. Plotting devices may be operated manually in vector, or if controlled digitally they can plot either in vector (like the flat-bed plotting machine), or in raster (like the automatic raster plotter).

In a plotting instrument, the paper or other substrate may be passed against the plotting head one line at a time, by rotating it on a drum, or drawing it under the plotting head, and the head is then 'stepped' forward to the next position. Multiple heads may plot several lines on each pass.

In a cathode ray tube, the electron beam, which excites the phosphors to produce a transient image, may be driven in vector, thus 'drawing' upon the screen, or in raster, as used in the continuous 'pictures' of the normal domestic television.

Machine printing processes may be truly simultaneous if the flat inked surface is pressed against the paper, but the more common rotary method of printing, in which a cylinder rotates in contact with either another cylinder or a flat bed, actually has a rolling line of contact. Such methods should properly be described as near-simultaneous, as a short time must elapse before the whole image surface is covered (see Fig. 71).

Mode of image formation should not be confused with speed of output, although simultaneous would appear to be faster than sequential. A sheet of photographic film may be exposed simultaneously to initiate the image formation, but the subsequent processing, although simultaneous at each step, takes time. On the other hand, some drum plotters, operating sequentially in raster mode, can produce images at a very high speed.

## Analogue and digital control

Although it is customary to refer to conventional cartographic production as 'manual', this has only come about because much of the initial generation of the required image components is carried out by hand, or by using manually controlled instruments. Unfortunately, this also suggests that maps are created solely by manual drawing methods, which is far from the case. There is a limited sphere of cartography in which a cartographer actually draws the entire map completely, and then has it reproduced by printing, but this type of cartographic practice is rare, and can only achieve a very limited graphic image. Most maps are produced through a combination of manual and reprographic methods, and often the two are combined in image production. Therefore, it would seem preferable to classify conventional methods as analogue, in order to distinguish them properly from digital methods, in which control of image formation is exercised through a computer, operating in a sequential mode, even though this may be at high speed.

In both analogue and digital production, the complete map image may not appear until the end of the sequence of operations used to construct it. However, the end product must be a visible image. Digital information itself is not 'visible', and the maps produced digitally may appear either as transient images displayed on a screen, or as 'hard copy' produced by a printing or plotting device. In all cases these final images are themselves formed by deposition or the use of photoreactive materials.

## IMAGING OPERATIONS

### Generation

As a map is not simply a reproduction of something

that already exists, its basic components have to be created in the first place. Different components, such as grids and graticules, names and boundary lines, have to be constructed or assembled, and for many maps a great variety of source information has to be put together in graphic form. These original or initial images can be generated by various means. Traditionally, this was the particular province of manual methods of construction; drawing in ink and pencil, and engraving. For a long period in the development of modern cartography, image generation was limited to what could be produced by hand, and any map was fundamentally the product of a skilled cartographic craftsman.

Today most maps are produced through generating images by a variety of means. Both manual and reprographic processes are used in analogue production, and increasingly there is a combination of analogue and digital methods for image generation. For example, a nautical chart may have the graticule and border produced through a digital plotting machine; the linework of the chart may be drawn or scribed by hand; lettering and figures will be composed initially by the reproduction of typeset material subsequently added to the originals manually; and tints for land, foreshore and shallow water colours may be produced by first creating a mask of each area manually and then using this to construct the required image reprographically.

It is also characteristic that in many cases the final image required is not produced in one operation, but is carried out by first constructing one element, and then changing or modifying this into something else. Many of the images that appear on a modern map could not be produced by manual methods alone, and therefore the whole production system uses a combination of different methods and operations for different cartographic requirements.

### Duplication and copying

There are many operations for which a duplicate or copy of an image is required. The most common ones include the transfer of an image to another material; reversing the image form from negative to positive and vice versa; changing the image position from right reading to wrong reading; producing a guide image on another piece of material, possibly with a superficial layer for further processing; and making a duplicate for record purposes or safe keeping. The controlling factor in duplication is that an identical image, or facsimile, is required, without any changes to its dimensions or detail. Therefore, it is very important that there should be no distortion introduced by the duplication process, or any loss of quality. This puts a premium on contact copying, using reprographic materials. Although duplication can be attempted manually, by tracing, this is impossible to carry out perfectly, and in fact it is a general rule in cartographic production, that once an image has been generated it should only be duplicated by reprographic methods.

The type of image affects the means used for duplication. Very fine lines or tints obviously require reprographic materials with a sufficiently high resolution. Continuous-tone images, such as orthophotographs or manually drawn hill shading can only be reproduced effectively by conventional photography, as other reprographic materials do not have sufficient tonal resolution.

## Combination

Because virtually all multi-colour maps are produced in both colour-separated and image-separated form, the combination of images is a common requirement. Although this may be delayed until the final stage of production, frequently it takes place after the elements of any one printing colour have been generated, and it may be operated both to combine different images and at the same time make modifications to them. For example, a particular printing colour for a map may contain line, lettering and tint elements. Different methods will be used to create these initially, so they will exist on three separate pieces of material. The next stage may be to combine them into a single image.

Reprographic materials and processes dominate combination. The different characteristics of these materials are exploited, particularly with regard to changes in image form. For example, if the components for one printing colour exist as a set of negatives, then it is relatively simple to combine these into one positive by photographic means, making use of the fact that in normal photographic processing, image form changes at each individual photographic copying stage. On the other hand, dichromated colloids can be used to combine three separate positives into a single positive, although it is a rather slow process. More complex procedures can also be used to carry out the combination of both negative and positive images at the same time. It must be noted that in image combination of this type using reprographic materials, the images to be combined are exposed successively, in order to maintain image quality.

## Separation

Because most multi-colour maps are produced by the colour-separated method, image separation is not such a common requirement in cartography as it is in most forms of commercial graphic art. Colour separation requires the extraction of the three primary printing colours from a multi-colour tonal original, usually with the addition of an extra black. This can be done by reprographic means, using photography and filters, or by electronic scanning. The process is central to standard multi-colour reproduction for printing, and consequently figures largely in all accounts of general graphic arts reproduction. It is only on rare occasions that any part of a map is initially produced as a multi-colour tonal image, which would have to be colour-separated before printing.

Where the separation of an image is required, it is usually part of the production process. For example, the response of ordinary photographic film is to relatively short wavelengths. Therefore, if a guide image in pale blue is printed down on a piece of material, and subsequently part of this is drawn up in black ink, contact copying to ordinary photographic film will result in the omission of the blue image. This makes use of the spectral sensitivity of reprographic materials to control image formation selectively.

## Changes and modifications

It is frequently the case during cartographic production that changes or modifications to existing images have to be carried out, in order to produce the desired characteristics, or because of the requirements of the standard printing processes. The principal types of change involved are geometrical transformation, scale change and changing image structure.

Generally speaking, changes to the geometry of an image can only be achieved by constructing a new image, rather than by modifying an existing one. This is one area where reprographic methods are of little use. Map detail on a different projection can be compiled visually by a cartographer, making the necessary modifications to shape and position by eye, and indeed in the past virtually all derived maps had to employ this means, although it was always slow and unsatisfactory. If the information is held in digital form, then geometrical transformation can be done much more rapidly and efficiently by computer processing. Affine transformations, such as translation, rotation and reflection in one axis, are frequently employed. Again, most of the processes of generalisation are in fact carefully controlled modifications, but these necessarily take place prior to the technical production.

Scale change is also often necessary in the preparatory stages of map production. It is only likely to be part of the main production process if the original images are constructed at a larger scale than the final map required, and therefore have to be reduced at some point to the desired scale. This is invariably done by photographic means in analogue map production.

The conversion of solid to tint, and of continuous tone to half-tone, are common cartographic operations, which lead to a change in image structure. Area tints are usually produced by first constructing a solid mask to define the areas required, and then converting this image into the required percentage tint. As continuous tone cannot be reproduced directly by lithographic printing – except in special circumstances – the normal practice is to convert continuous–tone images into equivalent half-tones by photographic processing, using half-tone screens. The same effect can be achieved by high-resolution digital scanning.

## Colour addition

Although the basic production of a map normally takes place using achromatic images, there may be particular stages or operations where coloured images are required. These do not necessarily have to be equivalent to the final colours of the map, but may be added solely for the purpose of either visual or reprographic discrimination. The chief processes in which the addition of colour is likely to occur include the compilation of the map manuscript, both for multi-coloured and monochromatic maps; the use of coloured guide images, both in production and revision, and the production of colour proofs to demonstrate the final appearance of the map, or a close approximation.

Colour addition can therefore take place manually or by reprographic methods. In a complex map, the compilation itself will be colour coded in order to assist in the differentiation of different map elements. This can be done easily through the use of coloured inks, although the density of these needs to be considered in relation to subsequent reproduction achromatically. Coloured guide images are usually produced using reprographic materials with dyes incorporated in them, or by making a stencil and then applying a dye to form the image. Blue is the commonest requirement, but sometimes in the execution of revision a red image

may be introduced to distinguish the unmodified detail from the corrections or additions.

Colour proofs may be required at several stages in the production of a multi-colour map. Although the final proofs may be produced using a standard printing process, reprographic methods offer a wide range of possibilities for the introduction of colour. Few of these produce perfect colour facsimiles of the printed colour, and some of the most useful are of limited value as proofs of specific colour appearance, but they serve a purpose as intermediate proofs for the checking of completeness and detail.

## Multiplication

Although any method capable of duplicating an image of a given type can be used to produce multiple copies, reprographic methods can only serve the function of multiplication by repeating the entire image-forming process for each copy. Generally, each copy needs a sensitised surface. The production of a large number of identical images is usually carried out by means of machine printing, although some reprographic processes and plotting instruments can be used if only a very small number of copies is required. Such printing methods are dominated by deposition, and operate in a near-simultaneous mode. The most suitable method for producing the final output is strongly affected by the total quantity involved, the format of the image and the nature of the colours required.

# BASE MATERIALS

Although some simple maps can be constructed on a single piece of material, cartographic production is generally characterised by the use of many separate materials for different images, all of which are required to fit together correctly. Consequently, considerable demands are placed on the base or supporting materials on which images are constructed. Although no single material can be said to satisfy all needs, in recent times there has been a considerable improvement in the provision of base materials suitable for cartographic work. Apart from improving the quality of maps, better materials add considerably to the speed, ease and efficiency with which cartographic operations can be undertaken.

The ideal base material should be dimensionally stable, durable, transparent or opaque as required, receptive to inks and sensitised coatings, sufficiently thin in gauge to facilitate tracing, and flexible. Materials used in the past rarely satisfied more than one or two of these requirements, and modern cartographic practice has been revolutionised by the introduction of sheet plastic materials.

## Stability

Dimensional stability is important in two respects. For maps that may be used directly for measurement, absolute changes in image size will be important. For most maps, especially multi-colour maps, the problem is more one of relative size than absolute size, as the relative sizes of the various components that make up the map will affect the correct fit of the images and the register of the different colours. In addition, although lack of stability is seen primarily as a problem for map users, it can introduce annoying problems during the production of a map from the point of view of the cartographer and other technicians. Although accuracy for its own sake is frequently condemned as impractical, it is in fact far easier and quicker to produce an 'accurate' map in the dimensional sense, than to struggle with separate pieces of material that do not fit together correctly.

For a long time in the past the only stable transparent material was glass, and the only stable opaque material was metal, or a surface such as paper supported by being laminated to a metal plate. The latter is still occasionally used for constructing images that are best produced by using a paper surface, the most obvious examples being some types of drawing, such as hill shading using an airbrush. Today, however, most cartographic and reprographic materials are available on plastic sheets, which are sufficiently stable dimensionally for all practical purposes.

## Transparency and opacity

If an opaque material is used for the original images, then the image can only be transferred to another surface by first copying it on to a transparent medium by using a process camera. This introduces the problem of maintaining the correct dimensions. When the only stable materials were opaque, being metal plates or highly compressed paper, this photographic stage was inevitable, and therefore working at a larger scale did not place any additional demand on the production system. However,

images produced on transparent or translucent base materials can be contact copied by a variety of means, without the use of a camera. Such contact copying avoids distortion during the copying process. In addition, by working at the final scale of the map, rather than a larger one, the actual amount of work is reduced. Therefore, with modern transparent materials, there is no reason to carry out production work at a larger scale unless this is demanded by some special problem in the organisation of the production.

## Receptivity

In order to draw with ink on a surface, it must be possible to deposit the ink smoothly, but with sufficient adhesion to form a permanent and durable image. Although paper surfaces are receptive to both pencil and ink, a clean ink drawing requires a firm, well-compressed surface, such as Bristol Board or its equivalent. Whereas for sketching or tonal drawings a slightly rough grain may be desirable, for fine-line ink work a smooth surface is essential.

The base materials used for sensitised reprographic coatings need to be stable against wetting and common chemical agents, as they are likely to be processed at some stage. They must also allow the surface coatings to adhere satisfactorily. Porous substances are generally unsatisfactory in this respect, and paper can only be coated with a solution if it is supported on a stable base.

## Gauge and flexibility

A translucent material used for tracing should be thin enough to avoid any possibility of parallax, but not so thin that it does not lie flat and provide a firm surface on which to work. A matt surface is generally necessary for drawing. Materials that are matt on both sides will normally provide a better working surface, as there is less tendency for the material to slip on the underlying surface. For reprographic purposes there is also an advantage if the supporting sensitised materials are thin, especially if more than two pieces of material are being exposed at the same time, as this reduces the space between surfaces introduced by the thickness of the base material.

Flexible materials offer many advantages in practical use. Large format images suffer less risk of damage while being handled and moved, and as in some kinds of processing machine the material is passed through a system using cylinders, it must be sufficiently flexible.

## Plastic sheet materials

The first plastics to become important for cartographic purposes were the vinyl chloride polymers, a large group produced for many different requirements. The early types were poor in stability and discoloured easily. The present materials are developed from co-polymers such as vinyl chloride – methyl acetate and vinyl chloride – maleate ester. The sheet materials are produced by rolling through a calendering machine. All vinyl sheets have a tendency to retain static, which attracts dust. The vinyl plastics are relatively stable thermally, and very low in moisture absorption. The sheet film can be dissolved by ketones, esters such as amyl acetate, and chlorinated hydrocarbons such as ethylene. These polyvinyl materials always suffered from a susceptibility to shatter along edges if carelessly handled, and this brittleness meant that they were never wholly satisfactory for cartographic originals which needed to be kept for a long time. Although they played an important role in the transition to the use of plastic materials, for virtually all purposes they have been superseded by the polyester materials that now dominate graphic arts where dimensional stability and durability are important.

There are many types of polyester materials, and the one which provides the base for most graphic arts purposes is polyethylene terephthalate. This is a condensation product of ethylene glycol and terephthalic acid. The monomer is heated to cause polymerisation. This material is produced in sheet form by extrusion. To make a film with suitable properties, it is heated and stretched in opposite directions, which causes plane orientation of the molecules. The manufactured product is sold under a variety of trade names. It has a very low coefficient of expansion, and a very high folding resistance, so much so that it is virtually impossible to make a clean fold in it. The film is tough, durable and impermeable. It transmits 90% of light in the visible spectrum, but cuts off sharply at the end of the ultra-violet. Unfortunately, its resistance means that it has not been possible to produce a dye that will penetrate it, and therefore all images lie on the surface. It is normally given a matt surface for drawing by chemical or mechanical treatment. It is available in both sheet and roll form, and in a variety of gauges.

Polyester sheet materials satisfy virtually all the requirements of cartographic production, for both drawing and reprographic purposes, and all the coated materials used in cartographic production are normally made available on polyester bases. This has the great advantage that all the images produced are on the same base material, eliminating any relative dimensional differences between different components. The combination of polyester materials and punch register systems has raised the general standard of cartographic production, and correctly fitted and registered images are now normally achieved.

# 7 PROCESS PHOTOGRAPHY

Photography is the most widely used reprographic method, and this is due to its capability and flexibility in many kinds of image formation. Process photography is essentially that aspect of photography that deals with the preparation of images for reproduction in print. It differs from studio photography in that the end product is not a photograph, but an image in machine-printed form.

Process photography can take place either through the use of a process camera, or by contact copying. Contact copying of course does not permit any change in scale, as there is no lens system involved, but it is more widely used in modern cartographic production than camera photography. Photography is extensively used for making a positive from a negative, and vice versa; for the combination of several negatives into one positive; for making percentage tints with negative masks; for converting continuous tone to half-tone; and for a variety of both line and tonal images in which special masking effects are required. Despite the fact that photographic materials and processing, especially with large format images, are relatively expensive, its versatility ensures that it retains its place as the most significant reprographic process.

## The photographic process

The basic principle of photography is that a material containing substances sensitive to radiation within the wavelengths of visible light, is exposed to radiant energy reflected from or transmitted through the subject being photographed. The light-sensitive material is affected in proportion to the amount of radiant energy received. Initially, a latent image is formed, at this stage invisible, and this is subsequently developed by processing in chemical solutions, to form a visible image. When this stage is completed, the unexposed, and therefore still sensitive, materials have to be removed by washing. Many stages are involved in the complete operation, and as the material is highly sensitive, these have to be carefully controlled.

## THE PHOTOGRAPHIC EMULSION

The light-sensitive layer of the photographic film or plate is conventionally referred to as the emulsion. It consists principally of light-sensitive materials, silver halides, distributed in a colloid, gelatin. The silver halides mainly used are silver bromide, silver iodide and silver chloride. Of these, silver bromide forms the main constituent of most photographic emulsions, as it has the highest speed of reaction to exposure. The halides consist of fine grains, distributed as evenly as possible in the colloid. The gelatin has to be sufficiently strong to adhere in a thin layer to the base support, and sufficiently transparent to allow radiation to reach and affect the sensitised material. It is normally applied to the film on top of a substratum, and in addition the back of the base material is coated to absorb the incident light that might otherwise be reflected from it into the emulsion during exposure.

## Resolution and speed

Because of the nature of photographic materials, the properties of resolution and speed of reaction are connected. The larger the grains of silver halide the faster is the emulsion response. If these could be evenly distributed then resolution would present few problems. But the emulsion grains tend to occur in clumps or groups, and high-speed films can exhibit a 'grainy effect' (Fig. 46). Speed of reaction is most

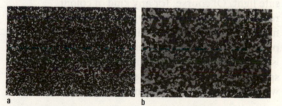

a                              b

46. Enlargements of the 'grain' of different photographic emulsions (reproduced in half-tone)

important in daylight photography, or the photography of moving objects, as the exposure must be very brief. With artificial illumination and powerful sources of radiation, speed is much less important, and therefore the emulsions used for reprographic work are essentially fine grain and relatively slow. In practice these fine-grain emulsions have a high enough resolution to reproduce any type of image likely to be encountered in cartographic work.

## Spectral sensitivity

The silver halides are basically sensitive to radiant energy at wavelengths in the blue, violet and ultra-violet parts of the spectrum. Spectral sensitivity can be extended by adding dyes during manufacture. If the sensitivity is extended into the green part of the spectrum then the emulsion is known as orthochromatic (Fig. 47). If it is further increased to include the red as well, making it sensitive to the whole of the visible spectrum, then it is known as panchromatic.

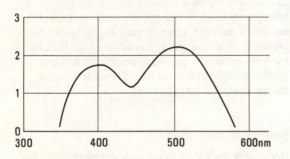

**47.** Spectral sensitivity curve, orthochromatic emulsion

Although most cartographic processing uses achromatic images, orthochromatic films are preferred, as they have shorter exposure times than blue-sensitive films, and can be used under a red safe light. Of course, for any kind of colour reproduction it is necessary to use panchromatic film, but this is rarely the case in cartographic production. The spectral sensitivity of any emulsion is usually described by means of a wedge spectrogram, which indicates graphically the wavelengths to which the emulsion is sensitive.

## Contrast and density range

The latent image formed on exposure is converted into a visible and permanent image by development, and the resultant image is composed of metallic silver, which appears black or grey. The degree of opacity of the image is therefore dependent on both exposure and subsequent development. In a photograph of black lines on a white background, the areas unaffected by radiant energy (corresponding to the black lines) will have all the unexposed silver halides removed by washing, and only the translucent base material and gelatin coating will remain, thus forming a negative image.

Transmission is the ratio of transmitted light to incident light, and is expressed as

$$T = \frac{I_t}{I_o}$$

Opacity is the reciprocal of transmission, as it describes the extent to which the material interferes with the transmission of light. Transmission is always less than unity, and is normally expressed as a percentage. Opacity is always greater than unity.

The degree of opacity of an image is normally measured as its density ($D$), which is the logarithm of the opacity. Therefore,

$$D = \log_{10}O = \log\frac{I_o}{I_t}$$

So that, for example, if $T = 50\%$ (½), then

$$D = \log\frac{2}{1} = \log 2 = 0.3$$

The use of logarithmic values is convenient in expressing rates of change graphically.

## The characteristic curve

The characteristic curve is a graph (Fig. 48) that shows the relationship between exposure ($E$) and the resultant density in the negative photographic image. The normal method of composing this is by

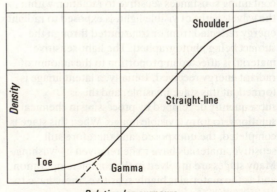

**48.** Characteristic curve

the use of a step wedge containing a range or scale of different densities between maximum and minimum opacity, arranged as a series of 'steps'. This represents the range of densities that might occur in a continuous-tone photograph. Density is plotted on the vertical axis, and the logarithm of exposure (step-wedge density) on the horizontal axis. If both axes are scaled in equal units, then if the negative produced corresponded exactly to the series of step-wedge densities, the curve would be a straight line at 45°. If the curve was a straight line at any other angle, then there would be an equal rate of change between exposure and resultant density.

In a normal characteristic curve usually three distinct parts can be distinguished; the toe of the curve, where increase in exposure gives little or no increase in density; the 'straight-line' portion of the curve where the rate of change is constant; and the shoulder of the curve, where again any increase in exposure leads to little or no change in density. The curve indicates that very small and very large amounts of exposure do not produce the anticipated effect on density in the negative.

The relationship between exposure and the resultant density depends on both the length and angle of the straight-line portion of the curve. If the straight line is extended to meet the base, then the tangent of this angle is expressed as gamma (γ). A gamma of 1.0 is equal to an angle of 45°. Gamma shows the relationship between the densities in the subject and the densities in the negative. If the gamma is high, then a small change in exposure leads to a large change in density. If the gamma is low, then a large change in exposure makes a small difference in density. Gamma is a measure of contrast, and this is different to density range, which is the difference between the maximum and minimum densities in the subject, or on the negative. A long, steep, straight-line portion indicates a large density range in the negative, which may be quite different to the density range of the subject.

The gamma of a negative is affected by development, and by varying the duration and method of development, different gammas can be produced using the same emulsion and exposure. Agitation in development generally leads to an increased rate of development, and therefore a higher gamma.

## Contrast and emulsion type

In process work, there are basically two different requirements. For the reproduction of discrete images, which are typical of most cartographic operations, high contrast is needed, as the objective is to make the maximum difference in density between image (transparent on the negative), and background (opaque on the negative). Therefore, an emulsion with a steep straight-line curve divides the negative densities into essentially transparent and opaque parts (Fig. 49). For the reproduction of continuous tone, the objective is in the first place to represent the tonal values in the subject by equivalent density values in the negative. This requires less contrast, as the intermediate densities between maximum and minimum should be retained. Note that in this respect a half-tone image, being composed of discrete dots or lines, needs a high contrast emulsion for sharp reproduction.

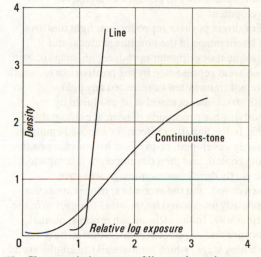

**49.** Characteristic curves of line and continuous-tone emulsions

## SPECIALISED EMULSIONS AND PROCESSING

### Positive reversal

In normal photographic operations, the exposure of a positive image yields a negative, and conversely the exposure of a negative image leads to a positive. Although this has many advantages, it is also desirable in some cases to be able to make a positive directly from a positive. As this is the reverse of the normal procedure, it is usually referred to as positive reversal.

Solarisation reverses the image formation through

using a high degree of over-exposure, so that the silver recombines with the bromine in the exposed non-image areas, which then cease to develop. Normal development follows for the positive image areas, which have received much less exposure.

The Herschel effect is more commonly used to produce direct image reversal. The principle is that a latent image can be destroyed by exposure to long wavelengths. In practice this type of film is chemically fogged during manufacture, and therefore if developed without any further exposure would produce a completely black image. Thus it behaves in the opposite way to normal photographic emulsions. Long wavelength exposure destroys this ability to form an image by development only, whereas a further exposure to short wavelengths restores the ability to produce an image on development.

For direct positive reproduction, light destroys the latent image in the non-image areas, and subsequent development produces an image only in those areas represented by the positive image, which have not transmitted or reflected any light.

Reversal images can also be produced by modifying the processing of normal photographic films. In the Sabattier effect, the exposed emulsion is partly developed, re-exposed without the positive being copied, and then development is completed. The partly developed non-image areas are desensitised, and the second exposure causes the previously unexposed image areas to develop in the normal way. In the etch–bleach process, normal exposure and development are followed by a bleaching stage, which removes all the emulsion and gelatin in the exposed and developed areas. Subsequent exposure then produces the positive image, which is isolated on a clear film base.

## Photolysis

Exposure of a photographic emulsion to intense light brings about the decomposition of the silver halides, and the finely divided silver particles appear visually as a faint coloured image. Often referred to as 'secondary' images, the colour depends upon the degree of over-exposure and the type of emulsion. Prolonged exposure of this type requires a high-density negative for effective copying. The resultant colours may be grey, brown, pink, violet or yellow.

## Dye former development

Although the silver halides normally produce an achromatic image, special dye former developers can change this to a coloured image during development. Two methods are possible. In the first, development with a dye is followed by bleaching to remove the silver image; in the second, a finished photographic image is bleached to remove all the developed silver, and subsequently redeveloped with a dye former and bleached again.

## Diffusion transfer

Although in the normal photographic process a positive is usually made by exposing a negative of the image in a separate operation, the two processes can be combined by diffusion transfer. The negative layer contains a silver-halide emulsion. The positive layer is not light sensitive, but the gelatin contains development nuclei which can react with unexposed silver in the negative. Both layers may be incorporated on a single supporting film, or the negative latent image may be brought into contact with a separate sensitised sheet to produce the positive.

Exposure to a positive image means that in the exposed non-image areas of the negative the silver halides are developed into silver, which does not affect the positive layer or film. The soluble silver salts in the unexposed (image) parts of the negative can react with the positive layer. Development occurs when the two layers are brought into contact, and they are subsequently peeled apart.

The best known example is probably the Polaroid/Land process. After exposure the double layer is drawn between rollers, releasing developer from tiny vesicles. Continuous-tone images can be reproduced.

The process has been extended to colour reproduction, by adding blue, green and red sensitised layers. The alkaline processing agent is a thin layer between the negative and positive components, and again this is ruptured under pressure to release developer. In the most advanced system, the processing agent contains both alkali and titanium dioxide. The titanium dioxide forms a white pigmented surface against which the coloured image can be seen, and therefore there is no physical separation of layers. The final image intensifies under exposure to light.

## Developments in photographic films and processing

The manufacture of photographic films for the

graphic arts industry is a highly sophisticated technology, and many different types of film are provided for different purposes. The main variations depend on the choice of base material, the speed of reaction to exposure, the degree of contrast required, and subsequent use of the photographic image.

For reproducing discrete images (usually referred to as line images), most manufacturers provide high-contrast orthochromatic films, on either a clear or a matt support. For cartographic work, where dimensional stability is important, polyester-based films are invariably used.

Major developments in the emulsions and processing have taken place, partly to satisfy specialised needs, and partly to meet increasing competition from other types of photoreactive materials. Conventional photographic processing is relatively slow and expensive, and processing has to be carried out under safe light conditions. Consequently there has been a rapid development of automatic processing units (which have the advantage that the silver can be retrieved); contact copying films which can be used in daylight; much improved direct positive films for positive – positive copying; and wash-off films in which both the unexposed emulsion and the gelatin layer are removed from the base.

Rapid access processing refers to the reduction of development time and complication in an automatic processing system. In the conventional form, specially devised emulsions are developed at higher temperatures in processor units, reducing the processing time to about one to two minutes for each image. In a different version, developer is contained within the photographic emulsion, which remains inert until brought into contact with a special activator. This process can operate at lower temperatures, and has the advantage that the efficiency of the developing solution is not changed during the operation of the processing machine.

## PHOTOGRAPHIC OPERATIONS

### The process camera

The process camera serves three main purposes. If the original image is on an opaque material, then it must be photographed by reflected light, and this can be done using the camera. If a change of scale is required, then the lens system of the camera makes

this possible. And if conversion to half-tone is being carried out using a glass crossline screen, this is also a function of the process camera.

As they frequently have to deal with large images, process cameras are often large instruments. They are generally classified by the dimensions of the maximum image size that can be produced. Although various types and configurations are available, the basic arrangement is that the copy to be photographed is held on a copy board or in a transparency holder, and is illuminated by reflected or transmitted light; the lens system is mounted in a light-proof bellows connected to the camera back, which usually has a vacuum support to hold the unexposed film. In most cameras there is also a device for bringing the glass cross-line screen into operation if required.

In order to control both scale and focus, two of the three elements have to be movable. In the gallery camera, the copyholder is fixed, and both the lens and camera back can be moved along a track. In the more common dark-room camera the camera back is fixed in the wall of the dark room, and both the lens and copyholder are movable along the track (Fig. 50). Some cameras use a floor track, whereas others have the units suspended from an overhead gantry.

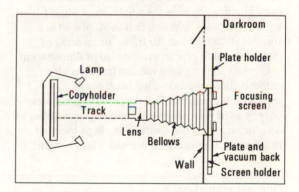

**50.** Plan of dark-room process camera

With large format images, the geometry of the entire camera system is highly important. The planes of the copyholder, lens and camera back containing the film to be exposed must be absolutely parallel, whatever movements are made. Consequently, large process cameras have to have an extremely robust and rigid construction in order to maintain correct geometrical relations. Large cameras are usually horizontal, but smaller cameras are sometimes arranged vertically.

In order to focus and scale the image, the camera normally has a ground-glass viewing screen, which is subsequently moved out of the way to bring the unexposed film into position for exposure. In modern semi-automatic cameras, all operations are carried out by the operator in the dark room through a control console.

## Contact copying

Contact copying with photographic materials is widely used for many operations in which a change of scale is not involved, including both line and half-tone image production. It is normally carried out with a vacuum frame and a source of illumination, operating through transmitted light. In a vacuum frame, the plate glass cover can be lifted, and the translucent image and unexposed film inserted on top of a black rubber blanket. Large frames are normally loaded in the horizontal position, and then swung to the vertical position before exposure, but a vertical arrangement is also possible. After loading, a partial vacuum is introduced by means of a pump, and the purpose of this is to bring about absolute contact between the image and the unexposed light-sensitive material. For large images there is a problem of obtaining even illumination over the whole image area. Either a powerful point light source is used, placed at a suitable distance from the frame, or an array of lamps is used to give an even spread of illumination.

Contact copying with vacuum frames is also extensively employed with other types of reprographic material, but the process camera can only be used with silver-halide photographic film, as no other material has a sufficiently high speed of reaction to be of practical use with a lens system.

## Line reproduction

Discrete images are often referred to as 'line' images, and as in many cases most of the map information is carried by line and point symbols, the reproduction of such images is important cartographically. The objective is a sharp distinction between the image detail, which should be transparent on the negative, and the background or non-image areas, which should be opaque on the negative. The edges of all detail should be sharp, without any intermediate densities being present.

The reproduction of line images has been assisted by the introduction of lithographic or 'lith'

emulsions, so named because reproduction by lithographic printing requires an image consisting of discrete elements. Both line images and half-tones need to be sharply differentiated in order to reproduce well on the printing plate, and subsequently in print. The characteristic curve of a lith emulsion has a short toe and a high gamma (Fig. 49). A special developer is used, which can produce very high densities, and development is infectious, as the initial reaction results in accelerated development of adjacent grains. Both exposure and development have to be carefully controlled.

## Continuous tone and half-tone

Because the printing processes used for the printing of maps – predominantly lithography – can only reproduce discrete images, a continuous-tone subject can only be reproduced in print by being converted to half-tone. This is highly important in general graphic arts, where the half-tone conversion normally accompanies or follows the colour separation process. The principal cartographic applications of half-tone are the reproduction of continuous-tone images such as hill shading and orthophotographs. Continuous tone can be reproduced photographically, using a suitable silver-halide emulsion, and a continuous-tone negative of the original is often made as the first step in half-tone production. The breaking up of the continuous-tone image into discrete elements is done using half-tone screens, of which there are basically two types: the glass cross-line screen, and the vignetted contact film screen.

## Half-tone screens

A cross-line glass screen consists of a series of fine parallel lines placed at right angles, dividing the screen into tiny apertures (Fig. 51). Light can only penetrate the screen through these apertures, and the resultant image has a 'dot' form (Fig. 52). The size of the 'dot' formed on the negative in the camera depends primarily on the intensity of light

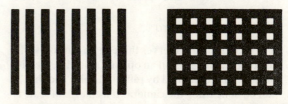

**51.** Diagram of the structure of line and cross-line glass half-tone screens

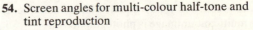

**52.** Dot formation (enlarged) produced by a cross-line half-tone screen

reflected from or transmitted through the copy at any point. Light areas of the copy will reflect most of the incident light, and therefore form large 'dots' on the negative. Conversely, dark areas of the copy will reflect little incident light and form only a small dot through the screen aperture. The screen has to be used in a process camera, and is placed at a distance in front of the negative film held in the camera back. The distance of the screen from the emulsion is critical, as a sharp dot must be formed. According to the diffraction theory, the screen acts as a diffraction grating.

The structure of the vignetted film screen is quite different. Instead of a grid or grating of opaque lines, the screen consists of a regular series of graded 'dots', which vary in density from centre to periphery (Fig. 53). The resultant dot size in the half-tone negative again depends on the intensity of light reaching any point on the screen. A high intensity will penetrate most of the screen dot, whereas a low intensity will only penetrate a small point. The vignetted dot screen is used in contact with the emulsion of the negative. With both types of screen the resultant half-tone image should reproduce the different tones of the original subject, and the 'dot' structure is below normal visual resolution. In practice, perfect reproduction of the tonal values by the half-tone reproduction is virtually impossible to attain, and good half-tone reproduction photographically can only be achieved by complex exposure and processing operations.

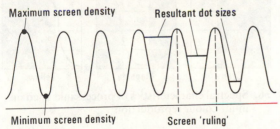

**53.** Vignetted dot contact film screen

The resolution of the half-tone screen is controlled by the frequency of the lines or vignetted dots of which it is composed. This is described by the number of lines per centimetre or inch. Those used for high-quality reproduction normally have 48, 54 or 60 lines per centimetre, although finer screens are available.

A range of screens, identified by number of lines per centimetre and inch, is as follows.

| | |
|---|---|
| Lines per cm | 25 30 34 40 44 48 54 60 70 80 90 100 120 |
| Lines per inch | 65 75 85 100 110 120 133 150 175 200 225 250 300 |

For rectangular screens, formats range from 12 × 17 cm to 80 × 100 cm. Circular screens will cover the same range of rectangular sizes.

## Screen angle

In multi-colour reproduction, where the tonal values of each colour are represented by half-tones, the angular position of the different coloured images is critical. If two or more coloured half-tones are superimposed, there must be an angular separation between them. If this is not done, interference patterns called *moiré* may be introduced. The angular separations are based on 30° and 15°. Where the standard process colours are used, a typical system is to place the black at 45°, the yellow at 90°, cyan at 75° and magenta at 105° (Fig. 54). Other trichromatic combinations, using red, blue and yellow, follow the same principle. Placing the different half-tones at the correct angle is achieved with the glass cross-line screen by rotating it in the holder in the camera back, the screen itself being circular. With vignetted screens used in contact,

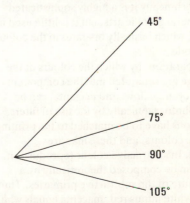

**54.** Screen angles for multi-colour half-tone and tint reproduction

either the image being reproduced must be placed against the screen at the correct angle, or more commonly, sets of identical screens at different angles are used.

## Half-tone processes

In printed reproduction the density range that can be achieved is limited. For example, in lithographic printing the maximum density range is usually about 1.4, due to the fact that a thin ink film is produced, and glossy papers with high reflection are usually avoided. The consequence of this is that although the original subject may have a much higher density range, it is not possible to reproduce this in print. To make a good reproduction the density range has to be restricted during the half-tone operation.

The normal practice in using the crossline screen in the camera is to produce the half-tone or screened image in one operation. With the vignetted dot screen it may be carried out in two stages. First, a continuous-tone negative of the original is produced, using a suitable emulsion, and then this is contact copied through the vignetted dot screen on to a lith type film to produce the half-tone. The advantage is that the density range can be adjusted on the continuous-tone negative. In order to improve the reproduction of very light and very dark areas of the original, multiple exposures may be used.

## Colour reproduction

The expression 'colour reproduction' is normally applied to the operation whereby a multi-colour tonal original is converted into a set of separate colour half-tones for reproduction in print. Therefore, it involves both colour separation and half-tone. Technically it is a highly sophisticated aspect of general graphic arts, but it is little used in cartography, which basically operates in the colour-separated mode.

Colour separation, by which the colours of the original image are separated into the component primaries of cyan, yellow, and magenta, can be carried out photographically by the use of filters. The separations have to be matched to the printing inks for these colours, and these inks are standardised. In subtractive colour reproduction, any colour can be composed theoretically by a suitable mixture of the subtractive primaries. Thus cyan equals white minus red; magenta equals white minus green; and yellow equals white minus blue. If the multi-colour image is photographed on

panchromatic film through a blue filter, then the resultant negative records the blue light reflected by the original. This separation negative is converted into a positive image on the printing plate, and this positive will be the converse of the negative; therefore, it will print yellow. Similarly, the green light record produces the magenta positive, and the red light record produces the cyan positive (Fig. 55).

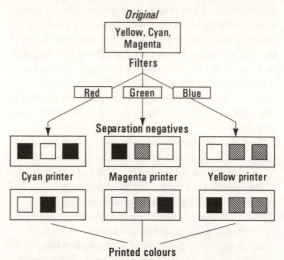
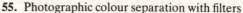

**55.** Photographic colour separation with filters

Because the printing ink pigments are not spectrally pure they do not match exactly the separation images produced by the photographic colour filters (Fig. 56). In addition, as the images are converted into half-tone, these half-tones are not a perfect match for the tonal values of the original.

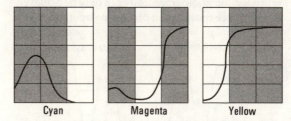
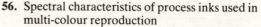

**56.** Spectral characteristics of process inks used in multi-colour reproduction

Consequently, in order to improve the quality of the reproduction, colour correction is carried out. For photographic reproduction, this is done by using photographic density masks, which are positive tonal images made from the separation negatives (Fig. 57). The density of these masks is controlled, so that they add to the densities on the separation negatives. The cyan ink absorbs red, but also a little green and blue, so it is necessary to reduce the

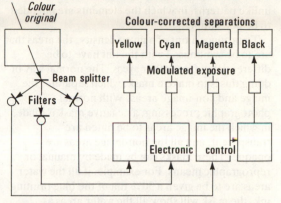

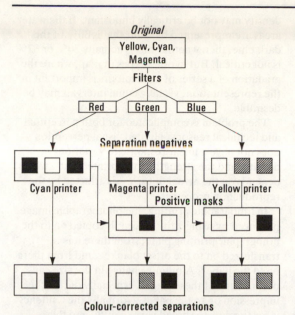

**57.** Colour correction by positive masking

**58.** Colour separation and correction by electronic scanning

amount of magenta printed in cyan areas. Yellow is spectrally the most efficient of the three process inks, but it absorbs a little green as well as blue. Therefore, the yellow should be reduced in magenta areas. In simple one-stage masking, the cyan printer is left unmodified; a reduced positive is made from the cyan negative and superimposed on the magenta negative to make the colour corrected positive. Similarly, a positive mask made from the magenta separation negative is placed in contact with the yellow separation negative to produce the yellow positive. This adds density to the negative in the magenta areas, reducing the yellow in those parts where yellow is also present.

In practice more complex systems are used, and for the very highest quality of reproduction, the half-tone dot sizes can be further modified by manual retouching, reducing the dot sizes where necessary by etching. These involved procedures were always time consuming and therefore expensive, and increasingly colour separation is carried out by electronic scanning.

## Electronic scanning

In the scanning method the image to be reproduced is scanned in a raster by moving a small light spot across it, point by point. This may be transmitted through the original or reflected from it. The reflected or transmitted light, the intensity of which will depend on the tonal value at any point, generates a small electric signal, which is then

multiplied (Fig. 58). This signal can then be modified, and so corrections can be introduced. If filters are added, then the process can produce colour separations, and signals from the other filtered records can be used for colour correction. As the image is scanned in a raster, it is also converted into a half-tone at the same time as the colour separation is carried out. The system can be operated in various ways; some will operate directly on the multi-colour image, whereas others can process colour-separated images produced photographically in the first place. In modern devices feedback from one image is used to carry out corrections to others. High-resolution raster scanning means that the speed and efficiency of the colour separation and half-tone process is increased, and high-quality colour reproduction is now normally achieved.

## Production of tint by photography

In map production, tints are used much more extensively than half-tones, as they provide a useful means of creating a range of area colours for area symbols, and can also be used to lighten line symbols. A tint is similar in structure to a half-tone, being composed of an array of discrete points or lines, but differs in that the same dot size or line gauge occurs all over the image. Therefore, making a tint requires a screen of the required density for each tint value. Tint screens cannot be used for half-tone reproduction.

In the printed image, a 10% tint is one in which the printed dots occupy 10% of any unit area, the remainder being 'white' space. So what is perceived is a combination of 10% printing ink and 90% white, giving the visual impression of a very light continuous colour. Tints are below visual resolution, unlike patterns, in which the elements are intended to be perceptible.

To produce a tint of a given density, the areas that are to be symbolised with the tint have to be differentiated from other areas. The normal way of doing this is to make a mask, which separates the image and non-image areas. With normal photographic processing, a negative mask is made, in which the image areas to be tinted are transparent, and all the non-image areas are opaque. These masks can be made by manual or reprographic means. For example, if all the water areas are to be given a 30% tint of the blue printing ink, the mask will show all the water areas as transparent, and therefore separate these from all other areas.

## Percentage tint screens

Tint screens are film screens used in contact. The screen resolution is measured by the number of lines of dots or lines per centimetre or inch, referred to as the screen ruling, like half-tone screens. Typical screen rulings are as follows.

| Lines per cm | 25 | 34 | 40 | 48 | 54 | 60 |
|---|---|---|---|---|---|---|
| Lines per inch | 65 | 85 | 100 | 120 | 133 | 150 |

The tint density or value is normally expressed as the proportion of incident light transmitted through the screen. Therefore, a 10% tint screen is a negative which transmits only 10% of the incident light, producing a pattern of very small opaque dots on the positive. In order to provide a range of different densities, sets of tint screens are normally manufactured in a series of percentage values. Although values based on 10% differences are common, very often more steps are needed at the lighter end of the scale. For example, the ByChrome Company manufactures a series of percentage screen tints with the following values (Fig. 59).

% 5 10 15 20 25 30 40 50 60 70 80 90

However, it is clear that such a series does not produce a set of visually equal intervals. In many cases, if only a single percentage tint is required to distinguish an area, the perceived percentage

density may not be critically important. If the water areas are represented by a tint that is 30% of the dark blue, then whether this is actually 30% or 32% is not critical. But in other types of map, where the gradation of a series of area colours is important in the representation, visually equal intervals may be desirable.

The problem is complicated for both perceptual and technical reasons. Human visual perception does not have a linear response to variations in density; and the strength of the printed tint – which depends on the actual dot size – is affected by the reproduction processes.

In lithographic printing, the photographic image, produced by using a tint screen, is copied on to the lithographic printing plate; from there it is transferred on to the offset blanket; and from there on to the paper. A minute change in dot size can occur at any of these stages, especially during the impression on to the paper. Normally the tendency is for the dots to spread very slightly, and this is referred to as dot gain. Therefore, the printed dot is not necessarily the same as the dot on the original tint screen, and any change in dot size, of course, affects the perceived density of the image. In order to compensate for this, tint screens can be produced which appear to have visually equal intervals in print (see Brown 1980). The Bychrome coordinated tint screens (ByChrome Company, Inc. 1986) have the following characteristics (Fig. 60).

| Code | Density | Transmission % | Density as % of solid on offset paper |
|---|---|---|---|
| ½A | 1.52 | 3.02 | 5.22 |
| A | 1.22 | 6.03 | 8.7 |
| B | 1.00 | 10.00 | 13.91 |
| C | 0.77 | 16.98 | 19.13 |
| D | 0.61 | 24.55 | 25.22 |
| E | 0.51 | 30.9 | 33.04 |
| F | 0.41 | 38.9 | 42.61 |
| G | 0.39 | 40.74 | 46.09 |
| H | 0.25 | 56.23 | 64.35 |
| J | 0.19 | 64.57 | 74.78 |
| K | 0.13 | 74.13 | 85.22 |
| L | 0.09 | 81.28 | 93.91 |
| M | 0.05 | 89.13 | 97.37 |

Tint screens are also produced in angled sets, as if two or more colours are printed in the same area,

**120 LINE**  5  5  10  10  20  20  30  30  40  40  50  50  60  60  70  70  80  80  90

**59.** ByChrome percentage-calibrated screen tints

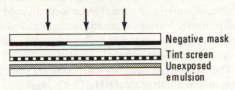

**60.** ByChrome code-controlled screen tints

they must be placed with the correct angular separation to avoid *moiré* effects.

## Photographic processing of tints

To produce the required tint photographically, the unexposed photographic film is placed face up in the vacuum frame; the required tint screen is placed face down at the correct angle; and then the negative mask is added, also face down (Fig. 61).

Negative mask
Tint screen
Unexposed emulsion

**61.** Photographic production of screen tint with negative mask

The object is to have the minimum separation between the unexposed film and tint screen itself, in order to produce sharp dense dots in the tint. On exposure, the light is transmitted though the mask, which controls the areas being exposed, and then through the tint screen, which forms the image. The resultant positive should have dots of exactly the same size as on the tint screen itself.

Although screens with a dot structure are the most common, line screens are also widely used, and indeed some organisations maintain that they are more easily controlled in reproduction than dot screens. Screens with a low resolution, that is with relatively few lines per centimetre or inch, can be used to introduce a perceptible effect. If the individual lines of which the tint is composed can be seen, then the tint takes on some of the characteristics of a pattern.

Although small-format tint screens (up to 50 × 60 cm, or 20 × 24 in) of the type widely used in the graphic arts industry, are relatively easily obtained, large-format tint screens for cartographic purposes are often produced in much smaller quantities, and are relatively expensive. Screens up to 120 × 150 cm can be obtained from suppliers such as Arnold Cook Ltd.

## Bi-angle screens

Screening a line image is not common in cartography, but it is very useful in particular cases, especially the reduction of a black line to appear as grey. Although the process works satisfactorily with a normal tint screen if the screen is at an angle of 45°, if the screen angle is close to the line direction, then the interruption to line continuity is quite noticeable.

A bi-angle screen is composed of two tint screens of different percentages combined at an angle of 30°. The percentage tint value is the addition of the two components. The screens are used in the normal way. Of course, it is desirable to use fine screens to minimise the visual interruption, and consequently screen rulings of at least 40 lines per centimetre (100 lines per inch) are always used.

# 8 NON-SILVER-HALIDE PROCESSES

Although the term photography can be used to refer
to any method by which light is used to form an
image, it is normally associated with silver-halide
materials. Even so, there is a large number of other
materials that will form images on exposure to
suitable radiation, and many of these are important
in cartography. Most of them are particularly useful
in dealing with line images, as they have little or no
capacity for continuous-tone reproduction. They
include processes that can be used to produce
positives directly from positives, for constructing
tints, for image combination, and for the addition of
colour. They are all characterised by the fact that
the image formation depends fundamentally upon
the radiant energy, and there is no equivalent to the
development stage used in silver-halide
photography.

## Ferric salts processes

These have been important historically, because
they provided a simple and inexpensive means of
copying large format line images. The ferric salts
react with other substances either to provide a
coloured image, or to make a sensitive layer
insoluble. The blueprint process uses a solution
containing potassium ferricyanide and ferric
ammonium nitrate, applied to a gelatin-sized paper.
Exposure to short wavelength radiation causes
reduction to ferrous salt, which reacts with the
ferricyanide to give a blue image. The process
involves image reversal, a positive resulting from the
exposure of a negative. After exposure it is washed
in water to remove the unexposed materials, and
then developed with an oxidising agent to produce
the deep blue colour.

The basic process is still used occasionally to
provide guide images in blue (blue keys) for
cartographic production. In this form, exposure is
followed by washing without the development stage
needed to produce a deep blue, as only a faint blue
image is required.

## DICHROMATED COLLOIDS

Dichromated colloids consist of a colloid, such as
gelatin, albumen or glue, dissolved in water, to
which a soluble dichromate is added. On exposure
to short wavelength radiation, the sensitised layer
hardens and becomes insoluble. Unexposed areas
can then be removed by washing, leaving the
resistant layer behind.

The organic colloids used for these materials were
mainly either carbohydrates or proteins. Gum
arabic, albumen and animal glues were all used for
this purpose. Various synthetic materials have also
been introduced, including cellulose acetate, resins,
vinyl polymers and polyvinyl alcohol.

The dichromates used as sensitisers are principally
ammonium dichromate, $(NH_4)Cr_2O_7$, and
potassium dichromate, $K_2Cr_2O_7$. The hardening of
the dichromates on exposure depends on a number
of factors; spectral sensitivity, dichromate
concentration, pH value, thickness and moisture of
coating, and dark reaction. The dichromates absorb
radiation with wavelengths in the ultra-violet, violet
and blue regions of the spectrum (Fig. 62). The
solutions gradually lose their sensitivity even when
kept in the dark, and therefore surfaces to be
exposed have to be coated shortly before exposure.

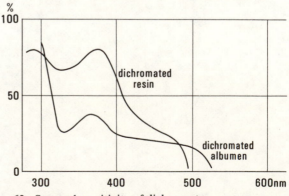

62. Spectral sensitivity of dichromates

In practice a thin layer of coating is applied to the surface of the support material, by hand, with a whirler or a coating machine, and allowed to dry. It can be handled quite safely in subdued light, as the material is relatively insensitive. Exposure has to be made to a powerful source of radiation, and exposures of several minutes are common with proprietary coatings. Because natural materials are involved, and because sensitivity is affected by local conditions of humidity, temperature, coating thickness, etc., it is not possible to standardise exposure times.

For a long period the chief use of dichromated colloids in the printing industry was for image formation on printing plates. They are now little used for this purpose, but as they can be applied as light-sensitive layers to plastic sheets, either opaque or translucent, they are still useful for a number of cartographic production requirements. There are basically two methods of image formation. If the sensitised layer is exposed to a line negative, the image areas are made insoluble, and the unexposed coating is then dissolved (Fig. 63) leaving a positive image. If a coloured dye is added to the coating, then the resultant image will be coloured. If exposure is made to a line positive, the non-image areas are hardened, and the image areas can be dissolved away, leaving a resistant stencil around the image. The unexposed coating is removed with a calcium chloride 'developer'. Dye can then be applied to the surface of the sheet, and will only remain in the exposed image areas. The hardened stencil is then removed with water.

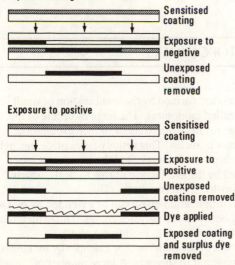

**Exposure to negative**

Sensitised coating

Exposure to negative

Unexposed coating removed

**Exposure to positive**

Sensitised coating

Exposure to positive

Unexposed coating removed

Dye applied

Exposed coating and surplus dye removed

**63.** Image formation with dichromates

The negative-working process can be operated with coatings applied by hand, the solution being wiped on to the surface of the sheet and allowed to dry. Positive-working solutions need to be applied with a centrifuge (whirler) or a coating machine. It is also possible to apply most coatings by using a coating machine, which is capable of rolling a thin layer of liquid or viscous material on to a supporting layer, either film or paper. These machines can be obtained in a variety of sizes. During the long period when dichromated colloid solutions were used to form images on lithographic printing plates, the necessary equipment was available, and therefore these operations on plastic sheet materials could be carried out without any additional equipment. With the virtual disappearance of dichromated colloid coatings for printing plates, this is no longer the case, and this is one reason why the negative-working system is more widely used.

The process will only reproduce discrete images. Sufficient exposure is needed to harden the colloid layer right through to the base, so that it is fully resistant. Exposure times are not critical, and over-exposure is common in practice. With plastic sheet materials, the image-forming process can be repeated as many times as desired, re-coating the surface and adding another image. Therefore, it can be used to combine a series of negatives to form one positive image, or to combine a series of positives to form one positive image. It can be used achromatically, with black dye, for example to construct a combination of a line and tint image. It is a simple means of producing large-format images in either achromatic or chromatic form, but tends to be rather slow. Whereas the dyes used with the positive-working system will actually penetrate the surface of polyvinyl sheets, this is not the case with polyester sheet materials, on which all images are superficial.

## DIAZO COMPOUNDS

Diazo compounds are complex unsaturated aromatic compounds, formed by treating an aromatic amine salt, such as aniline hydrochloride, with sodium chloride, in the presence of excess mineral acid. Such materials have the property of combining with other substances, such as phenols, to produce coloured azo dyes. For rapid image formation, this coupling reaction is brought about by creating an alkaline environment. Reaction is prevented in the applied coating by keeping it acidic. Exposure to radiation with a wavelength predominantly in the violet and ultra-violet destroys this coupling ability, and the release of nitrogen in the exposed areas brings about the formation of a colourless phenol.

All the diazo compounds are chemically complex, and many thousands have been investigated. The useful ones must be soluble in water in order that sensitising solutions can be prepared. The product resulting from exposure must be sufficiently clear and colourless to provide a 'background' without any density or discoloration. As there is no development of the image, all the energy required to destroy the coupling ability in the exposed areas has to be provided by the radiation source.

The spectral sensitivity of diazo compounds is mainly in the violet and ultra-violet, and for normal materials, composed mainly of benzene and naphthalene diazonium salts, 380 to 400 nm is the peak sensitivity (Fig. 64). Carbon arc, mercury vapour and gas discharge lamps are suitable sources of radiation.

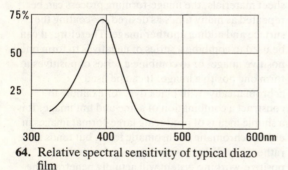

**64.** Relative spectral sensitivity of typical diazo film

The principal characteristic of diazo coatings is that a positive image is formed by the exposure of a positive. There are several different ways in which the coloured image can be 'developed'. In the semi-wet system, the sheet is coated with the diazo solution only, and the colour and coupling agent are applied to the exposed surface through a solution which is alkaline. The widely used 'plan printers' or dyeline machines, normally employ this method, which is easy to operate. The colour is controlled by the dye coupler used. In the dry process, the coating contains both the diazo compound and the dye coupler. After exposure, development of the image is brought about by releasing ammonia vapour over the material, usually by heating aqueous ammonia in an evaporation chamber, and allowing the fumes to reach the diazo. As the fumes are noxious, good ventilation is required, with the removal of the

fumes by means of a ducting system and fan.

As the process is positive-positive, only one image can be produced on the coated material. Diazo-sensitised papers and films are available, and those used for cartographic purposes are on polyester plastic bases.

In the standard diazo processing machines, exposure is usually carried out by passing the positive image and the coated material around a cylinder, inside which are one or more tubular lamps. Although exposure intensity is adequate, lack of full contact between the positive and the diazo material means that images with fine detail, which are common in cartography, may suffer from poor contact during exposure. One alternative is to expose the image in a normal vacuum frame, and then carry out the rest of the processing to produce the visible image in the relevant machine.

The chief use of diazo material is to provide large-format line image reproductions at relatively low cost. Diazo materials have good resolution, and properly operated the diazo process will provide high quality line, tint and half-tone copies. A limited range of coloured images can be produced, but each needs a separate piece of material. Negative-working diazo materials do exist. With the diazo-sulphonate group, the sensitised compound does not have the property of coupling with a dye, but it can be restored by exposure to ultra-violet radiation, so that the image is formed in the exposed areas instead of being destroyed. A fixing operation is needed to remove the unexposed material, and the image generally suffers from background discoloration.

## PHOTOPOLYMERS

Polymers are formed by the combination of small units, called monomers. The large molecule of the polymer results from chain or branch grouping of monomer particles. This activity takes place through the absorption of energy, and the reaction is initiated by the production of free radicals, which are short-lived intermediates. This photomultiplication effect can be extended by cross-linking between the polymer chains. The main effect of polymerisation is to modify the physical properties of the material, for example by making it more resistant to solvents. A large number of substances have this property of polymerisation to some degree, including polyvinyl acetates, polyvinyl esters, cellulose acetates and polyamides. Initiators include carbonyl, diazo compounds, halogen, and other dye compounds. Although some of these polymers are sensitive to radiation at about 300 nm, this can be extended to about 500 nm with other sensitisers. Unlike silver halides, the process is not assisted by chemical development, so the imaging process is completed either by dissolving away the unexposed material in a solvent (preventing further polymerisation by exposure to a different wavelength) or by heating. Many variations in image formation are possible, and photopolymerisation is likely to increase in importance as an imaging system.

Present applications of photopolymerisation include coatings on lithographic printing plates, and the production of coloured images by various means. The polymer layer can include a dye or pigment, or it can be characterised by a degree of 'tackiness'. In this form, pigment particles will adhere to the surface of the unexposed material, and exposure and polymerisation destroy this tackiness, so controlling image formation. Exposure and polymerisation can also be used with some materials to bring about changes in the adhesive properties of the material, and this also can be used to control image formation.

Photopolymers are also useful as an inexpensive recording medium that can be applied to large surfaces, for example large-format displays. Different materials have been produced which are sensitive to red and ultra-violet light, or an electron beam. Laser beams can be used to 'draw' on the film to produce an image almost immediately.

## PHOTOCONDUCTIVE PROCESSES

A quite different method of image formation is based on the fact that some materials are electrical insulators in the dark, but become conductors when exposed to light, or in some cases infra-red radiation. This variation in static charge retention can be used to form a 'latent' image, which can be made into a permanent visible image by the subsequent application of fine particles of pigment that are attracted to the latent image by opposite polarity. These methods are also referred to as electrostatic, electrophotographic, or electrostatographic.

### Electrostatic processes

Different applications of this basic method can be distinguished according to the way in which the pigmented image is formed. The image may be formed directly on sensitised material, or on a sensitised plate and then transferred to a paper or other surface.

In the xerographic process, a coating of vitreous selenium is applied to a support. This coated plate is passed under a charging grid, which applies a positive static charge to the surface (Fig. 65). The charged plate is then exposed to a positive image, either in contact, or by projection. The spectral sensitivity of the sensitive coating lies within the visible spectrum, being approximately the same as that of orthochromatic photographic emulsions.

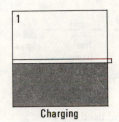
Charging

Charged plate

Exposure to image

Charge remains in image area

Charged image cascaded

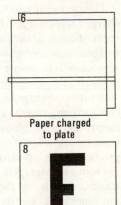
Paper charged to plate

Powder image on paper

Image heat fused

**65.** Image formation by the xerographic process

Exposure causes the static charge to disappear in the non-image 'background' areas, leaving a latent image of the copy on the surface of the plate. Negatively charged resinous fine particles are cascaded over the plate, and adhere only in the positively charged image areas. The image can be transferred by bringing a sheet of paper into contact with the plate, and giving it a positive electrostatic charge, which attracts particles from the plate to form a 'printed' image. The final image is normally stabilised by being heated (Fig. 66).

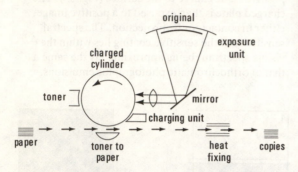

**66.** Schematic diagram of a xerographic machine

Charging

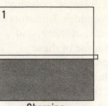
Charged plate

Exposure to image

Charge remains in image area

Liquid toner

Image fixed by evaporation

**67.** Image formation by the electrostatic process

In the direct method the material used is a photoconductive zinc oxide in a silicone resin, which forms a thin film on the supporting base, which may be paper or metal. The sheet is charged and exposed to the positive image (Fig. 67). Instead of cascading with powder, a liquid toner is normally used, in which fine particles are suspended in a liquid which can be removed by evaporation. In the direct reproduction system, each sheet is sensitised with this coating.

This change in conductivity of an electrically charged layer can also be brought about by heat. A sheet of material such as polyethylene is charged and placed in contact with the image being reproduced. Exposure to infra-red radiation then dissipates the charge in the exposed areas. Particles with the opposite polarity will then adhere to the charged image areas. This is an electrothermographic system rather than an electrophotographic one. Coloured images can be produced with liquid toners.

From a cartographic point of view, one disadvantage is that the available commercial machines have been devised primarily for office copying rather than high-quality reproduction. With these, large solid areas do not reproduce fully, suffering from a 'halo' effect, and resolution is comparatively low, so that fine tints or half-tones do not reproduce properly. Various special machines have been developed for cartographic purposes in the past, but the main line of current development is to make use of the electrostatic principle in a different way, by using it to create raster images.

Whereas the conventional electrostatic machines expose the whole image simultaneously, it is also possible to use a radiant beam to control image formation on the raster plotter principle. With these machines, either the presence or absence of a charge at any point in the raster will determine whether or not a drop of ink is deposited, or an image charge is given selectively to image points in the raster and subsequently developed by contact with liquid toners. The development of these machines has been important for high-speed monochrome plotting, and also for the generation of large-format multi-colour copies.

With a laser printer the original image is scanned, and then this is used to expose the image in raster format, point by point. If the photoconductive surface has been positively charged, exposure to the laser beam destroys the charge at any point, leaving behind a positively charged image. This is then made visible and permanent by the application of negatively charged toner. This can be transferred to paper and heat fused.

The laser copying system operates by deflecting the laser beam with mirrors at high speed to form a charged image on a selenium drum. An alternative method is to use ion deposition. In this a charged image is produced on a steel cylinder by firing ions through a charging matrix. Toner is applied to the cylinder, and then transferred to paper under pressure, in contrast to the heat fusion used with the laser printer.

In the Versatec machine, charges are applied through a programmed voltage to an array of 'writing' points, grouped in a stationery drawing or plotting head. These points create minute electrostatic dots on the paper surface, and the visible image is formed by contact with a liquid toner. A complete 'line' is generated at a time, and the paper is passed over the plotting head in steps (Fig. 68). Multi-colour plotters produce four colours by passing the paper through four times, using a different colour toner each time.

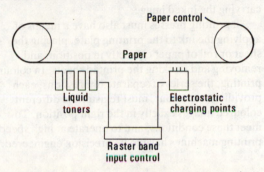

**68.** Schematic diagram of raster-line electrostatic plotting machine

In the continuous ink jet plotter, the minute ink drops are positively charged. If the charge is large enough the drops are broken up into a fine spray. Image formation is controlled by switching the charge on and off. In the 'drop-on-demand' method, fine capillaries hold liquid ink at a slight vacuum. A piezoceramic crystal is charged with an input voltage, and this 'squeezes' a tiny drop of ink on to the paper surface.

## OTHER REPROGRAPHIC PROCESSES

### Thermographic processes

Although thermographic processes have been developed experimentally over a long time, they are likely to become more important with raster plotting and printing devices. In the thermal transfer process (Fig. 69), a raster scan is used to melt tiny dots of coloured pigment on to film or paper, using the raster band method. The coloured wax pigments are carried on a separate sheet, and the complete multi-colour image is produced by transferring the colours consecutively.

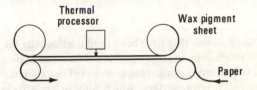

**69.** Schematic diagram of thermal transfer copier

## PHOTOCHROMIC PROCESSES

Photochromic materials change colour when excited by radiation of the correct wavelength and intensity, but revert to their image-less state when the radiation ceases. Therefore, they form transient images. They are not processed in any way, and cannot be made permanent.

If the photochromic materials are incorporated in polymer layers, the rate of image decay can be reduced, making the image slightly more stable. Photochromic glass uses inorganic compounds, and in some types the glass is progressively darkened by exposure to short wavelengths but reverts on exposure to long wavelengths.

The process can be modified so that a visible image is formed by the introduction of a dye, to create a permanent image. Being grainless they have a very high resolution of up to 1000 lines per millimetre, and this makes them useful as intermediates in microfilming. They can also be 'drawn' on by exposing them to a radiant beam, so producing a short-term visual display.

# 9 MACHINE PRINTING

The processes that have been used traditionally for the production of large numbers of facsimile copies have two characteristics in common. They all make use of a printing surface, which, although relatively expensive to produce in the first place, can be used to generate large numbers of duplicates. And they operate at high speed, using unsensitised materials on which to form the image, therefore producing large numbers of copies at relatively low cost per copy. Thus the cost per copy is lowest when large numbers are produced, and diminishes in proportion to the total quantity.

The principal printing processes can be classified according to two factors: the nature of the printing surface, that is the way in which the printing image is separated from non-printing areas; and the way in which this surface is brought into contact with the paper or other substrate, that is the impression. The printing surface may be in relief, intaglio or planographic form (Fig. 70). The machine operation may be classified as platen, cylinder or rotary (Fig. 71).

In a platen machine, both the printing surface and the paper support are flat surfaces. In practice this type of action is limited to relatively small sizes. In cylinder machines, the paper sheet is held on a flat bed, and the printing surface is on a cylinder that is passed across the flat sheet. In rotary machines the printing surface lies on a cylinder, and this rotates against another cylinder, the paper passing between the two. There are several variations on these basic principles, but in general a rotary action is required for high speed, as this has a continuous, instead of a reciprocating, movement. In order to form a clean impression, the correct pressure must be applied to bring the paper into contact with the printing surface carrying the inked image.

All printing presses must also have a means of applying the ink to the printing plate, placing the sheet or roll of paper correctly in position, and removing and stacking the printed sheets. In colour printing, the series of separate impressions, each providing one colour, must register the different coloured images exactly in the right position. To meet these conditions, and to operate at high speed, printing machines have to be precision engineered.

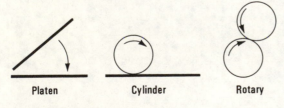

**70.** Types of printing surface

**71.** Types of impression

## RELIEF PRINTING PROCESSES

In this method the printing image is raised above the non-printing areas on a metal plate. It is one of the earliest printing systems, and has been of major importance for the printing of text, hence its common name of 'letterpress'. Originally the characters were cut into wooden blocks, but this was soon superseded by metal type. In modern practice a graphic image is transferred to the surface of the metal plate and protected by a 'resist'. The surrounding metal is then etched away with acid. Although for a long period simple line maps, mainly as illustrations in books or journals, were also produced in this way, the method is now rarely encountered.

# INTAGLIO PROCESSES

## Engraving

In intaglio methods the image is recessed into the surface of a metal plate. Ink is then wiped on to the plate, pressed into the recesses, and the surplus removed. Paper is brought into contact with the inked plate under pressure, usually by rolling a cylinder across the sheet. This was the earliest method devised to obtain a copy of a large graphic image, and in the form of engraving it was used for several hundred years as the major map printing method. The engraved plate was produced by hand, the engraver cutting the image into the copper plate with sharp tools. It is distinguished by the fact that the original 'drawing' of the map was also the printing surface. Because the impression results from direct contact between the plate and the paper, the image must be 'wrong reading' on the plate. In later developments manual engraving was aided by chemical etching of patterns and symbols.

Engraving could produce a very fine line image, but this was not sufficiently durable to withstand many impressions. At a time when the total number of printed copies of a map was relatively small, both quality and speed of output were adequate. But continued use soon wears the plate, and there was no simple means of speeding up the printing process itself. The great increase in the quantity of printed maps which took place in the latter half of the nineteenth century revealed these deficiencies. For a good impression, the paper needed to be damped, and the consequent stretching and contracting meant that it was virtually impossible to print a series of colours in register. Normally the printed image was in black, and any other colour was added subsequently by hand. This also led naturally to a cartographic style in which a fine-line image was embellished by the limited application of area colour, and this manner influenced map design for a long period after engraving had been supplanted as a method of production.

## Gravure

This is an intaglio process, developed largely in order to reproduce coloured tonal illustrations without the effect of the half-tone screen. The plate is divided into minute cells, each of which is etched, the depth depending on the tonal value at that point (Fig. 72). It is the only printing process in which the

**72.** Ink-film thickness in gravure cells

ink film actually varies in thickness. After inking, the plate surface has to be cleaned by passing a blade across it. The printing surface is expensive to construct, and the process is normally used for long runs of high-quality colour reproduction. Although the individual cells are minute, they can still interfere with the continuity of a fine-line image, and in this respect the method is not really suited to maps, although it has been used for map production on occasions.

# PLANOGRAPHIC PROCESSES

In planographic processes, the distinction between image and non-image areas on the printing surface is formed in the same plane, or more correctly, almost in the same plane. Several methods make use of this, although they differ markedly in other aspects, including lithography, collotype and screen printing.

In the original form of lithography, an artist or lithographer drew the image on the surface of a flat piece of limestone (hence the name of the process). This greasy image was capable of repelling water, whereas the remainder of the stone surface was capable of retaining a thin film of moisture. In order to take a print, the stone was first inked and then damped. A sheet of paper could be placed on the stone and a roller passed across it, thus transferring the ink to the paper. In the original form, fine-line images were drawn on smooth stones, but it was soon realised that tones could be introduced by using a grainy surface, and drawing on this with a greasy crayon, the pressure exerted determining the density of the image at any point. Multi-coloured images could be produced by transferring a key or outline to a set of stones, and then drawing the appropriate parts for each colour. In this form (autolithography) the process became not only a means of reproduction, but also an artist's medium.

Although high-quality stones could be difficult to obtain, it was possible to produce large format images, and this was a great advantage when the process began to be used for maps.

## OFFSET PHOTOLITHOGRAPHY

The use of a stone block obviously limited direct printing to the flat-bed method. Lithography was transformed by two significant developments in reproduction and printing: the replacement of the stone by a grained metal plate, and the construction of the printing surface by the use of a photosensitive layer, instead of by hand. The introduction of a flexible printing plate meant that the printing surface could be wrapped round a cylinder, making rotary action possible. The use of light-sensitive materials meant that the map image could first be drawn, and then photographed, and this photographic negative could be used to expose a light-sensitive coating on the plate. The final refinement came about with the addition of an indirect printing action. Instead of transferring the inked image directly from the metal plate to the paper, it is first offset on to another cylinder covered by a resilient 'blanket', and from this transferred to the paper. The basic configuration of the printing unit is shown in Fig. 73. Ink is applied to the surface of the plate through a set of rollers; the damping solution is applied through a 'fountain'; and the impression takes place between the offset cylinder and the impression cylinder.

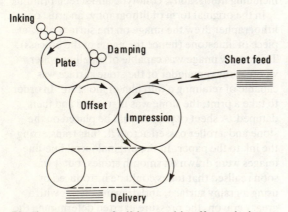

**73.** Arrangement of a lithographic offset printing unit

Modern offset presses may be single-colour machines, or may print up to six colours in sequence. The most common are probably two- and four-colour machines. In some multi-colour machines, two-colour printing units have a common impression cylinder, for which various configurations are possible (Fig. 74). The feeder unit controls either a sheet feed, or a continuous roll of

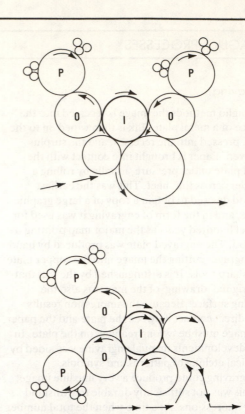

**74.** Two-colour printing units with common impression cylinders

paper known as a 'web'. Grippers position each sheet against front and side guides, and at the right moment insert the sheet correctly.

Apart from the technical problems of producing a perfect inked image at high speed, it is clear that there are physical limitations to any printing press. Space must be allowed along the sheet edge for the grippers that lift and move the sheet, so that the printing area is always less than the actual sheet dimensions. Thus there is a maximum sheet size, which can be printed on any press, as the plate circumference must remain constant. It is obviously uneconomic to use large presses for small-format sheets. Modern presses may print at up to 10 000 impressions per hour, but most high-quality work is carried out at speeds of about 5000 to 8000 impressions per hour.

The rate of production has to be related to the amount of time that is needed to set up the press so that it is ready for printing, known as the 'make-ready'. For a multi-colour press, for which several printing units have to be adjusted for correct colour and register, this may take several hours. It follows that the process is uneconomic if only a few sheets are to be printed, as the cost per copy would be exorbitantly high. Therefore, the high-speed rotary press is ideally suited to the production of many thousands of copies. For a printing run of half a million, the cost per copy is only fractionally more than the cost of the paper.

The economics of printing exert their own influence on map production, especially commercial production. Organisations that carry out their own printing are likely to acquire presses suited to their general requirements, but once installed, the need to keep the machine operating profitably exerts an influence on production planning and the flow of work. Multi-colour machines are too expensive to stay idle. In addition, both large sheet maps, and those containing folded sheets in book sections, will be affected by the maximum sheet size that can be printed. For example, the normal starting point for the planning of an atlas will take into account the page size, which will result from the folding of a sheet of a given printed format.

The need to print relatively large numbers of copies at one time has also affected the provision of many specialised maps. Unless a relatively large market can be assured for a commercial publication, there is little possibility of the venture being undertaken. Many ideas for useful maps have foundered on this economic fact. In this respect the development of a wide range of colour copying machines may in turn realise the possibility of a greater range of published specialised maps at a reasonable cost.

## Printing plates for offset lithography

For a long period the sensitised layer on both lithographic and letterpress printing plates was provided by dichromated colloids. Because of the dark reaction these had to be coated before use. Variations in coating thickness and sensitivity to changes in humidity also meant that standardisation of processing was impossible. Zinc plates were widely used, and these were generally 'grained' individually.

Printing plates for offset lithography are now dominated by pre-sensitised anodised aluminium plates, most of which have a photopolymer or diazo-sensitised coating. The most common are photopolymer and diazo–resin compounds, and they are available for either negative or positive working.

In negative-working plates (where the image is formed by exposure to a negative), the photopolymer layer reacts to actinic light by becoming insoluble. Exposed to a negative, this

insoluble resin layer also becomes distinctly hydrophobic, which assists the separation of image and non-image areas. The unexposed, unhardened non-image areas on the plate are dissolved away either in water or a special solvent (Fig. 75). These are subtractive plates, in the sense that the coating is removed from the non-image areas. The modern versions of these plates have a very high resistance to wear, and printing runs of over 1 000 000 impressions have been achieved. Diazo-coated plates are usually additive. After exposure, the insoluble coating in the exposed (image) areas has to be reinforced by the addition of a resinous developer. Generally speaking the unexposed coating is removed and the resin applied in one operation.

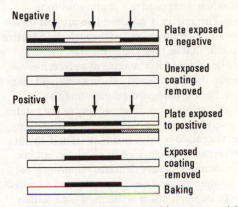

**75.** Negative- and positive-working pre-sensitised printing plates

Positive plates of the subtractive type also use diazo-sensitised polymers, derived from naphthalene compounds. Exposure to actinic light causes the exposed (non-image) areas to become acidic, and they can be dissolved away in an alkaline developer. Baking at a temperature of 200 to 250 °C improves the resistance of the image surface, and a plate that would normally print for about 40 000 copies can be used for up to 400 000 by this method.

Standardised graining and coating during manufacture also means that press plates can be processed automatically, rather than by hand, and automatic plate processors are now widely used.

The question of whether the plate image should be made from a negative or a positive affects not only the plate-making process, but also the preparatory stages in map construction. As the image on the printing plate has to be 'right reading', the negative or positive from which it is made must be 'wrong reading', in order to obtain full contact. Generally speaking, making plates from negatives is easier in terms of processing, but can be more difficult if several images occur on the same plate, and they have to be correctly positioned. This 'layout' on the plate is much easier with positives, as the non-image areas are transparent, whereas on negatives they are opaque. In addition, many positive-working plates have a visible image after exposure (although this is of little significance with automatic processing), and generally deletions are easier to carry out.

## Collotype

This is one of the few printing methods capable of reproducing continuous-tone images without a half-tone screen. The printing surface is a layer of gelatin, which is exposed to a continuous-tone negative. The gelatin is 'hardened' in proportion to the amount of light transmitted through the negative at any point; that is, this is controlled by the negative densities. After soaking in a glycerine–water bath, the unhardened gelatin swells and takes up water, but the hardened gelatin remains receptive to printing ink. The swelling of the gelatin causes a series of minute 'wrinkles' or reticulations to form; the raised points remain moist and do not print, whereas the slightly lower points accept ink.

The plate image is very delicate, requiring great skill in production, and can only withstand a limited number of impressions. It is still used occasionally for fine art reproductions, but is not really suited to the high-speed production of large numbers of copies. It has been used experimentally for the production of orthophotomaps.

## Screen printing

In this process a negative stencil is formed on top of a fine mesh, or screen, and ink is forced through this under pressure. Simple versions of 'printing' by this method are well known. Like gravure, this results in the printed image being composed of tiny points, or cells, but in screen printing the pressure produces a thick ink film, and the tiny points of ink will merge in dark areas. In the most advanced industrial practice the stencil is produced on the screen by reprographic methods, exposure making a resistant stencil in a light-sensitive coating, and the inking operation is semi-automatic (Kers 1978). Modern screens are made from synthetics such as nylon, or stainless steel.

Screen printing is widely used for posters, and for printing on metal and plastic. It has been employed for map production, despite the difficulty of reproducing fine lines through a screen, and can be considered as a possible method for the colour printing of a small number of copies.

## PRINTING PAPER AND INK

### Paper

Although maps can be produced on various materials and surfaces, the most familiar product is still the conventional image of printed ink on paper. The different types of paper depend on the nature of the materials from which they are composed, and the processes used in manufacture and finishing.

There are many kinds of fibrous material. Some are specially suitable for high-quality paper; others are used because they are cheap or plentifully available. They include materials processed specially to make paper, and also waste reclaimed from other materials, such as rag. The main natural fibres used in paper production are leaf fibres such as esparto and sisal; wood fibres from both coniferous and deciduous trees; grass fibres, derived from cereal straw; and bast fibres produced from flax, jute, hemp and ramie. The most important ones, in relation to printing papers for maps, are those produced from esparto, wood and rags.

Esparto is not particularly strong, but it is bulky and opaque, and can be made into a flat, smooth sheet. It is generally regarded as the best basis for high-quality lithographic printing. Normally its lack of strength is remedied by blending it with a little sulphite-bleached wood pulp, usually about 15%.

Wood is the most widely used basic material, and it can be of high or low quality. Mechanical wood pulp is usually made from spruce, whereas chemical wood pulp is made from conifers such as pine and deciduous trees such as poplar.

One of the principal factors in determining paper quality and type is the fibre length. Whereas chemical wood pulp usually has fibres of about 5 mm in length, cotton and linen rags have much narrower and longer fibres, up to 40 mm. Long fibres provide the strongest and most durable materials, but the short, bulky fibres make the best paper surfaces. Much paper making is therefore a compromise between these two qualities.

In their natural state, a mat of vegetable fibres would tend to be very absorbent, poorly coloured, and have an uneven surface. Because of this, other materials are added during manufacture, including dyes, sizes and starches. These are included to seal the fibres and make them more wet-resistant. China clay or titanium oxide is used to 'load' the mat of fibres, filling the spaces between them, to make a smoother surface. There is a limit to the amount of loading possible, as it decreases the strength of the finished paper.

For the printing of fine detail, the smoothness of the paper is important. Smoothness is not the same as gloss, which increases reflectance from the paper surface, but can also interfere with legibility by annoying reflections.

## Paper manufacture

The manufacture of paper is a lengthy process, in which the 'slurry' of wet fibrous material is gradually converted into a long roll of paper, by moving it along a travelling web or 'wire'. Gradually the proportion of water is reduced, and other materials are added. With the movement of the wire, the fibres gradually settle into the 'machine direction' of the paper. If the finished paper expands or contracts, the change is greatest parallel to the fibres, as they swell more than they increase in length. The paper is also stronger along the grain than across it. More water is removed under pressure, and the whole mass is dried by being passed between heated drums. Then it is usually compressed and given a smoother finish by being 'calendered' between rollers, which makes the paper more compact, firmer and smoother. This normal calendering produces a paper with a 'machine finish'.

In addition, the paper may be coated to produce a very fine, smooth surface. Such heavily coated papers are usually known as 'art' papers, and are often used for high-quality illustrations, but they rapidly lose their surface strength and smoothness if they are handled. Papers for more durable products, including maps, are never given a heavily coated finish, as this would deteriorate rapidly in use, and decrease the strength of the paper. Most map printing papers are good quality litho papers, basically a mixture of esparto and wood pulp, with a machine finish.

## Paper characteristics

From a printer's point of view, the characteristics of paper of most concern are printability, opacity, surface strength and stability. To make a good printed impression the paper surface must be smooth and firm, but the paper slightly elastic in

order to withstand being passed between cylinders under pressure. Unless the surface is smooth and even, it is difficult to obtain an equal impression over the full width of the sheet. It must also be at the correct pH, as this affects the drying of the ink.

Opacity is most important when the paper is to be printed on both sides, but it is essential in order to provide good reflection from the surface. Surface strength is necessary because high-speed printing subjects the paper to a good deal of mechanical stress, and lack of surface strength may result in fluffing (the pulling away of fibres by the ink), and picking (the separation of pieces of coating from the fibres).

As paper is susceptible to variations in humidity, which affect its dimensional stability, control of both the humidity of the paper and the ambient conditions under which it is stored and printed is important. If sheets are kept in large stacks, then a change in humidity can lead to curling at the edges. A sheet that does not lie flat cannot be properly controlled during printing. Air conditioning in both paper storage and press rooms is highly desirable, but in some cases temperature control and the use of humidifiers where necessary is adequate in a temperate climate.

## Paper dimensions

Paper sheet sizes are defined by substance and format. The paper weight and thickness is expressed in grams per square metre (gsm), and the sheet dimensions described in accordance with international paper sizes. The international A sizes are as follows:

| Code | Dimensions in mm |
|------|------------------|
| 2A ........ | 1189 × 1682 |
| A0 ........ | 841 × 1189 |
| A1 ........ | 594 × 841 |
| A2 ........ | 420 × 594 |
| A3 ........ | 297 × 420 |
| A4 ........ | 210 × 297 |

## Synthetic papers

Despite the many disadvantages of paper as a basis for high-quality colour printing, with its dimensional instability and limited resistance to wear and wetting, it continues to be widely used for map production, no doubt because it is available in a variety of types and finishes, and because of its relative cheapness. There are many instances when more resistant materials are clearly desirable for maps, especially when they are to be used under adverse weather conditions. Synthetic papers are available for this purpose, but despite their greater resistance to wear they can also have disadvantages, either in printability or cost.

There are basically two types. Synthetic fibre materials are manufactured in normal paper-making machines, and closely resemble ordinary paper in surface texture and feel. Any dimensional change is very small, and resistance to both abrasion and wetting is very high. They can be printed on with normal printing press settings and inks.

Thermoplastic films, primarily of the polyester type, are cheaper to manufacture, but generally require special inks and press settings, and can be affected by some of the chemical solutions used in the printing industry.

For many purposes the synthetic papers obviously do not provide a sufficiently great advantage to the consumer to justify their additional cost, but some map producers, especially national mapping agencies, issue topographic series or special sheets using synthetic papers.

## PRINTING INKS

Like paper, ink is often taken for granted, but the conditions under which a clean impression is produced on a sheet of paper, and transferred to a pile without being affected by the following sheet, are technically complicated. Modern high-speed printing is dependent upon the availability of inks that print quickly, adhere firmly, produce permanent images and dry rapidly. Although the complete drying of the ink may take several hours, the surface of the ink must be stable enough to resist any displacement in a few seconds.

Printing inks contain fine particles of pigment suspended in a vehicle or carrier, to which other ingredients are added to speed up drying and give better bonding to the stock. Both natural and synthetic pigments are used, but in general the spectral characteristics of most inks are far from ideal. As inks act by subtractive colour, in multi-colour printing the inks must be sufficiently

transparent to allow the maximum reflection of light from the paper surface. In order to maintain the selected colour, and to match it in subsequent printing, inks are produced to standards, such as BS (British Standard) specifications.

### Printed colour

Reflection, transmission and absorption are involved in the apparent colour of an ink layer on paper (Fig. 76). Some incident white light may be reflected from the surface of the ink film. Of the light penetrating the ink film, some will be reflected; some transmitted to the paper surface and then reflected through the ink layer; and some absorbed by both paper and ink. In the lithographic printing process, the splitting of the ink film between plate and blanket, and blanket and paper, helps to produce a thin ink film, leading to good transparency. But the superimposition of a solid layer of ink over another is avoided, as this increases surface reflection and internal absorption, reducing the proportion reflected from the paper surface.

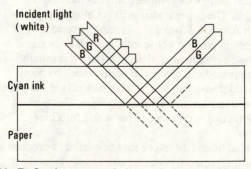

**76.** Reflection, transmission and absorption of an ink layer on paper

Unless the standard process colour inks are being used, suitable colours can be selected from the samples produced by the ink manufacturers. These usually consist of a book or set of printed slips, demonstrating the range of ink colours. When selecting colours from such samples it is important to realise that what appears to be a clear, strong hue when seen in a relatively large area will not have the same level of contrast when it appears as a fine line. The safest method is to place a piece of white card over the printed sample, leaving only a fine line exposed, and use this as a basis for judging the suitability of the ink for the particular map element.

**PART THREE**
# CARTOGRAPHIC PRODUCTION

# 10 ORGANISATION AND PLANNING

## GENERAL CONSIDERATIONS

Once a particular map has been decided upon, then it is possible to begin to plan the actual production. This has to take into account several sets of factors. Assuming that the map is required by a map author, or an external client, general discussion will have produced agreement on the overall nature of the map, and the required end product. If the work is being produced commercially, the client may then want a full cost estimate, possibly with a fixed delivery date. If only the general concept of the map has been decided, then the cartographic organisation may have to consider the amount of information to be assembled, the availability of source material, etc., in order to arrive at some understanding of the total informational input. The external factors are those that control the circumstances under which the map is to be produced. Invariably these impose limitations – in total cost, in time, in required format – and normally therefore in the amount of graphic elaboration that is possible.

The cartographic task is to solve the problem of producing the required map in accordance with the client's or author's wishes, making full use of the available resources, and overcoming any technical difficulties. At this point, therefore, the theories and principles of cartography are applied to a particular map-making task.

The technical organisation of the map has to take into account the nature of the end product, and break down the total production into a series of stages that will accomplish this efficiently. The extent of this technical planning will depend on the complexity of the map. A simple black-and-white illustration at small format may need no more than a discussion between the map author and the cartographer, and it is unlikely that questions of time will be important. But external restrictions on colour and format will strongly influence the final map. On the other hand, a large format road map, representing a considerable investment by the publisher, will require extensive organisation and planning, involving several departments and many people. Despite the great diversity of map products, there are certain characteristics of map production that apply in general. These concern both the nature of cartographic production, and the need to carry out the work in a particular order.

## GENERAL CHARACTERISTICS OF CARTOGRAPHIC PRODUCTION

Cartographic production is characterised by being both colour separated and image separated. Although the whole information may be assembled in a single compilation, a separate image has to be produced for each printed colour, and the production is organised to achieve this from the beginning. In this respect cartographic production is fundamentally different to most graphic arts production, in which a complete coloured image (a picture or photograph) is made first, and then colour separated by photographic or electronic means.

In addition, map production is image separated, because different methods are used to construct different types of image. This can apply to even a single-colour map, such as a black-and-white map. The line symbols may be drawn or scribed; the lettering may be produced as a separate original; and a percentage tint – constructed by photographic processing with a tint screen – will also require a separate original. Subsequently all three are combined to form a single image, from which the printing plate is made.

### Register and fit

Keeping the correct relative position between the different map images involves two distinct tasks. In multi-colour printing, the different coloured images must obviously be correctly positioned in relation to each other. This is referred to as register. Unfortunately there is no accepted term in the English language for the parallel problem, which characterises the image-separated mode of map production, in which the different images within one printing colour also have to fit together correctly. This is often referred to also as 'register', but whereas lack of register is a printing problem, incorrect fit occurs during cartographic production.

For example, on the blue printing 'plate' (generally used as a synonym for the blue-coloured printed impression), the drainage lines and shorelines may be in a dark blue line, and are likely to be produced by drawing or scribing: the blue tint, probably about 30% of the dark blue, will be produced separately with a mask and tint screen; the lettering on the dark blue will also be produced separately by mounting the prepared names on an overlay. These three separate images have to be combined, and they have to fit together correctly. Any errors in this image fit should be detected immediately, but if they exist they cannot be put right subsequently during printing. It would seem appropriate to employ the two terms, register and fit, to refer to these different aspects.

Correct register and fit are now normally achieved, due to the use of stable materials, contact copying and correct practice. For this a punch register system is widely used, in which all the materials used in the production are punched with a pair of slots or holes, on at least one edge, and fitted together wherever necessary by placing the separate images on a punch register bar, which has 'pins' that fit the holes exactly. Simple small-format systems are comparatively inexpensive. Large and more expensive systems can be applied to the entire production, right through to the positioning of the final images on the printing plates (see Kers 1980). These are generally more suitable for the production of large-format sheet maps. If such a register system is used, then of course the punching of all materials for the production work must take place before any operations are carried out on them. Apart from ensuring correct fit, and at least the possibility of correct register, they also assist reprographic operations, as it is far easier for a photographer to fit separate pieces of material together under conditions of safelight or subdued illumination than to do so by the visual inspection of register marks.

## Scale of work

Another factor that influences the entire production is the decision about scale of work. If the map originals are to be drawn on an opaque medium – in some circumstances a common practice – then at some stage they have to be transferred to a translucent base, so that they can be copied on to printing plates. If a photographic stage is necessary, then scale change can be introduced at the same time. For pen-and-ink drawing, this working at a larger scale offers advantages in execution; the specification of symbols is not so demanding, which can ease the task for the cartographic draftsman. On the other hand, photographic reduction is not so simple or straightforward as it might appear, and is more expensive than contact copying. Contact copying with stable materials avoids any problems with fit and register. Working at the same scale as the final product also means that the actual images are smaller, and can be produced more quickly than those that are larger in scale, and have to be reduced.

If a large working scale is adopted, this has consequences throughout the production sequence, from compilation to final image. With modern equipment and materials, and stable translucent base materials, it tends to be the exception rather than the rule, and should only be considered as a suitable working method where it can be seen to yield advantages for a particular product.

## Quality

Establishing and maintaining a proper quality of work is important for all cartographic production. Quality depends on several factors. These include correct information, proper use of graphic representation methods, and the production of a final image that has sharp, clear details with the different colours correctly fitted and registered.

Reference is often made to 'accuracy', but this term is difficult to define as a whole. It can be used (and usually is used) to refer principally to metrical accuracy – that is the correct location of detail. With high-quality map production, this should be within 0.2 mm for points and lines. But this can only be based on the nature and quality of the original information, which can be much more difficult to assess. For a basic, large-scale topographic map or chart, the locational accuracy of measured points or lines can be checked against the original survey information, and indeed usually is. For a small-scale derived map, the locational quality will be a function of the adequacy, consistency and completeness of the source material, the care taken in compilation, and good generalisation. This also depends on consistency of treatment.

Technical quality will depend on ensuring that all images are exactly according to the specification; that all lines and areas on a positive do not transmit any light during copying, and similarly that all dense areas on negatives are opaque; that the edges of lines and letters are sharp; and that all images are correctly fitted together. With screened tints, great care is needed to ensure that the tints are free from blemishes, such as small marks or holes that can easily result from dust.

# PRODUCTION ORGANISATION

Before the technical stages of production can begin, a certain number of preparatory operations must be undertaken. First of all the map must be compiled or the information assembled into the form and format in which it is to appear. Normally this involves constructing the metrical framework (the map base), consisting of either graticule or grid, before assembling the information for the complete map or map sheet. In the case of a special-subject map, the first stage is to construct the reference framework (compilation base) that will control the assembly of the special information. For many maps the compilation of the information for the required area, scale and format will then be the next major task.

Once the information has been put together, the map has to be designed. At this stage each symbol is specified so that it can be technically constructed. If the map is technically complicated, the next stage is to prepare a flow diagram to explain exactly what images have to be made, and in what sequence, in order to complete the map production. This cannot be done unless both the compiled information and the symbol specification are available. Therefore, the flow diagram translates the map design into a series of specific tasks in production.

If the map cost has to be estimated, or produced to a fixed date, then the compilation and the flow diagram are used as a basis for cost estimation. This has to take into account the total amount of work involved in producing the required images, and the cost of all the technical operations, such as making photographic negatives and positives. With this information, a production schedule can be worked out, showing how long the various production operations should take, and fitting them into the overall work of the organisation, or arranging the total staff input so that the deadline for production is reached.

If the map then goes ahead, the technical production begins. The various images are produced in accordance with the symbol specification, based on the compilation, until the whole map is complete. Normally at least one proofing stage is built into the system; very often a proof is required as soon as the images for each colour have been produced; sometimes a final colour proof is needed to show the client before the machine printing takes place.

Each of these stages is followed by checking: the basic framework or compilation base; the complete compilation; each individual image as it is produced; the assembled images at proof stage; and again before the plates are released for machine printing. Given the amount of information on many maps, and the breakdown of the production into a large number of separate pieces and stages, checking at each stage is absolutely vital.

## Construction of the map base

Once the production of the map has been planned, the first step is to prepare the information. This usually takes place in two stages: first the construction of the basic framework of grid and/or graticule lines, sheet border and neatline; and second the compilation of the map detail. There are many variations on this procedure. With some simple illustrative maps there may be no graticule or grid, and the topographic outline becomes the reference base. If so, then this needs to be compiled first at the required scale, and with the correct degree of generalisation. For topographic maps, the grid is usually constructed first, if it is to be included either on the map or in the map border, as the coordinates of the projection intersections are normally expressed in grid coordinates. Whether the map is basic or derived, the construction of this framework or outline lays down exactly the locational base of the map, to which all other information is in turn fitted.

## Compilation

The compilation stage consists of assembling all the required information in the form of a compilation, or a plot. For a basic map, this may be taken directly from ground or air survey plots, but generally these have to be incorporated into the overall sheet area and framework. It may be possible to use the survey plots directly, but this is unlikely unless the map is of a very small area. For derived maps, the compiled information is likely to be taken from more than one source, and being at a smaller scale than the source maps, will involve generalisation. Although maps may be the main source of information, other numerical or text sources may also be used. For special-subject maps there will be a topographic reference base, functioning as a locational framework for the detailed representation of the special subject.

For new maps, it is always preferable to complete the compilation and check it thoroughly before any production starts. In addition, the finalisation of the design specification cannot be undertaken properly until the information to be represented can be studied, and this is difficult to do without a complete compilation.

## Checking

Once the framework and the compilation of detailed information have been completed, it is essential to carry out a thorough check. No map is error free, and the more complex the map the more likely it is to contain errors. This checking must be done systematically, not by casual inspection, if necessary by making a list of items and working through it. At this stage both errors and omissions can be rectified easily: if discovered at later production stages they can result in a loss of time, and can interfere with the quality of the production. It is particularly important to check what appears to be obvious, as major items in the map content are often overlooked in the pursuit of small details.

## Symbol specification

Unless the map is being produced to a standard specification, the design then has to be formulated in specific detail, by providing a symbol specification and detailed sheet layout. For an individual map, it is better to create the final design specification after the compilation has been made, so that problems in representation become apparent, and symbols can be imagined in the context of their actual disposition on the map. The design specification must be expressed in measurements, or by reference to standards, and should be so arranged that technical staff can understand what is wanted without any confusion (Fig. 77).

The normal procedure is to deal with the symbols in groups, according to their dominant colour, dividing them into physical and cultural features respectively. Line gauge and form, point symbol

**77.** Examples of symbol specification

dimensions, area patterns and colours, all have to be described in specific terms. For an important and complex map, it may be desirable to construct an experimental section of the map to proof stage, and examine this, before undertaking the whole production. The detailed layout of marginal information is usually prepared at the compilation stage, along with the name arrangement, so that all the text material can be dealt with at one time.

# PRODUCTION PLANNING

In any organisation involved in production, the flow of work and the assignment of tasks to staff depends on planning and control. All map-making organisations have to arrange their production in relation to available staff and technical facilities. Therefore, the stages and operations in production must be broken down and dealt with as specific processes.

If maps are being generated for internal or public needs, it may not be necessary to produce exact cost estimates for each job, but any commercial organisation must also be able to use its planning system to estimate costs and then monitor performance. Estimating and costing depend on being able to identify every operation in the production system, in order to calculate the time required for completion, and the cost of materials and processing. In addition, the costs of plant, equipment replacement or improvement, and administration, have to be borne by the production income.

Production planning for cartographic purposes involves three main aspects or stages. First, the technical planning of the map must specify exactly the individual operations needed for that product, the types of image and processing, and the correct sequence. This is normally achieved through the production of a flow diagram. Then the heads of departments can estimate the actual amount of time required for their staff to execute particular tasks, and the cost accountants can estimate how much specific items such as photographic processing or plate making will involve, combining these with man-hour estimates. This will eventually lead to a full cost estimate. Finally, the planning staff or management can then translate this into a programme of working time, which will indicate how many or which staff will be engaged on the map production at particular periods, making due allowance for unworked time, vacations and fixed holidays. This translates the production time into specific calendar dates, with 'deadlines' set for the completion of individual parts of the whole job. A map which may take several months to complete, and which will involve many people in different departments, therefore requires a complex organisation, to ensure efficient production and where necessary delivery to the client at the agreed time.

Within this overall planning there are both objectives and constraints. Both underestimation and overestimation must be avoided, especially in competitive commercial work. The technical production system chosen for the map must be adequate in quality, but not excessive, and should be both technically and economically efficient. Methods need to reflect the capacities/preferences of staff, as well as theoretical principles. The methods used in a large production scheme should be those that can be performed to acceptable standards as a matter of routine, and not dependent on the specialised skills of a few people. Assumptions have to be made about work rate and productivity. If the planned time is exceeded, and the date of delivery fixed, extra work can only be input through overtime or additional support from outside staff, both of which will lead to extra costs. And as in most organisations employing skilled staff and expensive equipment, the total capacity of the organisation tends to be relatively fixed, the desirable objective is to maintain a continuous flow of work. Where this flow of work depends on obtaining a regular input of work from clients, this is an added complication.

For a smaller organisation that depends on other firms or departments for some of its production processing, for example printing or half-tone photography, the planning schedule must also take into account the availability of such services, and their costs.

Planning and organisation is therefore a combined function of both production and specialised planning staff. Efficient operation depends on full and proper consultation and discussion of all problems, and proper exploitation of the knowledge and skill of the staff. Anticipation of problems, so that the production schedule is not affected by unforeseen events, is the key to good planning.

## FLOW DIAGRAM

The construction of a flow diagram to describe the technical organisation of a map will vary with the nature and complexity of the map itself. A small black-and-white illustration will rarely require an extensive planning operation, especially if the type of work and final product is familiar to the staff involved. A four- or six-colour sheet, or an atlas, will require extensive planning of technical details before the rest of the estimating and costing procedures can even begin. It is in this field that the cartographers need to have a good understanding of the possible technical methods that may be used, their relative advantages and disadvantages for the particular organisation, and their relative costs. With a new map, it can happen that very small changes to the initial design can lead to a significant cost reduction. In other cases, overall costs are set initially, and the whole of the design specification and technical planning are conditioned by the fact that the work must be accomplished within these limits.

Most cartographic organisations evolve a system for constructing a flow diagram, and all such systems have certain elements in common (see Geodætisk Institut 1986). A comprehensive flow diagram system for cartographic purposes is described by Shearer (1982). Flow diagrams can serve several related purposes. In the initial planning of the map, when factors of scale, content and representation are being considered, as any element is designed it is inevitably followed by the question 'How is this to be made?' The answer to this can be outlined by constructing a flow diagram which will show how the proposed map could be produced, and this can then be used for further consideration of the design.

Sophisticated graphic representations tend to have their consequence in more complex and therefore more time-consuming production tasks, and if the map is to be produced to a fixed budget, then the graphic representation has to take this into account from the very beginning. What may be graphically desirable may turn out to be technically too expensive. After the form of the map has been set, then the full flow diagram is produced to organise the production in a technical sense, and this in turn can be examined by the heads of technical departments to ensure that the methods and sequence specified are technically correct, realistic and efficient.

In nearly all cases, there is more than one way of producing a given map element. A line image drawn in ink produces a positive: a scribed line image produces a negative. There may be advantages or disadvantages with each, depending on the nature of the work. The method chosen must be appropriate to the product in terms of quality, technically correct and preferably the most economical for the particular organisation. Scribing may give a higher quality line than pen-and-ink drawing, but pen-and-ink drawing is perfectly satisfactory for many purposes. But if the map has five separate sheets with dense contouring, there is little doubt that in same-scale work scribing will be both better and faster. So the suitability of methods is judged according to the type of image and product. Such judgements cannot be made unless the cartographer is fully aware of the existence of different methods, their relative advantages and disadvantages, and the capacities of the organisation within which he works. Adopting a different way of producing an image, or carrying out a process, will often involve the special knowledge and skill of technical staff in several departments. Such discussions cannot operate effectively unless all those involved have a basic knowledge of the methods and problems of the other fields. Rigid adherence to particular methods often means that work is done uneconomically or inefficiently. No organisation can constantly change its working practices, or replace its technical equipment, but developments in equipment and materials often lead to minor improvements in routine tasks, or to the inclusion of a process that had been thought previously to be too difficult or uneconomic.

## Flow diagram elements

In order to be useful, the flow diagram must employ a standardised set of signs to represent the technical characteristics of the images to be made, and the processes and operations to be carried out. Each individual image, or each operation on an image, is most easily represented by a square (Fig. 78). The corners can be used to describe the image characteristics, which are image position (top left), image form (top right), image type (bottom left) and nature of operation (bottom right). If necessary, the

type of material to be used can be added to the bottom right diagonal: the transparency or opacity of the base material can be shown by the bottom line; and the nature of any end product can be shown by adding to the bottom and right-hand lines. A number can be placed in the centre, so that additional notes or explanation can be made. In some cases this numerical sequence can be used to indicate the order of image construction.

Lines connecting the boxes show the sequence of operations and processes for which the images are to be used. For example, for a contact copy, the line connecting a negative with a positive copy includes the sign for contact copying placed across the line. Copying with a tint screen uses the same basic sign but adds information about percentage tint, screen ruling and screen angle. Photography with a camera employs the copying sign, but with a diagrammatic lens added. A separate process is distinguished from a further operation on the same material, and the use of an image as a reference employs a different line form.

By this means, all the types of drawing and reprographic material, and all the operations of drawing, scribing, stick-up, masking, contact copying, direct positive copying, image combination, positive masking, and so on, can be clearly described and identified. The sequence will show, for example, that before a percentage tint is made photographically, the necessary mask must have been produced by manual or other means.

It is unlikely that many organisations will have such a variety of work that all the items in the full system are actually used. Even so, there would be a considerable gain to the cartographic community as a whole if everyone was familiar with, and employed, a standard flow diagram. This would remove ambiguity and confusion, and if flow diagrams were always included in descriptions of maps and production methods, a much clearer understanding would result. Considering the highly ambiguous state of the English language in respect to graphic arts and reprographics terminology, this improvement in communication is long overdue. It would also make possible a truly international exchange of technical information.

## Arrangement of the flow diagram

The positioning of the boxes and connecting lines may be organised so that each box is placed with sufficient space to show the necessary details, without making the diagram unduly large. Normally the operations for each printing colour are arranged

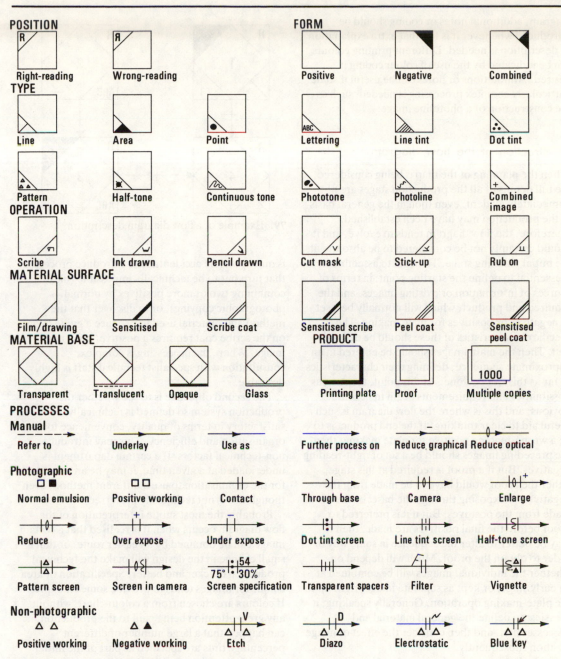

**POSITION**

Right-reading | Wrong-reading

**TYPE**

Line | Area | Point

Pattern | Half-tone | Continuous tone

**OPERATION**

Scribe | Ink drawn | Pencil drawn

**MATERIAL SURFACE**

Film/drawing | Sensitised | Scribe coat

**MATERIAL BASE**

Transparent | Translucent | Opaque | Glass

**PROCESSES**

**Manual**

Refer to | Underlay | Use as

**Photographic**

Normal emulsion | Positive working | Contact

Reduce | Over expose | Under expose

Pattern screen | Screen in camera | Screen specification

**Non-photographic**

Positive working | Negative working | Etch

**FORM**

Positive | Negative | Combined

Lettering | Line tint | Dot tint

Phototone | Photoline | Combined image

Cut mask | Stick-up | Rub on

Sensitised scribe | Peel coat | Sensitised peel coat

**PRODUCT**

Printing plate | Proof | Multiple copies

Further process on | Reduce graphical | Reduce optical

Through base | Camera | Enlarge

Dot tint screen | Line tint screen | Half-tone screen

Transparent spacers | Filter | Vignette

Diazo | Electrostatic | Blue key

**78.** Elements of a flow diagram

in a vertical column, with any necessary connections sideways to other printing colour images. The numerical sequence may indicate the order or production to a limited degree, but it does not always make clear that some operations cannot be started until after others have been completed. The alternative arrangement uses more space, but places the boxes, or rows of boxes, in a time sequence, each horizontal level being arranged so that it

indicates the sequence. In such a diagram any operations on the same level can be carried out at the same time, and those linked vertically must follow in a specific order. Although theoretically the diagram could even be related to a real time scale, this would tend to extend it to unrealistic proportions. For large cartographic tasks the flow diagram can be used as a basic for network planning.

Despite the advantages of using a standard

diagram, additional notes or coding should be introduced whenever it is felt that extra explanation or description is needed. Different printing colours can be indicated by the use of colour coding if desired. Annotations or notes can be useful if some particularly complex processing is needed, such as the construction of a photoline image.

## Construction of the flow diagram

When the planning of the map is being considered, the full details of all the production stages are not immediately evident, even though the general basis of the production may have been established. Therefore, the flow diagram tends to evolve, and it should certainly not be considered to be absolute at the initial planning stage. To begin to assemble it, it is essential to define the starting point, in terms of sources of information or existing images, and the required end product, which will normally be a set of negatives or positives for plate making. The specific characteristics of these should be defined first. Then the main images should be entered in an approximate sequence, defining their characteristics so far as this can be done. At this point, the various possibilities and requirements begin to become obvious, and this is where the flow diagram is such a useful aid to clear thinking. If the end product is to be a wrong-reading positive for plate making, then the preceding images should be a set of right-reading negatives. But if a proof is required at this stage, either the proof would have to be made from the negatives by exposing through the base, or the proof made from the positives. But if it is preferred to proof before the final positives are made, then it may be better to alter the sequence in some way in order to obtain the proof. Much will depend on whether the individual images will be combined at an early stage, or kept as separate images right up to the plate-making operation. Generally speaking, it is easiest to define the general material and processes first, and then consider the effect of image position subsequently.

Figure 79 gives a simple example of how three image elements are described and connected. Square 1 represents a wrong-reading scribed line negative; square 2 is a right-reading stick-up of lettering; this is contact-copied to produce a wrong-reading negative (square 3); both 1 and 3 are wrong-reading negatives, so they can be copied successively on to another piece of film to produce 4, which is a right-reading combined line positive.

The minimum requirement for the flow diagram is that it should be technically correct. Surprisingly, it

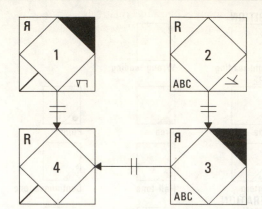

**79.** Example of a flow diagram description

is quite possible accidentally to introduce processes that turn out to be technically impossible (like combining two or more positives by normal photographic copying), or to discover that the method and material used to produce a guide image on the scribe coat requires a positive, not a negative, image. Where the processing is complex, consultation with specialist technical staff is highly desirable.

The second objective is to see whether the production system so defined is technically satisfactory in terms of quality, convenience for the organisation and efficiency. This may introduce non-technical factors. If a certain department is underloaded at a given time, it may be an advantage for the organisation to use a different method, even though this is not technically preferred.

Probably the most subtle interpretation of the flow diagram occurs when it is realised that a certain image can be obtained by an easier route, or that a small change in the design will make the technical production quicker and better. Specification of area colour for images tends to provide some examples. If colours are chosen from a colour chart, without any great attention being paid to the production, it can happen that a large number of different percentage tints are specified as part of the initial design. In practice, the difference between any two tints of the yellow is likely to be imperceptible, or of minor importance in the finished map. Time and material can be saved if the total number of masks, tint exposures and combinations can be reduced. Although perhaps this may make little difference in a single map, if it applies to a whole series, or a number of pages in an atlas, the cumulative effect can be considerable.

Like almost every other aspect of cartographic production, most of the planning, preparation or construction takes place in a relatively straightforward manner. But each map tends to provide some particular and intractable problem, whether it is in the information, the design, or the production. As with other problems, very often an answer can be found by taking a different starting point, and questioning the apparently 'obvious' processes that have been adopted. If a particular sheet only has a few names on it, these could be mounted directly on the line positive instead of making a separate overlay, plus negative, plus combination. The longer method may be quite correct where the stick-up has a great deal on it, but this does not mean that the routine should be followed regardless of the quantity of work on any particular image. Using a large sheet of strip mask in order to produce a tint for one lake is highly extravagant and time consuming. The ability to notice and enquire further into such questions shows how the flow diagram can bring to light the details of the production method before production is undertaken. It is also a very good way of learning about technical processes and their employment in actual map production. What matters is that the objectives are achieved, not that certain routine practices are followed. In this sense it is a valuable corrective to any study of methods in themselves.

## Estimating

The estimation of the time needed to complete any piece of work carried out manually must be related both to the difficulty of the operation and the total amount of work. Although the flow diagram describes the nature of the image to be produced, it does not contain any information about the quantity of work involved. The estimation of cartographic operations is most difficult where the total work is mainly dependent on the amount of detail in the image, rather than on processes and materials. This applies particularly in compilation and in all drawing office tasks.

Estimation is based on experience. It is essential that experienced members of staff should be able to state in advance the amount of time that a certain task should require. This must apply to compilation work as well as drafting, as in some types of map the compilation may actually involve more time than the construction of the images. For this it is essential that all staff are required to estimate, so that estimating is regarded as a necessary activity: and that the comparison of estimated time and actual time should be recorded and issued to the staff concerned.

For the standard drafting operations, estimating is normally the responsibility of the chief or section leader. The variation between one map sheet and another has to be taken into account, and most organisations producing sheet maps or charts classify the different sheets or charts in the series according to the quantity of work involved. Some sheets of lowland areas may have few contours on them; sheets of highland areas a great number. Standard line work, either drawn or scribed, can be considered in terms of total line length, but it is obviously more time consuming to construct an image with many different symbols, than a large image that is relatively uniform. Different types of drafting, such as double-line road casings, point symbol stick-up, or name stick-up, can usually be rated as either a given length per hour, or so many items per hour, different figures representing the difficulty of the work. For large production programmes, multiple regression analysis can be used to establish good estimates (see ACIC 1964).

All estimating has to allow for non-productive time, checking and corrections. It is clear that some operations will need more detailed and longer checking than others. It is fairly easy to ensure that on a scribed contour plate, all the contours plotted have been scribed. It is more difficult to check that all the figures on the contours are correct, or all the place names on a large sheet are correct. Failure to allow for checking and corrections usually has the consequence that work is not released on time, which in turn puts back all the ensuing operations, which have already been programmed in advance for other sections or departments.

## Costing

The total anticipated cost of a map product will involve the cost of hours of work, materials and processing. Hourly rates are known for the various staff, and technical departments normally have fixed costs for standard operations, based on job records. Therefore, the cost of making a photographic positive copy from a negative of a certain size should be a known factor, because the actual cost has been calculated from the records of previous operations. These standards have to be modified in relation to increases or changes in materials and supplies, rates of pay, or methods of processing. They can only be established if full job records are kept and analysed, so that the production organisation should be able to state at any time the expected cost of any routine operation. Costing of materials and processes needs information about the actual dimensions of the image, as this will affect both materials and processing time.

In addition to the direct production costs, allowance has to be made for all overheads. Different organisations employ different methods for allocating these. Some use a standard percentage added to every costing operation: others allocate them in relation to individual departments and procedures. Whereas the current costs of administrative staff salaries, rent, services and so on are known, allowances for future equipment or re-equipment are difficult to estimate, especially in any period when technical innovation and change proceed rapidly. Expensive equipment has to be justified on the basis of actual use, and few organisations can afford to obtain and maintain every possible item of equipment that might be of some use on some occasion. Small organisations often find it more economical to purchase special requirements and services from other suppliers or agencies than invest in equipment that will only be used occasionally.

## Production schedule

Once the job has been specified, costed and agreed, the planning reaches the stage of an actual commitment to a period of time, during which the work must be completed. This may be calculated simply on the basis of the estimated time needed for all the operations, or it may have to work backwards from the delivery or finishing date. If this is the case, then the whole operation will have been considered

in this light from the beginning. If necessary, more staff are allocated to the task, or some work is placed with outside agencies or individuals, or the job as a whole has to be reconsidered. When there is extreme pressure to comply with a delivery date, the maintenance of standards can be a difficult issue, and yet the future success of the organisation may well depend on the standards it maintains. This is where proper planning, and the anticipation of all relevant factors, is so important.

The production plan itself is likely to consist of some type of chart, which shows the expected progress of the work in real time. Depending on the type of work, specific orders for parts of the total job may then be issued to the departments concerned. These job description details may accompany the various operations as they progress. Provided that all staff have to state specifically what has been performed within the days and weeks of production activity, it should be possible to monitor actual progress against the plan. If necessary, action may be taken to rectify any delays. Nothing is more inimical to satisfactory progress than presenting a department or section with an unexpected demand for an output of work that was not anticipated.

Most cartographic production is still typified by high labour costs, despite the great progress made on the technical side. The search for quicker and better methods of production is related to this. If labour costs are high, then efficient planning is vital. This puts a premium on good management at all levels, and the ability to make maximum use of the available resources.

# 11 COMPILATION

## Compilation and sources of information

The compilation stage is normally the first occasion on which the whole of the map information is assembled. Consequently, it is also the point at which any deficiencies and inconsistencies in the source material are likely to be revealed. Although this may seem to be most relevant to small-scale maps derived from a variety of different sources, it is also the case that even specialised maps encounter problems. Although adequate source material is either known or believed to exist, it is the process of constructing the compilation that either verifies or contradicts these assumptions.

Limitations of source material, or disparities between different sources not only complicate the task of assembling the information, but are often crucial in deciding whether a given method of representation is possible. For example, a decision may have been made to use hypsometric colouring on a small-scale map of a continental region. It is known that general small-scale maps exist which should provide the basic information. The hypsometric colour scheme is designed theoretically to provide the required series of vertical bands. The first compilation step for this element in the map information is to assemble the delimiting contours. It then transpires that different source maps have used different contour intervals, and that some of the desired contours do not occur on all of the source maps. Either the additional contour information has to be produced by interpolation – a risky business at small scale using disparate sources – or a different approach adopted to relief representation.

## TYPES OF COMPILATION

Although the methods used to assemble the map information are essentially the same for all types of map, particular compilation tasks will vary according to map type. The greatest difference is between basic maps, produced from a survey specially undertaken for the map or series, and derived maps, produced at least in part from existing maps. Although basic maps are usually associated with topographic series, many specialised maps are also basic, at least so far as the special content is concerned.

The type of compilation will also depend on the complexity of the map. A small illustration may require no more than an outline, and perhaps some notes referring to particular needs or problems in production. A detailed multi-colour map, which will contain many thousands of pieces of information, will commonly consist of three pieces of material: the basic line compilation; an overlay with text and name information; and a colour guide that indicates the colour composition applied to areas, all carefully registered together. These are produced separately in order to minimise interference, especially if the line compilation needs to be copied on to drafting or reprographic material as a guide.

The importance of compilation in the map production as a whole also depends on map type. It is least complicated and most straightforward when the compilation is an assembly of survey plots generated specifically for the map. Conversely, it is most complicated in the production of a derived map covering extensive areas, using information from many disparate sources, which may themselves vary in projection, scale, content classification, date and reliability. In these circumstances the compilation of the map is likely to occupy the major part of the total production time.

## Sources of information

In theory, small-scale derived maps of large parts of the Earth's surface should be based on an examination of all the best available source maps, including the small-scale maps produced in individual countries. In practice this would be extremely difficult to achieve, and in most cases prohibitively expensive. Several factors account for this.

Large map collections are few in number. Although there are many libraries, both public and institutional, few of them have more than local map collections, and usually maps cannot be borrowed. Even with large collections, examination of the maps they hold is difficult without actually visiting the collection and working through both the catalogue and many maps. Although a number of publications refer to newly published maps and atlases, and classify them, getting hold of a copy of any one map may prove difficult, especially if the map deals with a specialised subject. Few catalogue lists give details of such things as contour vertical interval, which may be very important in judging the possible value of the map as an information source for compilation.

For many countries, topographic maps are virtually unobtainable, and large and medium-scale topographic maps are restricted, even though they are known to exist.

Therefore, many cartographic organisations tend to build up a collection of maps and atlases relevant to their particular needs. Such collections tend to concentrate on sheet maps produced by official national mapping agencies and major international agencies if they are obtainable, and upon other published maps and atlases issued by cartographic publishers with a high reputation. Once this collection grows, then organising and maintaining it becomes important, and specialist map curators may be employed for this purpose.

For small cartographic organisations the problem is more difficult. Both checking the flow of map publications and searching for particular maps takes time, and many hours can be spent in a fruitless search for information.

For small-scale coverage of the Earth's surface, most cartographic publishers also make use of their own published maps, often in the form of world or regional atlases. Once produced, this basic information tends to be maintained by constantly correcting, improving and up-dating it as new or better information is obtained. As inevitably published maps go out of date, checking for detail changes is an automatic first step in using them as compilation source material. Information about significant changes to important elements such as state and town names, international boundaries and so on, is often reported in the press before any maps are published showing the new information. Therefore, such news is also collected as a guide to map corrections.

In many cases, the topographic detail (drainage and relief) shown on a relatively old map is unlikely to have changed in any significant way. If such a map is a suitable source in terms of scale, projection and general content, it is likely to be used as the main compilation basis. The cultural information, which is liable to change much more rapidly, is then carefully checked and up-dated. The evaluation of the age, quality and reliability of source material may itself involve considerable study (see Bond 1973).

For initial information, some journals, such as The *Cartographic Journal*, contain sections listing recently published maps. Many national mapping agencies will offer information about their products and their current availability on request. Some map distributors also publish detailed lists, such as *GeoKartenbrief* (issued by Internationales Landkartenhaus), classified by area and subject, and these are very useful in initially appraising the available information. A general survey of cartographic sources has been published by Hodgkiss and Tatham (1986).

# CONSTRUCTION OF THE LOCATIONAL BASE

The basic reference framework of the map may consist of either selected graticule and/or grid lines, together with their systematic display in the map border, or no more than a limited topographic outline, usually distinguishing the land/water division, and probably accompanied by the main international or internal boundary lines. The former provides an absolute or relative locational base with a coordinate system: the latter operates on the assumption that the map user will 'place' the map information by recognising major geographical entities and figures.

## Grid and graticule

Many large and medium-scale maps use the grid as the major reference system, even though graticule positions are shown in the map border. In this case the sheets in a series are likely to be arranged on grid lines, providing a rectangular map sheet. This is much the easiest to construct, as frequently graticule intersections of selected latitudes and longitudes will be calculated and plotted in grid coordinates, and much of the survey data for the control of position will also be provided directly in grid coordinates.

For those map sheets using graticule lines as the main reference basis, grid positions may only be shown in the map border, although in some types of map, especially military maps, a grid is likely to be included on the face of the map as well. With graticule sheet lines the neat lines may be slightly curved at medium and small scale, and will vary in dimension according to latitude and projection. Where a national grid exists, it is still likely that the coordinates of the graticule intersections will be provided in grid coordinates, and therefore it is necessary to use a grid for construction purposes, even though the lines themselves may not appear on the face of the final map.

For small-scale maps, which will include the graticule only, various methods may be used to construct the graticule to scale, depending upon the projection. With the increasing importance of digital construction and compilation, there is little doubt that a computer plot, or even a list of coordinate positions in $x$ and $y$ is the most efficient way of obtaining the information in a form which makes it relatively simple to construct. Although some projections can be constructed by graphical means, plotting in a grid is the most straightforward. For compilation purposes, it is a great advantage if a projection constructed for a new map has a dense network of meridians and parallels, as these will be of assistance in compiling the detail correctly, even though only a few of them will be retained on the final map. One disadvantage of taking a projection directly from an existing map is that very often it is difficult to calculate the position of a closer network of projection lines, or even a different selection of principal meridians and parallels, as the intervals marked in the map border may not be adequate for this purpose.

## Grid construction

The need to include more detail for construction purposes can apply to other compilation tasks, as the compilation is a means of assembling the information, not just an imitation of the final map. For all scales and map types where locational information is important, a precise construction of the grid is essential, as this will control the accuracy with which all other coordinated points can be plotted. Grids are normally constructed by using either a coordinatograph or a gridding template. The coordinatograph can be used to plot lines or points at any interval, whereas the punched template, which is precision engineered, is limited to

a fixed set of equidistant points. But the template avoids the introduction of any errors in measurement, which are possible with the coordinatograph. Templates with 4 and 5 cm point spaces can be used for a variety of scales. The 5 cm template can be used for plotting grid intersections suitable for scales 1 : 10 000, 1 : 20 000, 1 : 100 000 and so on. At 1 : 25 000 scale, 1 km on the ground equals 4 cm on the map, and therefore the 4 cm template can be used for any multiple of 1 : 25 000 as well.

A point plotting device fits into any hole in the template, thus ensuring that the pricked point is exactly centred in each hole. It is essential that the template should be placed on a flat, smooth surface. After the grid intersections have been plotted it is normal practice to ring them lightly in pencil, in order to make them easier to identify, and then to mark their values in $x$ and $y$ coordinates. Intermediate lines must then be constructed by measurement from the defined grid lines if required, and measured points are plotted by measurement from the grid lines.

Rectangular coordinatographs are most widely used, as most of the survey data are provided in this form (Fig. 80). The instrument consists of a carriage on the $x$ axis, a rigid beam which moves freely along a rail on one side of the plotting surface, or with large instruments on both sides. The position of this gantry can be controlled by scales set against the rails. Along this beam another carriage is able to move in the $y$ direction, and is also positioned against a scale set on the beam. This carriage holds the plotting head, in which a pricker point, pencil or pen can be mounted. The measured point is located by using a viewer with magnification. Usually the carriage is brought to the nearest scale measurement by hand, locked in position, and then finely adjusted with a vernier scale. The overall plotting accuracy is generally within plus or minus 0.03 mm.

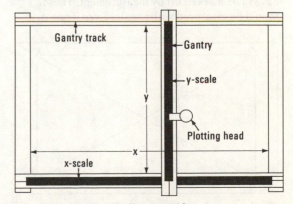

**80.** Rectangular coordinatograph

Coordinatographs range from small portable types to large instruments with their own plotting tables, from 70 × 100 cm to 150 × 200 cm.

The polar coordinatograph operates by bearing and distance, in which case the measurements are taken from an origin (Fig. 81). It is suitable for survey measurements based on direction and distance, but is of less general usefulness than the rectangular coordinatograph.

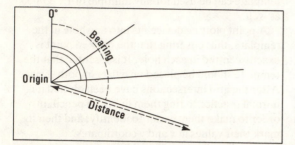

**81.** Polar coordinatograph

## Geometrical construction

In an emergency, if a gridding template or coordinatograph is not available, it is possible to construct a grid by hand, although this is never as accurate as one that is machine plotted. The initial step is to produce two axes at right angles. Ordinary drawing instruments such as set squares are not sufficiently accurate for this purpose. Several methods are possible. The simplest way is to draw diagonals from the sheet corners (Fig. 82). The point of intersection is the centre of the plot, and from this centre arcs can be drawn with a beam compass to intersect each diagonal. If the points marked by these intersections are joined, the lines on opposite sides will be parallel, and the adjacent lines perpendicular. The required grid intervals still have to be marked off by measurement. These should be measured with a scale. Even if one is

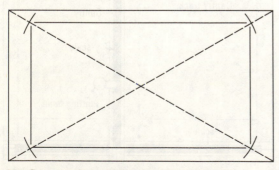

**82.** Construction with intersecting diagonals

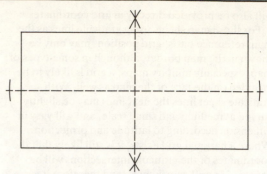

**83.** Construction with bisected centre line

incorrect, the error will not be accumulated, as will happen if any 'stepping' with dividers is used.

Alternatively, the central axis can be located by measuring half the width of the plotting sheet from one edge (Fig. 83). This line can be bisected by swinging arcs to each side, from each end of the initial line. With the two axes centrally placed, and perpendicular one to another, any other measurements can be made from one corner to obtain the desired grid intervals.

## CHANGING SCALE

A common requirement in compilation and construction is to change scale. When maps are being used wholly or partly as source material, the normal practice is to use larger scale maps, and therefore this information has to be reduced in scale to be included in the derived map. Even with basic map series, the survey or field plots are very often at a larger scale than the map, and the information needs to be reduced accordingly. Direct scale change can be carried out by both optical and mechanical methods. It becomes more complex when the scale transformation is combined with generalisation, which may be necessary in some cases.

The detail on an existing map or plot can be transferred to a new compilation at a different scale by using the equivalent grid square or graticule method. A series of grid squares on the map to be constructed is made equivalent to grid squares on the original map or plot (Fig. 84). These grid lines must be sufficiently close to facilitate the visual judgement of relative position. Detail is plotted by noting the points at which grid lines are intersected, and the linear shapes in between are approximated

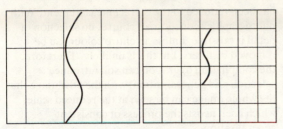

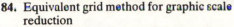

**84.** Equivalent grid method for graphic scale reduction

by visual inspection and judgement. If there is no difference in projection, and if the grid lines are close together, this can be done effectively.

With small-scale maps based on the graticule, then the polygons enclosed by meridians and parallels can be used in the same way although these may comprise curved lines (Fig. 85). On published maps they are rarely frequent enough on the map to make graphic scale change possible without additional lines being constructed. As noted above, this can be difficult on small-scale projections, where the meridians and parallels are neither regularly spaced nor at a constant distance. If the published map source is not a working copy, it may be necessary to construct the 'grid' on a transparent overlay. If a long line is being plotted (such as the outline of a continent), it pays to locate the line at a number of points along its course, so that these can act as a check on correct position during the progress of the work. It is surprisingly easy to make a mistake and carefully plot the line in the wrong squares!

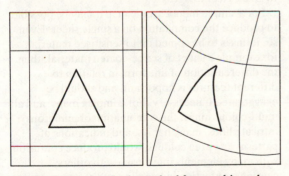

**85.** Equivalent graticule method for graphic scale reduction

## Optical methods

There are several instruments, some devised specially for cartographic work, which will project an image on to a screen through a lens system, so

that the image can be enlarged or reduced. The differences between them arise from the degree of scale change possible, the size of image that can be projected, and the accuracy with which the movements of lens and copyboard can be made in order to achieve correct scale and sharp focus. They can also project the image horizontally, upwards or downwards (Fig. 86). Control of movement may be through a simple chain drive with hand wheels, but the better instruments have machined drive shafts, and the most sophisticated ones are operated through electric motors. They also bring the image automatically into focus in relation to the scale change. Usually two or more interchangeable lenses are provided, making possible a scale change of up to five or seven times.

Of the different arrangements, the table with the image projected from directly underneath is the most simple. Other types make use of a mirror between the object and the projected image. Either the copyboard is horizontal, and the image is projected on to a vertical screen, or the copyboard is vertical and the image is projected on to a horizontal screen from above. The latter normally allows a

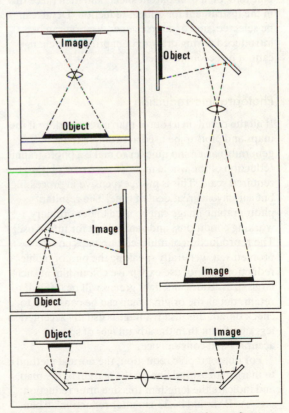

**86.** Different arrangements for scale change by optical projection

greater projected size, and a better working position, but of course the projected image must not be interrupted by interposing a hand between the projection and the image surface.

All these devices are basically similar to cameras, the projection screen being equivalent to the ground glass screen in the process camera back. Indeed, some instruments combine the functions of optical enlarger/reducer and that of a simple camera. Although setting the scales for reduction or enlargement will give the approximate scale ratio and focus, it is more satisfactory to construct an equivalent grid at the required scale, and make sure that the grid lines on the source map fit exactly to the projected image. Most of the smaller instruments can only deal with a limited field, and it is often necessary to construct the compilation by a series of separate reductions fitted together.

Although the projected image may be sharp, and would appear to give a good basis for accurate reproduction, in practice the reduced image lines, especially those in light colours, can be difficult to follow. For this reason it is quite common to make an extraction trace of a particular map element in black ink on a transparent sheet, and then place this in the instrument for optical reduction. Detail can be selected, and of course generalisation can be introduced during drawing, but projection change cannot be accommodated.

### Photographic reduction

If all the detail on a source map is required, or if the maps and plots to be used are complex, it is generally easier and quicker to make a photographic reduction of the whole map, or even part of it, to the required scale. This is more expensive in processing, but can save a great deal of time. The resultant photographic image can be used under ordinary working conditions, and can be kept for future use. The reproduction of multi-colour maps poses problems, as generally speaking the photographic reduction will not use expensive colour film at such large sizes. Blues and light greens will be difficult to retain, but as the original map can be consulted to check detail, this disadvantage is usually accepted as less significant than the advantages of speed, accuracy and convenience.

For photographic reduction, the normal method is to mark a measured line on the margin of the map, and indicate the length of this line after reduction; e.g. 'A–B should be 33 cm after reduction'. Then the photographer will adjust the projected image on the camera screen until the line on the map margin has been reduced to the correct measurement. With photographic reduction, of course, no selection of detail is possible, unless an entire colour can be removed by filters, but this is unlikely. Therefore, the compilation operation can still introduce selection and generalisation, with the advantage that the whole image can be seen at the reduced scale, which itself assists the process of consistent generalisation.

## COMPILATION PROCEDURE

The compilation contains all the information to be presented in the map; it defines all the locational detail; and it serves as the basis for all the ensuing production. Consequently it is an important document, and should be treated as such. Although there may be a few simple maps which can be compiled, designed and produced as a single operation, the procedure is only likely to be successful within strict limitations, and is unlikely to produce anything more than a repetition of something already familiar. In general a compilation should be made, even for a simple map, so that the design and production can be considered, and especially so that detail can be checked before production is commenced. It should always be made in ink, using coloured inks where necessary, in order to give a clear image.

For a simple monochrome map it may be possible to produce the compilation on a single sheet. Even so, if this is to be copied on to sensitised material, for example a sheet of scribe-coated material, then the differentiation of line form in relation to different features is important, and should be exaggerated if necessary. For complex maps, and all multi-colour maps, the combination of numerous intricate lines, many names, and indications of patterns and area colours, would produce confusion, with many elements interfering with others, so causing a reduction in legibility and certainty of identification. It is essential that all the features on the compilation can be clearly distinguished. Therefore, for such maps a multiple complication is used, consisting of two or three separate sheets, carefully registered together.

The line image normally controls the locational information; it contains linear features, outlines and the location of point symbols. Names, the arrangement of which must take into account the other map information and its representation, are produced on a separate translucent overlay. And area colours and patterns are usually indicated on another overlay, using the line image as a guide to area outlines and edges. Alternatively, for area colours, a copy of the line compilation can be made, and then this can be coloured up to indicate the various categories of area colour and pattern to be used in the production. The name arrangement should be done preferably as an overlay on both the line image and the area colour guide, so that the placing of the names can take both these elements into account.

## Order of work

The line image is produced first, as this will contain all the information of the reference base (neatlines, map margin, grid and/or graticule, control points, etc.). In some cases, bounding lines must be included, for example for vegetation or land use areas, even though these will not appear on the final map. Because the cultural features are located in a physical environment, the physical features are normally compiled first, and the cultural information added subsequently. For example, a boundary line may be defined as running along a watershed; the representation of the watershed by the generalised contours will control the position of the boundary as shown on the map. Similarly, cultural features placed in relation to surface waters must have these elements defined first. The usual order is drainage and shorelines (hydrographic features); contours and spot heights; buildings and settlements; roads and other transport and communication features; other linear features such as transmission lines; boundaries; and areas of vegetation and land use. The details of procedure will vary according to the type of map and the relative frequency of different topographic features.

With special-subject maps, the topographic reference base is compiled first, and then the special information, which will be the main map content, is assembled on this basis. There is a distinction between the base map, which may be appropriate in detail to the compilation of the special information, and the reference information to be included on the final map. These considerations are dealt with more fully in the chapters on different types of map.

It must be stressed that the compilation is not a manuscript imitation of the final map. It is constructed to display the map information clearly, so that it can then be produced according to the symbol specification. The colour coding and symbol forms used on the compilation do not need to be equivalent to the symbol specification used in production. If the coloured line compilation is subsequently reproduced as a guide image, then any distinction in hue will be lost; but the map features must still be distinguishable. Therefore, variation in line form is vitally important. Line gauge should approximate to that specified for the feature symbol, as otherwise complications that could affect the line simplification may be introduced. If in question, lines should be slightly thinner, but never thicker. Light blue and green coloured inks should be avoided if possible; if they are essential, they can usually be reinforced by the addition of a certain amount of black. In order to make a clear distinction there is no reason why contours should not be in red on the compilation, and 'green' outlines shown in purple, both of which will have a higher contrast and reproduce better than either light brown or green.

## Colour guide

This component identifies all the area symbols that are to be constructed separately from the line image. As most areas will have bounding lines, and as these will appear on one or more of the coloured line images, they have to be produced first, and the area colour symbols fitted to them.

The colour guide is superimposed on the required line images and the different area symbols and patterns indicated according to some code. This need not correspond to the specified colour of the particular area symbol. Although it may be convenient to use green to show all the areas represented by a solid green on the map, a pattern of lines of any colour will be sufficient, provided that this coding is explained on the colour guide or the production instructions. Generally, line patterns are quicker and easier to produce. Coloured inks are more effective than coloured crayons, although these can be used if desired.

Where area colour is complex, and especially where many colour combinations are used, it is usually necessary to make a production guide to describe the solids, tints and patterns in terms of their production components. For example, if the woodland is to be produced in a green made by a combination of 30% blue and 50% yellow, and if the map includes a 30% blue, which is also applied by itself to the water areas, then the production guide would show the areas to be included on the mask for 30% blue and the areas which form part of the green in one unit.

### The names overlay

In order to ensure that all names and figures are legible, and to avoid confusing the details of the line image, it is normal practice to make a separate overlay of all place names, figures and text information, unless this is very limited. This compilation operation is related to the production processes, as the names, figures and any text information in the form of title, explanatory notes and marginal information will usually be produced by photosetting, separately from the line images themselves. Very often names and figures will appear on different colours on the map. To make certain that none of these interfere with one another, they should all be compiled on the same overlay, and then the different colours distinguished by underlining or marking them to indicate the colour on which they will appear. By using the line compilation and referring to the colour guide, it is possible to judge the placing of names in relation to the topographic features, and to take into account marked colour contrasts in the map which will affect name legibility.

## NAME ARRANGEMENT

Names can be a significant component in the total map information, and in some kinds of map the selection and representation of names is a major aspect of the compilation work. The treatment of names within the total map content is important for both accuracy and legibility. Poorly positioned names are easily referred to the wrong location, and can be either misleading or incorrect. Poorly arranged names can be difficult to read, can cause confusion in interpretation and can interfere with other map information. As names always have a high priority in map information, very often the success of the map will be strongly affected by the legibility and appearance of the names.

Names may be positioned adjacent to, alongside, or within the feature, depending upon the nature of the representation and the relative size of the name. Point symbols and small areas often have common characteristics in relation to names, as if the name cannot be placed within an area feature it must be positioned alongside. Again, linear features such as estuaries may contain the name if the area is sufficiently large, but smaller rivers shown as single lines must have the name positioned alongside the linear feature.

As names take priority over other information, to the point where lettering is never interrupted, the obvious objective in name arrangement is to cause minimum interference to other map information and at the same time place the names for correct reference and ease of use. The problem is most difficult in monochrome maps, as lines, outlines and patterns may have to be interrupted to include names. In multi-colour maps, names in the darker hues can usually be superimposed on lines or areas with lighter colours, provided that there is sufficient contrast, but strong contrasts in background colour under a name should be avoided for the sake of legibility.

### Names applied to points or small areas

On many maps the names applied to points or small areas are the most numerous. As in the English language the reading order is from left to right, the ideal position for a name is immediately to the right of the symbol, so that the map symbol and the first letter of the name are adjacent. This provides the most rapid identification for the map user. If this is not possible, for example if the symbol is itself on a line or a strongly marked feature, positioning to the left of the symbol is possible, but care must be taken to ensure that the last letter of the name is then immediately adjacent to the symbol (Fig. 87). Most

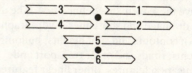

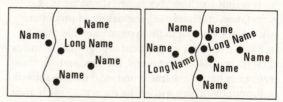

**87.** Placing of names for point symbols (and small areas)

point symbol names are aligned in relation to the reference system of the map, or its neatline, depending on projection and scale. At medium and large scales, names are usually kept parallel to the base line of the map. At small scales, and especially if the graticule has distinctly curved parallels, then placing may be parallel to the nearest parallel of latitude, and may vary slightly over the whole map (Fig. 88). For maps with concentric graticules, such as those of polar areas, special rules are applied (see Fig. 89).

If names, even short names, are moved from the horizontal position to a diagonal, then the rule is

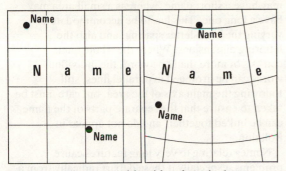

**88.** Name alignment with grid and graticule

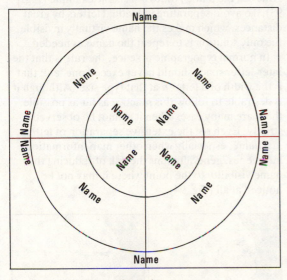

**89.** Name alignment for polar maps

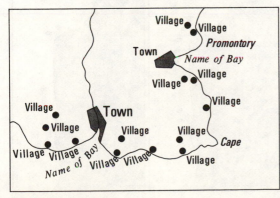

**90.** Name curvature

that the base line of the name must be slightly curved, and not straight. In areas where there are many names of point features in close proximity, this procedure often has to be introduced. Wherever possible, the name should be curved so that either the beginning or the end of the name is close to the horizontal (Fig. 90).

For a physical feature which also has a height, such as a mountain peak, the height value should be located correctly, even if this involves moving the name slightly out of position.

## Names applied to linear features

Although there may be many linear features on a map, the proportion named is usually much smaller than the point symbol categories. The most common are rivers. The name should be placed alongside the line, and follow its alignment (Fig. 91). As the

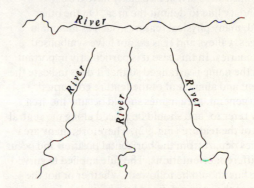

**91.** Name arrangement for linear features

section at which the name is placed can often be selected from a number of possibilities, positions near the vertical, or in a section of extreme irregularity, should be avoided unless there is no alternative. This therefore follows the same rule, that the horizontal position, or close to it, is the easiest for the map user. If the linear feature extends over a long distance, the name may be repeated as required. It should not be placed so that it is separated from the linear symbol by another feature. Features occurring on the map in a north–south orientation are the most difficult to deal with satisfactorily, but because the name must be aligned with the feature, it may not be possible to avoid near-vertical arrangements.

## Names applied to areas

In relation to scale, the area must be large enough

for the name to be contained within the area, and this frequently means that lettering size must be adjusted to the areal extent and the length of the name. As the reference to the feature must be clear, it is not usually possible to place the names of such features outside the area. For example, on a small-scale map, it is generally accepted that the name of a state cannot be positioned so that it lies in the territory of another state, although it can be placed in 'neutral' territory, such as the sea. Because area symbols vary so much in extent and shape, and because names of administrative areas of physical features are often compound names that may be quite long, the arrangement of names applied to areas is often the most difficult part of the name compilation.

There are basically two distinct cases: areas delimited by bounding lines, and areas that are located only by interpretation. Names of states and administrative divisions usually have a clear boundary line to delimit the area. On the other hand, many physical features, such as mountain ranges, valleys and seas do not have symbolised boundaries. In this case it is particularly important that the name is arranged so that it does indicate the extent and alignment of the feature concerned.

In general, area names should occupy the area they refer to, and should be placed along the central axis of the feature (Fig. 92). Therefore, many area names depart from the horizontal position and occur at different orientations. The rule applied to curved base line should be followed. Whether or not the edge of the area is defined by a boundary line, the most satisfactory arrangement is for the space between the first and last letter and the edge of the area to be approximately one and half times the inter-letter space. In order to make the name extend over the area, the letters frequently have to be spaced out, and indeed area names on maps are often referred to as 'spaced' names from the point of view of lettering and arrangement, although this is not always the case. It so happens that a large area may have a short name, whereas a small area may have a long one. This has to be accommodated by varying the inter-letter spacing, and also the lettering dimensions. Where the whole name consists of more than one word, the individual words can be arranged in parallel lines, still following the main axis of the area, but care must be taken to ensure that the separate parts of the name can be linked together, and do not appear in isolation.

Names relating to very long features cause problems, especially if these depart radically from a horizontal alignment. On small-scale maps of South America the name Andes is a classic example (Fig. 93). To avoid separating individual letters by great distances, which makes the name virtually invisible, the only solution is to repeat the name as needed.

In normal typographic practice, the rule is that the inter-letter space should never exceed one 'en'; that is the width of a letter *n* at that type size. Although it is desirable to follow this practice as far as possible, there are many cases when it cannot be observed strictly. Even so, the extensive separation of letters in a name, especially when other map information intervenes, generally runs the risk of reducing the name visibility to the point where it may not be noticed at all (Fig. 94).

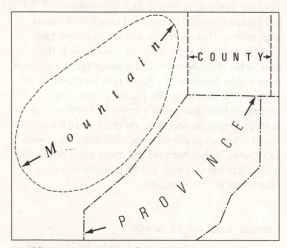

**92.** Name arrangement for area features

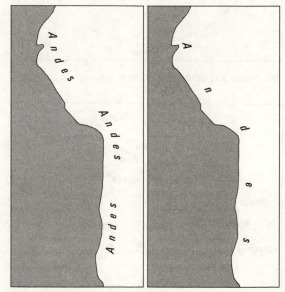

**93.** Names of long features

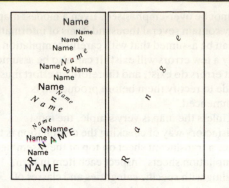

**94.** Spaced names in crowded areas

## Names in combination

If there are many names on the map, the usual procedure in name arrangement is to enter the area names lightly in the first place, avoiding point locations as far as possible, and then attempt to arrange the point and line names completely. In many cases it becomes necessary to adjust the position of some names in order to improve the arrangement of others. If all the point and line names are entered first, then very often it turns out that additional problems, which could have been avoided, are introduced for the lettering of the area name. (Fig. 95). Because of the many variables involved – correct identification, legibility, ease of reading, inter-letter spacing, lettering style and size – some fine adjustment of the original arrangement is normally inevitable. Names that intersect must be carefully watched, especially if they are about the same size and weight. Small names should be moved away from the alignment of larger ones, to avoid introducing interference. In crowded areas, it may help to spread the point symbol names outwards to avoid heavy concentrations in a small area. This fine adjustment makes a great deal of difference to the

overall quality of the map, and repays the extra time and trouble involved. Because of the infinite variety of shape, area and orientation of topographic features, it is impossible to adhere strictly to any set of rules, but in most cases a sensible arrangement can be made if the basic principles are followed.

## Lettering on the compilation

The arrangement of names on the compilation is usually done by hand, and it is the only surviving example of hand lettering in virtually all map production. Clearly, as the compilation does not imitate the final map, it is neither necessary nor desirable to attempt to reproduce the characteristics of topographic styles exactly, even if this could be done. However, it is important that the names inserted on the name compilation are correctly positioned, and occupy the same amount of space as the typeset lettering which will be produced. As the names then have to be listed prior to typesetting according to the lettering specification, it is also desirable if the main type variations – capital and lower case, Roman and italic – are followed on the compilation. This serves as a guide in listing names for setting. Those names that are to appear in a different colour (assuming that most of them will be in black), can be underlined lightly in a colour code, to indicate the image to be followed in production.

The lettering design is part of the design specification of the map. The compilation shows the content and its location. Once the name arrangement has been checked for completeness, and any omissions rectified, the names are then listed according to typographic style and size. On most photolettering machines, a different setting of the machine is required for each variation in size, and different masters are used for different type styles. Therefore, a complete list has to be made of all the names in a given style and size, (see example below), so that these can be set in one operation. The same basic organisation applies to digital typesetting.

*Univers 57 Roman capitals and lower case 1.8 mm*
Tarfside Fettercairn Auchenblae Gourdon Catterline
*Univers 57 Roman capitals and lower case 2 mm*
Edzell Lawrencekirk Inverbervie Johnshaven
*Univers 57 Roman capitals and lower case 2.4 mm*
Stonehaven Brechin Montrose Forfar Arbroath
*Univers 57 Roman capitals 3.5 mm*
DUNDEE
*Times Italic capitals and lower case 3 mm*
Firth of Tay

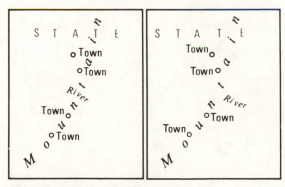

**95.** Adjustment to clarify spaced name

Every manual repetition may introduce error, so the names list should also be checked in detail. If there are many names in a single type category and size, then the cartographic draftsman may have difficulty in finding the typeset name when working on the name stick-up at the production stage. In this case the listing should follow a specific order, for example by listing names by grid square, in parallel lines from top to bottom. The name compilation is used to control the positioning of the names at the production stage. Any problems that have not been solved will become evident during production, which can waste time and delay progress. Therefore, it is always better to ensure that the name compilation and names list are correct before proceeding to production.

## CHECKING

Checking has been referred to several times in the discussion of compilation procedure. Its importance cannot be over-emphasised. Even a modest map may contain several thousand items of information. It can be assumed that with careful compilation, only a few errors will exist. It can also be assumed that errors do exist, and therefore an effort must be made to rectify them before production is commenced.

Unless the map is very simple, the most satisfactory way of checking the compilation is to place a translucent sheet on top of the assembled compilation sheets, inspect each item in turn, dealing with specific categories and classes of information, and mark the errors on this sheet. Casual visual inspection is not enough. If there are several hundred small point symbols on the sheet, it is very likely that one of them at least will have the name omitted, or that for one of them the name will have been transposed. Rivers running uphill, mountain tops that have turned into lakes, and even a grid line with a different figure at each end are not unknown. Errors in compilation should not be regarded as some sort of unusual disaster, but as a normal feature of map production. Once this fact has been recognised, then a series of checking procedures can be regarded as a normal part of the map production operation.

# 12 LINE IMAGE PRODUCTION

## The line image

The term 'line image' is somewhat confusing, because it is used in two different ways. The obvious reference is to any image composed of lines, and in many cases certainly linear features, outlines, lettering and point symbols are predominant. But in reprographic processing it also refers to any image composed of discrete elements. In this sense all images that can be directly reproduced by lithographic printing are 'line' images, because even tonal images have to be converted into half-tone or tint (that is, discrete images) to make them reproducible. As these require different methods in production, the first definition is used here.

In terms of production, the line image on a map will usually consist of three distinct elements: lines and outlines, lettering and point symbols. Patterns may also comprise line and point symbol elements, but these have to be constructed by different means. All the elements forming the line image for one printing colour eventually have to be combined. Image-separated production is simply a convenient way of producing the images in the first place.

In analogue cartography manual methods are used to construct both regular and irregular lines. Repeated symbols, such as point symbols or patterns, may also be produced by making use of prepared or pre-printed material, in order to avoid having to repeat small symbols that should be identical, especially if large numbers are required for the map. The general rule is to use manual drawing procedures where they are necessary, but to avoid using them for operations for which they are inefficient and slow. There are limitations to the resolution that can be achieved by manual processes. Very fine patterns, like identical very small point symbols, are impossible or extremely difficult to produce manually.

## POSITIVE LINE IMAGES

Positive images consist of opaque image elements on a clear or translucent background. Basic pencil drawing, painting and pen-and-ink drawing normally produce positive images. In cartographic practice, these images are achromatic, and the technical objective is to produce dense, sharp lines on a background that will either reflect or transmit light. For contact copying, the images must be on a sufficiently transparent base material for radiation to be transmitted during exposure. Little cartographic work is carried out on paper surfaces, as these are opaque (requiring process photography for image transfer), dimensionally unstable and difficult for erasure. Tracing paper is a thin paper made more transparent by being saturated in oils and then dried. It is unstable, fragile and difficult to work on if very much ink is deposited on the surface. Although cheap and available, it is only useful for either very small or very simple maps for which the original drawings will not be required for further use in the future. Polyester plastic drawing materials are now almost universally used unless there is a specific reason to employ some other kind of base material.

## Line gauge

In cartographic work, the line forms and dimensions are determined by the symbol specification. The objective is to produce the required lines according to the specification, at a quality which will ensure good reproduction. The specification of line gauge is usually expressed in tenths of a millimetre, and therefore the drawing instruments used must be able to produce the required line gauges. In practice this is difficult, no matter what type of drawing instrument is used. The choice lies between an open pen (with a single flexible nib); a ruling pen, with blades which can be adjusted with a screw; or a reservoir pen with interchangeable points, the points being fine tubes of a given dimension. The open pen is now rarely used, although it did have the single advantage that, as line gauge depended on pressure, variation in line gauge could be obtained while drawing a line. Ruling pens have advantages in that they can be set to a range of gauges (within the dimensions of the pen), but of course the correct line gauge has to be set each time the pen is used for a given dimension, and usually this can only be verified by drawing some lines and measuring them. Reservoir pens (often called technical pens) with interchangeable points have great advantages in speed of operation and uniformity of product, but the available line gauges then depend on the dimensions of the points provided, which may not include gauges needed for the particular map specification. For example, common line gauges in cartographic specifications are 0.1, 0.15 and 0.2 mm. With many technical pens, the manufactured point sizes are 0.13, 0.18 and 0.25 mm. This limits the range of line dimensions available.

### Line form

The chief requirements for cartographic symbols are continuous lines, multiple lines and interrupted line forms. Interrupted lines are normally produced by first drawing a continuous line and then breaking it as required, either by scraping on the positive drawing, or opaquing on the copy negative. Any scraper used for this purpose must have a smooth, slightly curved edge, so that the point is kept away from the surface of the material. Any attempt to draw interrupted lines directly results in uneven line thickness, rounded ends to line sections, and slow progress, as keeping a constant section length and interval requires great concentration.

The most common multiple line symbol is the double line road casing. It is virtually impossible to construct the two lines individually and keep them parallel. Double line ruling pens are available, but can be difficult to use. If the lines have to be produced in positive fashion, the most satisfactory solution is to draw the line as wide as the total symbol width, and then construct line casings from this by photographic processing. Combinations of different line gauges and different line forms in the same line symbol are often used on topographic maps. For example, one line may be continuous, the other interrupted, and the lines themselves may be at different gauges. It is far easier to construct these at the drawing stage by scribing.

Long straight lines are drawn with the aid of a straightedge. Although the ordinary scale or 'ruler' can be used for shorter lines, there is no guarantee that the edges of these instruments are actually straight. Arcs can be produced either with compasses, if they are small, or by a variety of additional aids for longer lines. Railway curves are

sets of plastic curves that range from small to large radii. They are particularly useful for some types of projection. If the curved line changes radius, french curves are used, the plastic curve being fitted against a section of the line.

## NEGATIVE LINE IMAGES

Negative line images are produced by scribing, in which a surface coating is removed from the base material by cutting, so that the line is transparent against an actinically opaque background. For this process, special tools and materials are needed.

Scribe-coated materials generally consist of a polyester film with a thin coloured coating that can be removed from the base. As often the scribing is done by tracing, the coating must be sufficiently translucent for the image to be seen, at least on a light table. As contact copying will usually be done to orthochromatic film, any coating that transmits only long wavelengths can be used. Most of the scribe coatings in general use are red, orange or rust. Green coatings can be used, but then any exposure must be made to photographic film or material, which is only sensitive to radiation in the 'blue' part of the spectrum. The coatings for scribing materials are basically coloured waxes, evenly applied over the sheet surface.

A great variety of scribing tools exists, but they can be divided broadly into two main groups (Fig. 96). The first group makes use of a fixed round point, the diameter of which controls the line gauge.

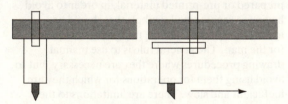

**96.** Round and chisel scribing cutters

These are pushed into the coating and remove it by sideways pressure. In the other type, the cutting edge is in the form of a chisel, and the line gauge depends on the width of this chisel blade. To cut an even line, the chisel point must be kept at right angles to the direction of movement, and is therefore mounted in a swivel. This requires less pressure to cut a clean line, but the movement of the eccentric swivel has to be anticipated as the cutter is pulled to follow any change in line direction.

Most scribing tools are mounted in a tripod. The cutting point is placed at one end, and the rear is supported by rotating balls, so that the instrument can be moved freely over the surface, but kept perpendicular to it (Fig. 97). It is obvious that to cut a clean line, the cutting point or edge must be in a vertical position. If it is tilted to one side, a ragged line will result. Other refinements include pressure-loaded instruments, in which the pressure on the scribing point is adjusted so that it cuts through the coating; and a magnifier mounted in front of the cutting point to help the draftsman to follow the line details.

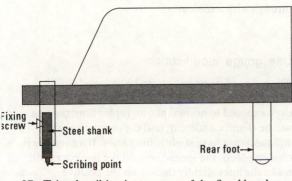

Fixing screw — Steel shank
Rear foot→
— Scribing point

**97.** Tripod scribing instrument of the fixed head (round cutter) type

## Line gauge

The line gauge depends on the dimensions of the cutting point, and therefore the cutters must be interchangeable. Standard gauges are provided by the manufacturers, and special gauges can be obtained to order. Most of the round point instruments use synthetic sapphires, and the chisel cutters hardened steel. One advantage of the scribing process is that the points and cutters are engineered to a fine tolerance, and therefore standardisation of line gauge is comparatively easy to obtain. Most of these tools are specially manufactured for cartographic purposes, in comparison with most reservoir pens, which are dominated by the needs of the general graphics markets.

## Line form

Multiple line forms are easily produced by using

scribing cutters with the required dimensions. Double and treble line scribers are available. Any combination of line gauge and inter-line space can be produced to order. As with ink-drawn lines, interruption is normally produced by scribing continuous lines first and then breaking them either on the scribe coat itself, by opaquing through the negative line, or by deletion on the subsequent photographic contact copy. Scribing has added considerably to the general use of multiple line symbols in map production.

## Guide images

Guide images or 'keys' can be added to both positive and negative materials. Although drawing and scribing are often performed by tracing, it is an advantage, especially for large images, if the information to be drawn or scribed is actually reproduced on the surface of the material. For ink-drawn images, a blue key can be added to the plastic film by simple 'blue-key' processing, in which a negative is exposed to a thin coating, and the unexposed parts washed off. The most common guide image is a copy of the line compilation. The draftsman then follows the lines relevant to the particular image, drawing them to specification. The blue image 'disappears' on exposure to sensitised materials which are only affected by short wavelengths.

For scribe-coated materials, guide images can be produced by using specially coated materials, in which there is a photosensitive layer on top of the scribe coating. Different materials can be exposed to either negative or positive images respectively, and they are either photographic or diazo based. Diazo-sensitised materials can be exposed to either a positive or a negative, depending on whether the guide image is wanted in positive or negative form. For working over a light table, many draftsmen prefer a negative image, in which the image lines are seen as translucent against a dark background. A simple way of adding a guide image is to apply a coloured dichromated colloid coating (usually of the type used for proofing from negatives); exposure to a positive image of the compilation or copy will result in a negative guide image on the scribe coat, and vice versa. The result is not very attractive, as the colour density varies, but it can be effective.

A continuous-tone image, such as an orthophoto, can also be exposed to scribe-coated material which has been manufactured with a continuous-tone photographic emulsion on the surface.

## Corrections and alterations

Special materials are available which can be painted on the scribe-coated surface, thereby replacing the scribe coat where it has been cut away. So a correction can be carried out by first painting out the incorrect section, and then re-scribing. Some cartographic draftsmen find that a greasy crayon can also be used to fill in small spaces on the surface, or to interrupt scribed lines. Even black ink can be used for this purpose.

## Positive and negative line images

Because of the great variety of needs in cartographic production, both positive and negative processes are useful. They have relative advantages and disadvantages. Ink drawing is less expensive in both materials and instruments. The images produced by a skilled cartographic draftsman are sufficient for most purposes. Additional material in the form of dry transfer or 'stick-up' images can be added easily to simple line originals. Even hand lettering can be included if necessary. Therefore, the medium is flexible. Maintaining a constant specification with very fine lines is difficult, requiring great concentration and control, and therefore tends to be relatively slow.

Scribing produces negative lines that reproduce well. It is easier to standardise than pen-and-ink drawing, and the production of fine lines to specification in a large image is quicker and easier. It is better for some line forms, such as double lines. The negative image is less flexible for other purposes. It is difficult to add lettering or patterns directly. Generally speaking, where large volumes of line work are required, and especially if many operators must be engaged at the same time, scribing makes it easier to control standards and maintain the specification of the line symbols. If only a small section of a line image is included in a particular map sheet, the use of pre-coated materials becomes extravagant.

If the scribe-coated material is based on a polyvinyl or polycarbonate film, the scribed lines can be dyed in black when the image is completed, and then the scribe coating removed with a solvent. This directly converts a negative into a positive. It is also possible to scribe the lines using a white coating, and then photograph the image (placed over a black background) with a process camera. The transparent scribed lines then appear as a positive image.

# LINE IMAGE PROCESSING

## Line copying

Using normal photographic materials, a negative line image can be contact copied to produce a positive line image, and vice versa. A positive copy can be made from a positive line image by dichromated colloids, diazo or positive reversal photographic film. A scribe-coated material with a sensitised coating can also be used to reproduce a scribed image in scribe-coated material. Exposure of the line image produces a resistant coating over the non-image areas, and then the lines are etched through with a solvent.

## Line gauge modification

The gauge of a line image can be changed slightly during contact copying. If a positive line image is over-exposed to normal photographic film, the line will be slightly undercut, and the contact negative will therefore have slightly finer lines. If a negative is over-exposed, the spread around the line edges will make the lines slightly thicker on the positive.

This variation can be controlled by the use of spacers. In normal contact copying, where an exact duplicate of the original image is required, it is important that there should be full contact between the two image surfaces. If contact copying is done with the image being exposed separated from the unexposed material by the thickness of the base film, then it is easier to modify the resultant line thickness (Fig. 98). This contact copying 'through the base' (or with an additional piece of thin transparent material, known as a spacer, interposed) normally means that a positive line will be thinner on the contact negative, and a negative line will be thicker on the contact positive. The two

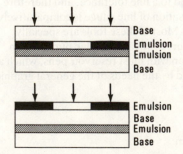

**98.** Contact copying and copying through the base

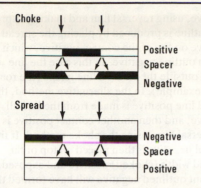

**99.** Choke and spread

images are often referred to as 'chokes' and 'spreads' respectively (Fig. 99).

The amount by which the line image is changed depends on three factors: the degree of diffusion in exposure; the thickness of any spacer; and the duration of the exposure. Both normal photographic film and positive reversal film can be used. If a positive is produced by exposing a negative to normal film, it is relatively easy to make a 'spread' positive, especially if the negative is separated from the positive film by a spacer. A diffuse light source will help to increase the amount of spread. If a positive is exposed to positive reversal film, then undercutting at the line edges will produce a 'choke', with a slightly narrower line. A negative exposed to reversal film (which will produce a duplicate negative) with additional spacing will produce a negative with slightly thicker lines (Fig. 100).

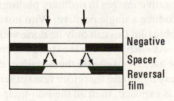

**100.** Spread negative using positive reversal film

## Scale-independent line modification

In the normal photographic procedure, the photographic reduction of an image on the process camera will reduce to all elements in the image in proportion to the scale change. If it is necessary to reduce the scale slightly, but maintain the dimensions of the symbols on the map, this can be done by using a special attachment, the Klimsch Variomat. This instrument is mounted in front of the camera lens, and contains a rotating glass disc, moved by a cam. The rays of light from the image are deflected through the disc, and therefore their exposure can be controlled independently of the overall scale reduction. If a negative image is photographed, the lines on the original image can be made thicker on the reproduction; and if a positive image is photographed, the lines on the resultant image can be made thinner. The variation in line thickness can be controlled by interchangeable discs and cams.

## Double lines and outlines

The double line of a road casing or the outline of an area can be produced by photographic processing using positive reversal film. As road symbols are frequently filled with an additional colour, the image for this coloured infill can be used if this is part of the map design. Otherwise a separate image of the desired road thickness would be needed. In the normal method for such a road symbol the two lines are drawn or scribed first, and then another line is drawn or scribed to represent the coloured infill. With reversal processing, the line for the coloured infill is produced first, and then this is used to produce the double line.

Two methods are possible. In the first, double exposures with filters of the line fill are made in order to make use of the properties of reversal film. In the second, a masking method is used.

To make a double line (or outline) image by taking advantage of the properties of reversal film directly, two exposures are required (Fig. 101).

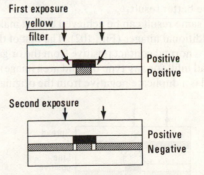

**101.** Double-line or outline image using positive reversal film

Either a positive or a negative of the road infill can be used, and this is normally a broad line image, between 2 and 4 mm wide. The first exposure is made through a yellow filter, and its duration determines the eventual width of the double lines. The longer the exposure, the thicker the lines. Development of the reversal film without any exposure would produce a black image. Exposure destroys this reaction. If a positive image is exposed first, then the area under the positive would develop out into a dense image, whereas the transparent background areas would revert to transparent. Prolonged exposure undercuts the line being copied, so that it would be slightly narrower if developed and completed at this stage. The second exposure is given without any filter, but using a light source strong in ultra-violet radiation. This restores the normal reaction of the photographic film, making the transparent background areas dense on development. But the first exposure has produced slightly thinner lines, and so there is a gap between the images produced by the two exposures. The result is a negative image, the lines representing the edges of the road infill.

Using a negative image of the road infill, the first exposure operates to restore the normal reaction along the image line, and would produce a dense image in the non-image areas, as these remain unexposed where they are concealed by the negative. In this case the original line becomes slightly thicker, as light creeps beyond the edges of the line on the negative during the prolonged exposure. The subsequent second exposure produces an image along the line, and again the line edges are revealed by the difference in width between the two images. The difference between the two methods is that with the positive system the two outlines are within the width of the infill, whereas with negative originals the two lines are outside the width of the infill. The latter is said to give the better result.

The same result can be achieved by first making two additional images (Fig. 102). The first of these can be a normal contact positive from the original of the road infill (or any type of solid area image). The second is a duplicate negative from the original infill

negative, using reversal film and making a spread. The outline is produced by placing the spread negative on top of the contact positive, which will give an outline negative. In this case the line width will be outside the original line image (the road infill in this example). In the alternative method, the normal line positive is made from the line fill negative, and then another contact positive is made on reversal film, using the choke method. If the original negative is then placed on top of the reduced width (choked) positive and exposed, the resultant outlined negative will have formed the lines inside the width of the original image.

## LINE COMBINATION

The combination of two or more line images is a common cartographic requirement. It can be carried out with either negatives or positives. Two methods are possible. Either a 'sandwich' of the line images can be exposed in one operation, employing simultaneous exposure of all the images; or the line images to be copied can be exposed successively, each one being exposed independently. The essential and important difference is that with the former, the second and other images will be separated from the sensitive film by the thickness of several pieces of base material, and will not be in full contact. With the second method, each image can be placed in proper contact, and correctly registered in position.

The most common requirement is the exposure of a set of negative images to ordinary photographic film, to produce a single positive. With normal photographic film this can only be done with negatives. As each negative is exposed in turn, the non-image areas, opaque on the negative, do not transmit any light, and therefore remain sensitive. If a positive is exposed, then all the non-image area of the entire film is exposed, and can no longer record any subsequent image.

The combination of positives is more difficult. It cannot be done using successive exposures on to positive reversal photographic film, as the exposure reverses the image formation process in all the exposed (background) areas. It can be done using dichromated colloid materials, as a sheet of translucent plastic can be coated, exposed to a positive, developed, dyed, washed off, and then re-coated for another exposure. The process is

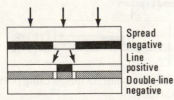

Spread negative
Line positive
Double-line negative

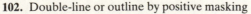

**102.** Double-line or outline by positive masking

effective but lengthy. Not surprisingly, there is generally an advantage in providing the separate images in negative form in cartographic production, as this makes their subsequent combination much more straightforward.

Combination of a set of negatives to produce a combined negative can be done photographically with reversal film, using the normal procedure of successive exposures.

# LETTERING

As the lettering on the map is normally produced by using typeset material, it is constructed by different processes to other components of the line image. Several different methods can be used.

Stencil lettering can be used, provided that the total amount of lettering is small, no small sizes are required, and the inadequate letter forms are acceptable. Letters that have enclosed loops or bowls have to be interrupted, and of course it is impossible to imitate the proper characteristics of type, such as serifs or thin and thick strokes. Although used occasionally on large-scale plans, stencil lettering serves little purpose in normal map production.

A modern version of stencilling makes use of an 'automatic' drawing system, in which master characters and figures are held in 'electronic' templates. Different lettering styles and sizes can be used. The device makes it possible to avoid the breaks in letters which occur with mechanical stencilling, and can also draw point symbols. Words are composed in a sequence and then plotted on the drawing surface. Although an improvement on ordinary stencil lettering, it is still an unwieldy method if large amounts of specified lettering have to be produced, and it does not properly reproduce the characteristics of type designs.

Letters can be assembled from pre-printed masters, the individual letters being either cut out and stuck down, or more commonly, transferred through pressure. Alignment and spacing still have to be controlled by the draftsman. It is virtually impossible either to obtain or to employ very small letter sizes. To deal with hundreds of names would be a very slow and expensive business. The chief use of the method is in simple black-and-white maps where only a few names are involved, particularly if the original image is drawn at a larger scale for subsequent photographic reduction. Titles for maps may also be produced by this method, especially if some style or size of type is wanted which is not available on the standard photolettering machine.

## Photosetting

The greater part of map lettering is now produced by photosetting. Photolettering machines have been developed especially for cartographic purposes. They operate on similar principles to the much larger machines used to set books and other large-volume continuous text. In such instruments, each letter, or 'character', exists as a negative image. Complete alphabets, together with figures and punctuation marks, are composed on discs or sheets. A different disc or film is required for each different type style. The instrument has a lens system which is used to project the master image at the required size on to photographic film, and the film containing the names and figures is then used, directly or indirectly, to construct the lettering on the map.

The advantages of such instruments are that lettering at suitable sizes can be produced, usually from 1 to 10 mm. The image is sharp and clear, and reproduces well. Virtually any type design can be provided, although each needs a separate master. The inter-letter spacing is controlled automatically on the better machines. Special characters, such as conventional point symbols (circles, squares and so on), can also be included on discs or slides made to order. On some instruments it is possible to set lines of text, fully justified, although the cartographic requirement for this facility is small.

For operations requiring high-quality reproduction, especially where the lettering is to be combined eventually with other images, output on to film in positive form is usually the most convenient. Some types of machine will provide an immediate print-out on either film or paper, but often they are restricted in size of lettering, and do not control spacing automatically.

In this respect map production has rather different requirements to graphic arts offices where lettering is mostly an addition to multi-coloured artwork, limited in quantity, and at comparatively large sizes. Some maps may need large volumes of lettering, but with a great variety of styles and sizes, which must be carefully controlled, and repeatable. Small maps may have few names and figures, but usually these will require a variety of styles and sizes. Type sizes ranging from at least 1.5 to 10 mm are necessary for cartographic work.

Although the typeset lettering image is often combined with line negatives, it is usually regarded as an advantage to produce the lettering in positive form in the first place. It is easier to read and handle, and any type of montage is difficult with negatives.

The setting of continuous text, or blocks of text combined with illustrations, is increasingly being done by electronic systems, in which the lettering, like the image, is composed of a very fine raster. The individual characters in the type face are held as digital masters. These systems are suited to all types of raster display device, and indeed one of the advantages of this type of composition is that it is obviously suited to computer display. Page composition is achieved by entering the desired content and then arranging it by controlled spacing and line movements.

With the approaching demise of cartographic photolettering machines of the optical–mechanical type, systems based on small computers will be increasingly important. Most of these operate by using the typographic characters in a type face converted to digital masters. There is no problem with obtaining the desired type size, and some machines will produce lettering from 1 to over 60 mm. Condensed, extended and italic versions can be produced through software control, in either positive or negative form. Output can be either right reading or wrong reading. The ability to control type width as well as height could be very important cartographically, especially for those type faces which have not been available in a condensed version.

Many of the current machines make use of laser technology for the output device that actually 'prints' the lettering. Typically, this consists of a helium–neon laser, which exposes the image on red-sensitised film or paper, through a raster pattern of horizontal scan lines. In the better systems the resolution of the raster scan is at least 0.01 mm. Another type of machine makes use of a cathode ray tube to direct an electron beam at normal photographic film or paper, also operating in the raster scan mode with a resolution of 0.02 mm.

## Montage

Once the names and figures have been produced on film or paper, it is necessary to mount them in position in their relevant colour images. This is normally referred to as 'stick up', an expression which does indicate the nature of the process effectively. If the amount of lettering is small, and

there is space to avoid other parts of the line image, then sometimes it is possible to mount the names and figures directly on the positive line drawing, or a contact positive made from the scribed negative. If a contact copy is to be made, then the film support for the lettering must be very thin, or good contact copying becomes difficult because of the thickness of the film.

For a long period this has been done by producing the lettering on thin film, called stripping film, and mounting the names by first preparing the film with a wax adhesive. Stripping film consists of a thin membrane with a sensitised emulsion, supported on a normal weight base film (Fig. 103). After exposure and development, the thin upper layer can be stripped away from the base, or pieces can be cut out first and then these are stripped away individually. If whole sheets of film can be exposed, assuming a sufficient volume of lettering, then the entire sheet can be coated with a waxy adhesive by removing the stripping film sheet and then running it through a waxing machine to obtain an even distribution of the adhesive. If small amounts only are needed, the waxing can be done by hand, by scraping a small quantity of wax on to the back of the film, either before or after removing it from the supporting base.

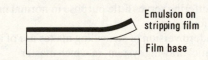

**103.** Stripping film

In the waxing process, the image position required, and indeed the method of exposing the film are all inter-related. If normal exposure is carried out, the result is a right-reading image with the lettering on top of the stripping film, on top of the supporting film base. To put adhesive on the back of the stripping film, to mount it in right-reading mode, the stripping film has to be removed from the base. Because it is very thin, it is difficult to handle and keep flat, and small sections will curl up rapidly. The film can be inserted in the photolettering machine so that it is exposed through the base, so producing a wrong-reading image. In this case, the adhesive can be applied to the surface of the stripping film before it is removed from the support. This is normally easier to control, especially when the names and figures are cut round before mounting in position separately. Conversely, a wrong-reading image can be constructed by exposing the complete stripping film in the normal way, waxing the surface before removing the stripping film from the base, and then mounting the lettering in reverse position.

A different method of producing the adhesive coating is to expose the lettering and figures on to ordinary photographic film first, and then to contact copy this positive on to a diazo-sensitised film that already has an adhesive coating on the back. This produces a positive from a positive, and again image position can be controlled by inserting the photographic film in the photosetting machine either facing or away from the exposure source. The adhesive-backed film is easier to deal with, as it is protected by a thin film on the adhesive side, which is removed before use. It is a thicker film, and not so transparent as stripping film itself. This means that although it is excellent for mounting the lettering, it is inadvisable to attempt to combine this image with a line image directly, as the thicker film causes lack of contact in copying, and affects exposure times as well.

Although many types of adhesive are in use, instant adhesives are quite impractical in cartographic work, as the positioning of names and figures usually has to be carefully adjusted before they are finally fixed in place. By using a wax adhesive, it is possible to move the whole name, or even cut partly between letters and move them individually in order to place the name on a curve (Fig. 104). On the other hand, the image is never really secure, as it is easy to disturb the small pieces of stick-up, and they also tend to accumulate dirt around the edges, leading to retouching on the subsequent contact copy. Therefore, it is always advisable when a large image has been completed, to contact copy it as soon as possible after checking, so that the risk of disturbance or damage is avoided. Once completed, it is difficult to tell whether a stick-up, which may contain several thousand separate pieces of material, has 'lost' an individual item, or if any of them has been slightly displaced.

If the stick-up is damaged after the contact

negative has been made, errors or omissions can only be detected by checking the negative against the positive on a light table.

## POINT SYMBOLS

Point symbols pose many problems in construction. Although the production of a number of regular symbols, such as small circles or squares, may seem to be straightforward, in practice it is a difficult part of the line image production. The critical factors are symbol size, identical shapes and sizes to specification, and reproduction quality.

A great variety of possible methods exists, which indicates the extent of the problem, and the diverse methods used to solve it. Large, regular point symbols can often be drawn or scribed in position, and therefore treated as part of the line image directly. The most difficult aspect is in the production of small symbols, especially when these are repeated in many places. Therefore, if individual isolated buildings are represented on the map by 1 mm solid black squares, the production method must be able to make numbers of these efficiently and correctly.

Methods include drawing in ink individually, using normal tools such as compasses where necessary; scribing, either with tools or with templates; photosetting and then mounting as part of the lettering stick-up; drawing a set of symbols at a much larger size, photographically reducing these, and then copying them on photographic or stripping film for the quantity needed; or dry transfer, adding the symbols to the positive ink-drawn image.

There is little doubt that dry transfer is an efficient method, provided that suitable symbols are available on the manufacturers' sheets of symbols, and assuming that the quantity is not so great as to make this method unduly wasteful and expensive. Prepared sheets of both geometric and iconic symbols are commercially available, but these tend to be too large for most cartographic work (Fig. 105). If the line image consists of an ink-drawn positive, then the symbols can be positioned and transferred directly to the same surface. They do not interfere with any subsequent contact copying. If the line images are produced by scribing, then the dry transfer image would be constructed on the same overlay as the lettering stick-up, being converted to a negative subsequently along with the names and

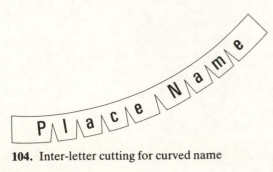

**104.** Inter-letter cutting for curved name

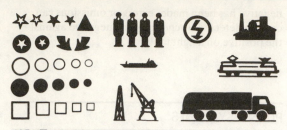

**105.** Types of pre-printed point symbols

figures. Some symbols can be produced directly by scribing, either by using the normal scribing tool and a template of plastic or metal, or by using special tools which, for example, will cut a square or rectangle of a given dimension.

If large numbers of symbols are needed for a standard series, the special production of tools or templates may be justified, and specific construction of the required symbols may be economic.

An effective way of producing a reasonable quantity of identical symbols at the same size is to draw each symbol carefully at a larger size, and after photographic reduction making a block of copies. This can then be contact copied on to stripping film, or even used to produce special dry transfer sheets, either by the use of prepared materials available from the manufacturer, or by buying the completed symbol sheets from the dry transfer manufacturer.

Some symbols can be included on the discs or slides used in photolettering machines, but often these have to be specially prepared. They can be adjusted in size, like the lettering, within the range of the machine, and quantities are produced by simply repeating them during the typesetting process.

Dry transfer is a clean and efficient operation. On the other hand, if the symbols are produced on film, and mounted in position as part of the stick-up, then the process of waxing or copying to an adhesive-backed film must be followed, exactly as with the lettering image.

# 13 PRODUCTION OF AREA IMAGES

The symbols applied to areas on a map may consist of solids, tints or patterns, or combinations of these. They may be in colour or achromatic. They indicate the type or class of the feature, or represent differences in value or quantity. Areal variations can also be represented by differences in tone, but the construction of continuous-tone images and their reproduction in print involves a different set of technical problems.

## SOLIDS

An area can be represented by simply filling it with the printing ink for that particular colour, or a combination of two or more inks. In this sense it is similar to a line image, but spread over an area. Although apparently simple, the method is of limited usefulness in most maps, and is technically more difficult than at first would appear.

In a multi-colour map, most of the printing inks chosen will have to be suitable for line images. They are usually too dark or too 'strong' to be used over large areas, as they would prevent the inclusion of other information, and would also tend to dominate the graphic design. Therefore, if used as solids they are restricted to small areas. It may be possible to use a yellow or a light green printed solid, as lines, lettering or point symbols in black, dark blue, red or brown would still be legible against the colour background, but even so only a light green could be used. In some specialised maps, in which a long series of area colours is needed, and no other information is present, the fully saturated (solid) colour may be used at one end of the scale, but such cases are exceptional.

### Positive images

Small solid areas can be drawn as part of the line image, by filling in with ink. The inks used for line drawing are devised so that they are suitable for fine lines, and flow easily in the pen. Such inks are generally unsatisfactory when spread over large areas. They tend to dry in irregular patches, and do not form a sufficiently dense image for subsequent reproduction. If the area is inked in, and then the sheet held up to the light, the lack of density will be obvious. Special inks do exist for use on some plastic base materials, but otherwise it is possible to employ the liquid opaque materials used by photographers for correcting negatives. But if large areas are involved, it is usually preferable to use a different method.

For a positive image, prepared materials can be used, in the form of strip mask (peelcoat). This consists of a transparent polyester plastic base, with a peelable surface film, available in various colours. Any colour that transmits only long wavelengths (red, brown, rust, orange) is suitable if subsequent exposure for copying is to be made to ordinary or orthochromatic photographic film, or diazo or dichromated colloid-sensitised materials, as these are only sensitive to short wavelength radiation.

A sheet of such material is registered in position, the outlines of the areas cut round with a sharp blade, and then the non-image areas peeled away. Cutting around curved edges can be assisted if a swivelling cutter is used. Thus the symbolised areas will be represented by the coloured film, and the non-image areas will be transparent.

This is a positive image, and can be treated just like any other positive image component. But if lines and point symbols in other parts of the map have been drawn in ink, or produced by dry transfer, the area symbol image has to be produced on a separate sheet of film, and subsequently combined, which adds to the total production work. In addition, if only a few small areas are needed in the map, the peeling away of most of the coloured film tends to be wasteful of an expensive material.

### Negative images

If the lines of the image are produced by scribing,

then it is possible to outline the areas by scribing around the boundaries, subsequently scraping away the scribe coating. Like ink drawing, this is only suitable for small areas. The strip mask peelable material can also be used to construct a negative image of the areas, and indeed this is a much more common practice, particularly if the line work is also in negative form. This still involves separate construction. The initial procedure for making a mask is the same, the area outlines being cut through the superficial film, but then the image areas themselves are peeled away, leaving them transparent against an actinically opaque background (Fig. 106). This negative mask can then be contact copied with other negative images to form a combined positive if required.

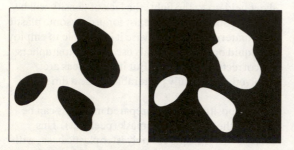

**106.** Positive and negative masks

# TINTS AND SHADES

Although coloured areas on a map may include a solid yellow or other light colour, the other printing colours are generally too strong to be used in the solid form. Therefore, their saturation is reduced and their lightness value raised by making tints. A tint decreases saturation, and by the exposure of more of the 'white' paper surface, increases the lightness of the colour. So pale tints can be made, even of the black. In this way, suitable area colours can be made from all the printing colours, and they can be combined in proportions to make combinations of colours as well. The production of a tint requires a mask of the defined areas, and a tint screen of the desired density.

The construction of a mask can be carried out in the same way as that described above for cutting masks for solids. The difference lies in the fact that the tint is made by exposure of the mask through a tint screen to produce the desired image. A mask can be either positive or negative, but negative masks are more widely used, as they tend to fit in

with the general production procedure, using conventional photographic materials for copying (Fig. 107).

**107.** Photographic production of tint with negative mask

## Controlling edges

The problem with the cut mask is that it is difficult to follow the outlines of the areas exactly. Most area symbols involving tints, or even combinations of solids and tints, are made to fit an enclosing line, and this line is also included, often being shown in a neutral colour, black or grey. Even in map designs where the bounding lines of areas are not shown, the coloured areas still need to fit together if they are adjacent. Minor variations between the outline and the edge of the tint will be revealed either as overlapping colours, or as white spaces. In addition, a map with very many different area symbols, like a geological map or a map of world vegetation types, requires a large number of masks, some of them with very irregular outlines. If the area colours are made up of two or more tints, the edges or outlines of these areas must be repeated exactly in two or more masks. This is both difficult and time consuming with manual mask construction.

Where the masks are complicated, and especially if the edges have to be repeated, it is possible to derive the mask outlines directly from the outlines of the areas. This requires a complete outline of all the areas, whether or not this is a component of the line image. In some cases, if area colours are bounded by the sea, or administrative boundaries, then the outline will in fact be composed of lines derived from several different originals. A combination of these may have to be made in order to produce the outlines of the areas on one image.

The process requires a pre-sensitised strip mask material, similar to the normal peel coat or strip mask, but with an additional superficial light-sensitive layer (Fig. 108). This is exposed to the line image. The developed image will protect the image areas, but leave the lines clear, so that the strip mask can be etched through with alcohol or a special solvent. This repeats the lines exactly, and cuts through the strip mask layer. Then the required areas are peeled away from the base, to produce either a negative or a positive as required.

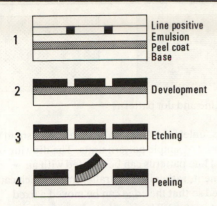

**108.** Pre-sensitised strip mask (peel coat)

In most cases the etched line image will also contain other lines in addition to those required for the mask areas. With a positive mask, all the unwanted non-image areas are peeled away. But with a negative mask the unwanted lines will have to be opaqued or filled in before the mask is exposed to produce the tint.

It is also possible to dye in the etched outline before peeling away the coating. This adds the line width to a positive mask, but subtracts it from a negative mask.

Pre-sensitised strip mask is particularly valuable if the areas are numerous, small and irregular, and if many masks have to be made from the same outline. Despite the additional costs of materials and processing, a great deal of production time can be saved, and the quality of the image noticeably improved.

## Tints based on raster processing

The more elaborate digitally based photolettering and composition machines are also capable of producing a wide variety of graphic forms as well as type. Although designed to 'fill in' outlined areas during composition on the screen, these could also be used to generate percentage tint screens directly. Both the ruling and the percentage density can be controlled to within fine limits (up to 117 lines per centimetre in one instance), and high quality output on film would be quite suitable for cartographic processing. If copied to an adhesive-backed translucent base, they could be applied to a map original in the same way as pre-printed patterns on adhesive film, cutting round the image areas and removing the surplus. Master screens generated at the required percentages could be contact copied to adhesive-backed material as required.

## Negative or positive mask

To some extent the choice of either negative or positive mask is a function of the general production planning, and the way in which the masks are integrated with the production system. The negative, or 'open-window' mask can be exposed to a photographic film, with the tint screen, to give a positive tint, and this can be done while the negative is being combined with other negatives in the same printing colour. Production of a tint with a positive mask can be done with dichromated colloid materials, using the positive-working system. As this requires an even coating, and therefore the use of a whirler or coating machine, the process is less favoured than in previous times, and the negative mask process is now used in most cases.

The choice of either negative or positive masks also raises the question of the effect of the width of the outline. For example, if the area outlines consist of lines with a gauge of 0.2 mm, then with a pre-sensitised negative mask the width of the outline will be added to each area. If the mask is peeled as a positive, the line width is subtracted from each area. It is possible to alter this by first peeling a positive mask, and then producing a contact copy negative of this, and this negative will be minus the line thickness.

## Screen angle

Where two or more colours are superimposed to produce another colour, the rows of dots or lines in the screened image must be placed at different angles, in order to avoid the *moiré* effect. These angles are standardised for the process printing colours, and other hues are placed at intermediate angles of 15 °. It is possible, if the image is small and the tint screen large, to place the mask at the right angle in relation to the screen, but for large areas this is usually impractical. Consequently, tint screens are usually made available in angled sets. So if nine or more percentages are desired, and each of these is provided at four different angles, a large number of separate screens will be involved.

## Colour charts

Most cartographic organisations either construct or make use of colour charts. These show in printed form the different percentage tints in individual colours, and also in combinations of two or three colours. The production of such a chart is described by Karssen (1975). Although many of these charts make use of the standard process colours, it is desirable to produce them in the printing inks commonly used by the cartographic organisation.

## Screen ruling

Tints should give the appearance of a continuous colour, and therefore fine screens are always preferable. However, some kinds of paper, including those commonly used for books, are not suitable for fine screens, which will not reproduce well. In such cases a slightly coarser screen, such as 40 lines per centimetre, will be more suitable.

## Shades

In contrast to a tint, a shade reduces the saturation and also reduces the lightness value of any colour. Therefore, the tint or solid has to be made darker. This can only be done in print by adding grey, or a percentage of the black.

Slight variations in shade can be very effective in controlling the balance between different hues. A very small percentage of grey or black is often enough to produce a noticeable difference. If a solid or tinted area is to be shaded, the same mask is used to produce a percentage tint of the grey or black, which in print is superimposed on the other printed colour. If tints of the black are used, they normally need to be very light, usually less than 10%.

## PATTERNS

Patterns, like point symbols, are apparently straightforward, but in practice pose many problems in production. They can be constructed by hand, drawing them in pen and ink, or scribing them in negative form. They can be produced by dry transfer, by the use of pre-printed material with adhesive backing, or by special pattern screens, using a mask as in tint production.

Patterns suitable for cartographic purposes usually consist of an array of small pattern units, equally spaced, and at a particular angle. To draw them individually means that the pattern unit, which may be a dot or a line, or an iconic symbol, must be repeated exactly; the spacing must be even; and the orientation consistent. Patterns such as these (Fig. 109) are difficult to produce manually with any

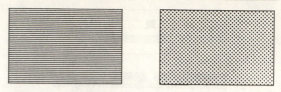

**109.** Line and dot patterns

success, unless the areas involved are very small and composed of simple pattern units. The drawing of parallel line patterns can be assisted with an instrument, rather like a parallel ruler, which can be adjusted so that the edge can be moved a fixed amount in steps. This controls spacing and orientation, but not line gauge or form.

Wherever possible, cartographic draftsmen prefer to use some type of pre-printed or standardised material, in which the pattern elements are already constructed. Such patterns are available commercially, either of the dry transfer type, or as adhesive-backed pre-printed thin sheets. Whereas a considerable range of patterns and 'tints' is available in the pre-printed adhesive-backed form, far fewer are available as dry transfer materials. Many of the line and dot patterns classified as 'screens', are in fact sufficiently coarse to be well above visual resolution. For example, the Mecanorma coordinated dot 'screens' with rulings of 27, 32 and 42 lines can be used as fine dot patterns. This also applies to the coarser coordinated line screens which are described in percentage terms.

These pre-printed patterns can be applied to an area in the usual way, either by rubbing down over the area and then trimming back to the outline, or by placing a piece of the pre-printed material over the area, cutting round the outline, and removing the surplus. Unfortunately many of these pre-printed pattern materials are limited in format (normally to a maximum of A4 size), and variable in quality. This makes it difficult to produce large areas of a particular pattern, although this tends to be an infrequent requirement.

There is no doubt that consistency and quality are best obtained by the use of pattern screens, if these are available. These are like tint screens, in which the pattern unit is regularly repeated, and they are exposed with masks to apply the pattern to the desired areas. Several types of pattern screen are available in a range of standard line and dot patterns from tint screen manufacturers. The difficulty is that the pattern desired for a particular map may not exist in either the pre-printed materials or the available pattern screens.

If only small discrete areas are required, then it may be possible to construct these as part of the 'stick-up' image for that colour. If they are extensive, so that they would interfere with linework or lettering, then it may be advisable to make a separate overlay of the patterns, and combine this subsequently with the other line image components.

Where patterns are used in large quantities, as in the production of a standard series, then it may be worth the cost of having the pattern screen constructed specially for the map specification.

Some patterns can also be produced with digitally based photolettering systems. The more sophisticated types, increasingly used for setting both text and graphics, can also generate a range of patterns based on the repetition of particular symbols or forms. These could be output as 'masters' and then copied and used like other stick-up material.

## Patterns by subtraction

Patterns can also be produced in negative form, by which the pattern is subtracted from the area colour, solid or tint, or one of the printed colours in a multi-colour combination. This is done by positive masking. The area colour or tint is constructed in the usual way, but the pattern is placed on top of the mask before exposure of the solid or the tint screen. This subtracts the pattern from the image, leaving it 'blank' against the solid or tint background (Fig. 110).

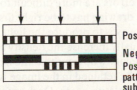

Positive pattern
Negative mask
Positive with pattern subtracted

**110.** Pattern subtraction by positive masking

If an area colour is made up from two printing colours, and the pattern subtracted from one of them, another visual variation can be produced. This type of pattern masking can be very effective, provided that the pattern elements are not too fine, as light or white lines against a light-coloured background require more contrast than the equivalent pattern in dark lines against a light background.

## Stipple

The term stipple is derived from the fine arts. A stipple created by an artist in painting is often obtained by literally sprinkling ink or paint from a brush, or dotting with a pen. In a sense it is a particular form of pattern, in that the pattern units (usually dots) are intended to be roughly of the same size and shape, but their arrangement is irregular. Some types of pattern symbol imitate this effect. For example, an irregular pattern of fine dots may be used to represent areas of sand, and a banded stipple effect used for lines of moraine. For small areas, and especially if a particular configuration or density of stipple is required, it can be produced most easily by drawing with ink. Pre-printed materials with stipple effects are also available (Fig. 111).

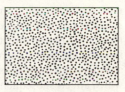

**111.** Stipple patterns

# CONTINUOUS TONE

Continuous-tone images are rarely constructed in cartography, and the only common example is the drawing of hill shading. The continuous-tone image has to be converted to a discrete image for reproduction by lithographic printing, and the usual method of doing this is by converting it photographically with a half-tone screen to produce a half-tone. A continuous-tone image may be a single achromatic image, or a complex multi-colour one. In all cases, the problems of construction are essentially different from other types of area image.

A continuous-tone drawing may be made with pencil, brush or airbrush. The object is to place a certain density of tone where required, and therefore the control of this density is the key to the drawing. With pencil, or pencils, this is achieved by controlled pressure and repetition, light tones being produced by a light pressure with a medium lead, dark tones by using more pressure and softer leads. In most cases at least two different degrees of softness are employed.

In brush work, the tone depends on the proportion of pigment to water, or the mixture of pigment and white in the gouache form. Diluting watercolour paints changes the transparency of the paint, and different levels of tone can be built up, in either monochrome or multi-coloured images.

Smooth gradations of tone are difficult to achieve with the brush only, and consequently the airbrush is used. This sprays fine particles of pigment in solution over the surface, and the density of the paint and the duration of spraying control the build-up of different tones. Generally speaking, a small-scale drawing with many small areas is most easily produced by pencil drawing, reinforced by some ink or brush work if necessary. Medium- and large-scale originals, or multi-colour originals, are easier to produce with the airbrush.

In both cases a suitable base material and a guide image are needed. For pencil drawing, the normal matt surface drawing film can be used, or the semi-opaque white variety. The translucent drawing film can be placed directly over the base information, which usually consists of drainage and contours, but the film itself has a slightly grey tone, which affects subsequent reproduction. Painting, especially with the airbrush, is very difficult on plastic film, even with a high degree of graining of the surface, as the propelled droplets of liquid tend to bounce on the surface. Paper surfaces are far better to work on, and the usual method is to employ a paper laminate of the aluminium/paper type, which is sufficiently stable and also has a receptive working surface. With opaque material, the guide image must be produced directly on the surface.

One of the best materials, although it needs special preparation, is a thick coating of white paint on a metal base. This can be given a guide image, but with the added advantage that it can be slightly etched into the surface, so that it does not disappear during the progress of the work.

## Airbrush

There are different types of airbrush. Those used for broad effects in artists' designs are often only suitable for producing smooth tones over rather large areas. Other types of airbrush have a much greater degree of control over the fine jet, and these can produce very detailed images. Most of the artists' airbrushes are driven by compressed air, fed through a control valve. Fine airbrushes require very low pressure, sometimes as little as 1½ pounds per square inch. Electric pumps with air reservoirs are also available, but these tend to be noisy in action, and it is difficult to control the pressure exactly to within fine limits.

The airbrush has a reservoir, which is filled with a small quantity of paint or ink, and in the most common type this is drawn into a fine tube under pressure and expelled through a nozzle. The area covered by the jet is partly a function of distance from the image, and partly a function of finger control on the airbrush. A skilled artist can build up large areas of even tone with transparent pigments, but in many cases it is much easier to use gouache watercolour paints, mixed with white to increase the lightness as required, which will form a surface of a constant tonal value, but with graded edges.

## Drawing procedure

With pencil drawing, the image can be built up gradually, working in lighter tones first, and then increasing the density as required. Areas are covered by light strokes of the lead, and then smoothed by the use of a stub, a small pointed paper cylinder. Tones can be gently graded out to edges with the stub, and areas of tone merged and softened. Heavy build-up of lead should be avoided, as this tends to introduce reflections during photography. The advantage of pencil drawing is that edges and small areas can be controlled carefully. The disadvantage is that large areas of even tone are difficult to produce. Even so, many of the minor variations in tone are likely to be reduced in the reproduction.

The application of ink or pigment to an absorbent surface means that errors are more difficult to deal with, and often the drawing has to be right first time. Again, the normal way is to produce the light tones first, and then gradually increase the density as required. It soon becomes obvious that the degree of contrast that can be retained in the image is largely a function of relative areas. Dark tones on small areas or bands need additional emphasis to make them stand out. It is difficult to maintain any sharp edges with the airbrush, which of course is intended to produce a soft effect, and therefore sharp transitions in tonal value, if needed for marked breaks of slope or changes in slope orientation, may be reinforced with a brush.

The other problem in execution is that it is impossible to avoid overspraying adjacent areas that should not be toned. Although on the land area these transitions are often unimportant, they have to be cleared from water areas, or any areas of the map that must not be shaded. This can be done either by applying a liquid masking fluid for protection – the mask being removed when the drawing is finished –

or by masking out the water areas during reproduction. If a mask for water areas already exists for the production of a tint, this is generally the easier method. Sheets of low-tack adhesive transparent mask are also useful for larger areas.

If the shading is part of a topographic map, then it is usually necessary to remove it from other symbolised areas and features, such as double-line roads and possibly built-up areas. Rather than attempting to avoid these edges on the drawing, a combined positive mask is made from the line images and masks, and the shading deleted from the half-tone in one operation during reproduction.

## Highlights

Those parts of the image that are not intended to receive any tone, or should have a very light tone, are the most difficult to deal with in reproduction. If the drawing is made on plastic drawing film, the greyish film will tend to reproduce a fine dot all over, equal to a very light tone. This has to be removed in some way. White can be added to small highlights in the drawing, either with a white crayon or white paint, but these additions tend to photograph with sharp edges, which is not what is wanted. The problem can be overcome by highlight masking, that is by drawing a tonal positive mask to cover the highlight areas. This is the converse of the shading itself, although normally restricted to the very fine highlight areas. In reproduction, this positive drawing is normally converted to a continuous-tone positive on film, and then this is placed over the continuous-tone negative before exposure, so adding density to the dark areas of the negative, which represent the light areas on the image (Fig. 112).

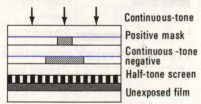

Continuous-tone
Positive mask
Continuous-tone negative
Half-tone screen
Unexposed film

**112.** Continuous-tone positive masking for highlights

If the original is drawn or painted on a white base, it is easier to obtain a clean photographic negative, and generally process photographers prefer such an image. However, the guide image is more difficult to deal with, as it is painted over in the course of producing the drawing. If a blue key is used, which is normal for an achromatic image, this will disappear on reproduction, but its obliteration during drawing may cause difficulties. The etched line, which can be viewed with side illumination, is useful in the sense that it is not obliterated during painting with the airbrush, but it has to be a fine image or else it will interfere with both the painting and the reproduction. A transparency of the guide image can be used to check the positioning of key features, by placing it temporarily over the drawing and checking the placing of the tones in relation to the interpreted slopes and drainage.

## Reproduction

The conversion of the continuous-tone image to half-tone is carried out by either of the methods of half-tone production described earlier. The drawing is placed over a white base if produced on a translucent film. Although the half-tone conversion may be nominally straightforward, it is one of the most difficult things to do well. The difficulty is in maintaining the correct tonal values, without losing contrast in either the highlight or dark areas of the image. Over-exposure will help to remove any residual highlight dot formation introduced by the half-tone screen, but then the reproduction of the dark tones is poor. Correct exposure for the dark tones will tend to leave a small highlight dot in the highlight areas.

If a glass crossline screen is used, normally more than one exposure is given. A short exposure to a white surface may be given first, in order to build up a little density in the highlight areas; followed by the main exposure for the overall tonal range. The half-tone negative may then be copied again, to produce a positive, and at this stage any additional positive masking can be introduced.

With a vignetted dot contact screen a continuous-tone negative is provided first, and then this negative is exposed through the half-tone screen. Any additional highlight masking is usually carried out at this stage.

For good reproduction, the tones in the drawing or painting must have sufficient density. If a pencil drawing is made on a very matt surface, like cartridge paper, the large spots of lead will be separated quite widely, even though the eye will resolve these into a single tone visually. But in photographic reproduction, the tonal density in such areas will be very uneven. Therefore, a relatively smooth drawing surface is needed.

## Multi-colour reproduction from a single original

Although the continuous-tone drawing is constructed as an achromatic image, and therefore can employ the full tonal range between white and black, the shaded image on the map will certainly be restricted so that its maximum density does not interfere with other detail. This is normally done by printing it in a grey or brown. The change to a lighter colour means that the tonal range is compressed, and contrast is reduced. There are several consequences to this. There is no point in producing an original image with very delicate and subtle tonal differences, as these will no longer be visible in the reproduction in grey. If the tonal range needs to be increased on the reproduction, in order to compensate for the loss of contrast, this can be done by extracting two versions of the tonal image by photographic processing, and adding one of these in a slightly stronger hue, such as a purple or dark blue.

The normal procedure is to obtain a full-contrast image to be reproduced in grey, and then to make a second image that contains only the darkest tonal areas of the original. This can be done by over-exposure, so moving the reproduction densities along the characteristic curve. If properly controlled, this can provide a subsidiary image. If this is then printed in a complementary but slightly darker hue, the dark tones can be reinforced and given more contrast (see Bantel 1973). As this is made from the same original image, it is better than the production of a separate drawing for this purpose.

In some cases, the only colour available for the shading is black. Although in theory this can reproduce all the intermediate greys, it is essential to control the reproduction so that the maximum density does not exceed about 40% of the black, otherwise a very heavy image will result.

## Multi-colour tonal images

On occasions, the shading may be combined with a colour impression of other terrain characteristics, such as vegetation and/or land use. Many small-scale maps of this type have been produced, the object being to give a 'natural' impression of the landscape. Therefore, the shading is itself produced in colour, and the drawing will also include other landscape colour and detail. The only way to produce this is by making a complete multi-colour original first, reproducing it through standard colour separation

procedures. These will certainly make use of the three- or four-colour process, either by photography with filters or by electronic scanning. The standard process colours must be included in the colour specification, as all colour separation is geared to the use of the process colours. The problems of obtaining the correct tonal balance still exist, as well as colour correction. This is probably the only case in normal cartographic production work where colour separation needs to be used.

For both achromatic and coloured originals, the printed image or images must also observe the rules for correct screen angles, which apply to half tones as well as tints.

# VIGNETTES

A vignette is a band of graded density along the edge of an area. It is most commonly used in cartography to provide a visual reinforcement of the land/water divide, without colouring all the water or land area. It is also frequently used on aeronautical charts to delimit areas; either areas of natural vegetation, such as forest, or restricted or controlled air space.

To produce a vignette, a negative mask is exposed through a half-tone screen, with the converse (positive) mask superimposed with an additional spacer. As light is diffused between the two mask edges, it diminishes in intensity away from the edge,

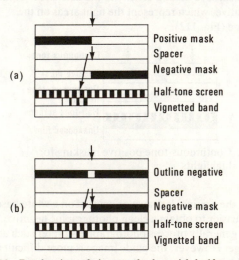

**113.** Production of vignetted edge with half-tone screen and masks

so producing a graded diminution of tone along a narrow band (Fig. 113).

An alternative method (Cuenin 1972) makes use of a negative of the outline in place of the positive mask. Opinions differ as to whether it is preferable to place the half-tone screen or the negative mask in contact with the unexposed film.

# 14 PROOFING

## PROOFING AND COPYING

When any colour-separated graphic image is being reproduced, it is necessary at some stage to assemble it as a single image, so that its correctness, completeness and quality can be checked. In many cases it is also necessary to supply a proof to the author or client, in order to ensure that the product meets with approval. As, in general, multi-colour maps are produced in the colour-separated mode, the actual appearance of the map in the proposed colours is not known until a proof demonstrates the images in colour, even if this is only approximate. Proofing is therefore an essential component of any multi-colour map production system.

The methods traditionally used for the proofing of multi-colour maps are all analogue methods, in which a negative or positive of each image is exposed to a sensitised surface, and this is used to produce an image in the required colour. The processes range from the proofing press, similar in principle to the offset litho machines, but normally using a cylinder and flat-bed system and simple manual feed, to proofs made by coloured dyes that only approximate the appearance of ink on paper.

Proofing overlaps the related need for the production of a small number of multi-colour copies, where this is in fact the full requirement for the map. In some cases, for example for internal circulation within an organisation, large numbers of printed copies are not the objective. Some specialised maps are likely to need only a small circulation. The difficulty has been the lack of a high-quality, large-format multi-colour reproduction system capable of generating a limited number of copies at reasonable cost. Despite the advances in reprographic methods, such a process is still not available. There is an obvious overlap between the need for proofing and the need for a limited number of finished copies, or a short-run print. A method satisfactory for one might well be the solution to the other. But whereas there are good proofing methods available for making one or two copies, the need to repeat the entire process each time means that such methods are still prohibitively expensive if some twenty-five or fifty copies are required. One important consequence of this is that in many cases desirable maps have not been undertaken at all because the quantity required has not been sufficient to justify lithographic printing, and the cost of making them by any other method has been too high.

The picture is further complicated because in the general field of graphic arts colour reproduction, electronic scanning is used both to convert to half-tone and colour separate, and the colour copiers based on this principle can also use the colour-separated raster data to produce a coloured image. Indeed, any maps treated as part of the 'artwork' in a graphic composition are likely to be reproduced in this fashion. This also creates a single copy that can be used as a proof, but the process is fast enough to be considered also as a short-run printing system. If the resolution of the system is sufficient for the proper reproduction of fine lines and tints, the process can be used for maps. Indeed, it is possible to make a colour proof of a map in the first place by one of the traditional reprographic methods, and then have this copied by a colour copier. Because colours can be modified, this second stage proof can be closer to the appearance of the printed image.

With rapid advances in technology, it is difficult to say where all these developments may lead in the future. However, it seems certain that the possibilities of producing small numbers of good multi-colour copies are likely to increase, and the elusive low-cost high-quality copying system may be devised. Such a development would be of great importance in extending the provision of maps of many sorts.

## REPROGRAPHIC PROOFING METHODS

### Types of proof

A colour proof may consist of either a complete

colour composite on one piece of material (superimposure), or a series of coloured images on transparent bases (overlay). Overlay proofs have the disadvantage that an assembly of separate translucent layers does not visually resemble a printed sheet, and if many layers are involved it is difficult to form any clear impression of those images that are at the bottom of the set. They do have the advantage that individual colour components can be extracted and changed without having to repeat the whole of the proofing process.

## Proofing functions

Because maps intended for colour printing by lithography are normally produced both image separated and colour separated, more than one proofing stage is usually required. The first proof stage is to check that the details of individual images are correct and complete, and that the various images can be both fitted and registered together. Although it may be an advantage if this check proof resembles the appearance of the final map – especially with experimental designs – the main objective at this stage is checking detail, not colour fidelity. A proof in colour does not necessarily have to be visually equivalent to the final product.

Before printing the full quantity required, it is normally necessary to have a proof, usually called therefore a pre-press proof, which is intended to ensure that the images have been correctly transferred and registered on the printing plates, and that the coloured image produced by the press run will be that desired. This proof of colour is the final check before printing, and for this purpose it is essential that the proof should match the visual appearance of the final map. In many cases, this pre-press proof has to be accepted by the client before the printing run takes place. Proofs made on a printing press are often referred to as 'machine proofs'.

There are many methods that will provide a proof in colour without matching the appearance of the final print, and comparatively few that will serve the purpose of a proof of colour. If a proof is made from the reproduction negatives or positives, but not from the printing plates, then it is still possible that the printed stock will not match the proof. Even so, many organisations prefer to have a good proof of colour before reaching the stage of making plates, as any errors at this point will result in new plates having to be made.

The most effective proofing system as a preparation for lithographic printing is the lithographic proof press, as this can use exactly the same paper, inks and plates as the production machines. But if any errors come to light at this stage, one or more new plates will be needed. It is common practice therefore to make a single proof copy by a reprographic process first, and then proof on the press subsequently, especially if several proof copies are wanted for the organisation or client.

One major difficulty with reprographic proofing methods, is that these are focused mainly on the general requirements of graphic arts colour reproduction. This field is dominated by process colours and relatively small formats, usually not more than A3 or even A4 in size. These may be sufficient for individual atlas pages, but most sheet maps and atlas sheet sections are produced at large formats, and frequently use more than the standard process colours. These differences in requirement tend to restrict the usefulness of proofing methods that would otherwise be suitable for cartographic purposes.

## Overlay methods

Most of these are based on diazo materials and are predominantly positive-working methods. Although the semi-wet diazo process can be used with diazo-sensitised sheets, most organisations avoid changing the liquid toner in the machine to different colours, and therefore the dry-development method is preferred. This has the advantage that the original positive can be exposed to the diazo sheet in the vacuum frame, giving proper contact in copying, and development carried out separately in the ammonia vapour unit. The colours black, cyan, magenta and yellow are generally available, and some other colours can be obtained, but none of them is a good match for printing inks. The only practical advantage of the method is that diazo copying is relatively quick and skilled labour is not required, so that large-format proofs are comparatively inexpensive. Overlay methods are most useful where the proof is essentially for checking purposes in the course of producing a standard map for which the specification is well known.

# SUPERIMPOSURE METHODS

## Dichromated colloids

Dichromated colloids, which for so long dominated the production of printing surfaces, retain their usefulness in cartography by providing a reasonably economic and straightforward method of proofing in colour. Coatings can be applied to plastic sheets, by hand, coating machine or in a whirler, and coloured images can be made from either negatives or positives.

With the advent of pre-sensitised lithographic printing plates, whirlers soon became obsolete, so that now wipe-on methods are more widely used. However, coating machines, which spread a thin layer on to any flat film or substrate, are much more efficient than manual application. This is particularly important with regard to the black. In general, there has been a decrease in the use of positive-working dichromated colloid coatings, which is in some ways unfortunate, as the dyed image is very durable, and the coloured dyes can be intermixed and diluted quite easily. They also provide a better black image than the wipe-on negative-working coatings. Even so, apart from the need to coat the plastic sheet for each exposure, the actual processing takes longer than that using negatives. In addition, as the method is mostly used for the first stage or check proof, and as at this stage in many cases the first set of image-separated components frequently consists of negatives, it can be useful to employ a proofing method that can work directly from the negatives.

Methods of making the proof are those described for dichromated colloid images. Exposure times vary for different colours and for different conditions of temperature, humidity and coating thickness. Unless the coating is complete and evenly distributed, some image areas will not form correctly. Either the smooth or the matt side of the plastic sheet can be used, and either polyvinyl or polyester plastics, although polyvinyl plastics are generally easier in practice. The coated plastic sheet can be dried with a hot-air blower, and handled safely in subdued daylight. The coated sheet, which of course contains the colour dye in the negative system, is then exposed in the vacuum frame, which hardens the image areas (Fig. 114). The unexposed coating is then washed off in water, and the whole process repeated for the next colour.

In the positive-working system, which requires two extra stages, the coating is first applied to the sheet in a whirler or a coating machine, and then dried. On exposure, the coating in the non-image areas is hardened, forming a stencil around the image, and the unexposed coating is removed with a developing agent such as a solution of calcium chloride. The appropriate dye is then worked over the sheet. It reaches the plastic in the exposed image areas, and the surplus is removed along with the

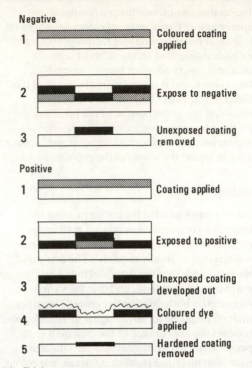

**Negative**

1 — Coloured coating applied

2 — Expose to negative

3 — Unexposed coating removed

**Positive**

1 — Coating applied

2 — Exposed to positive

3 — Unexposed coating developed out

4 — Coloured dye applied

5 — Hardened coating removed

**114.** Dichromate colour proofing by negative- and positive-working methods

resistant coating by scrubbing in water. If used on polyvinyl plastic sheets, the dye actually penetrates the surface of the material, making a very resistant image.

Relatively little skill is required to produce a proof, but the need to repeat the process for each image means that even the negative-working process is relatively slow. On the other hand, no special equipment is required, assuming that the organisation will have vacuum frames, light sources and sinks suitable for the type and size of work normally produced.

Dichromated colloid coatings are manufactured under a variety of trade names, but the processing of them is basically the same.

## Combined methods

The negative-working method often suffers from a poor black image, and as this is usually a critical element it can be a major disadvantage. One partial solution is to expose the black image to a wash-off photographic film, and then to develop and process this photographic image, which leaves behind a clear plastic base. Other colours are then added to this with the normal dichromated colloid method. The disadvantage is that the proof is on a translucent base, not an opaque white one, but the fine sharp black is a great improvement, especially where this is a significant component of the total information.

## ELECTROPHOTOGRAPHIC METHODS

The possibility of using coloured pigment particles, either dry or in the semi-wet form, has long been realised as a proofing method, and several machines have been developed for the purpose of colour proofing. However, the lack of a sufficiently large market for specialised cartographic equipment has meant that these machines were either developed in-house by major mapping organisations, or have had a rather uncertain and limited life as commercial products.

As normally the desired end product is a proof on paper, the indirect method of image formation is generally preferred, by which the electrostatic image is first created by exposure on to a plate, or master, and the charged particles then attracted to plain paper by opposite polarity. By controlling the charging sequence, images can be produced from either positives or negatives, although positive production is the most common. For reproduction on paper the dry method is preferred, and the magnetic brush principle has been adopted on one currently available machine. Although this machine will produce a proof up to 61 × 85 cm, this is still much smaller than the image size on large offset litho presses. Up to eight colours are available, and each takes a few minutes to produce. The colour quality is good, but the machine requires a comparatively large investment, both for hardware and proofing materials.

## PHOTOPOLYMERISATION METHODS

In many proofing methods, a coloured image is formed by transferring a colour on to a sheet of paper or plastic. They all involve some aspect of photopolymerisation. Dry transfer methods are obviously preferable for proofs on paper stock, and those that need liquid development are generally applied to white plastic base materials.

Although many methods exist under a great variety of trade names, one of the best known

examples is Gevaproof. The layer of coloured pigment is supported on a thin membrane, which is transferred to a sheet of white plastic stock, or other polyester film. The supporting membrane is then removed, and exposure made to a negative, using a strong ultra-violet light source. The exposed image areas are hardened in a special activator. Unexposed non-image areas are dissolved away in warm water. The final stage is a brief immersion in a solution of water and alcohol, followed by drying. A full range of colours is available, and the image quality is good.

The Cromalin system exploits another characteristic of some polymers, which is that they are tacky while unexposed, but lose this tackiness after suitable exposure. A layer of photopolymer is laminated by heated rollers on to the paper or other stock, the protective layer of polypropylene film being removed from the contact side. A positive is exposed and the non-image areas lose their tackiness. The covering film is then stripped away, and the required colour powder is applied with a pad, adhering only to the tacky image. The process is repeated for each colour. Finally, a clear lamination can be applied, followed by a longer exposure to give a shiny finish, or a matt toner can be used to give a matt finish.

The process can be carried out from negatives, but this requires a different type of photopolymer layer. As the sensitised material is supplied in rolls, the only restriction is the width of the material (68.5 cm), which is not large enough for some sheet maps. A full range of colours is possible, and the proof of colour is a good match for printed copy. Again, compared with dichromated colloids or diazo proofs, the cost of the machine and special materials is relatively high.

## PRINTED PROOFS

Printed proofs can be made on either a standard printing press or a special proof press. The proofing press normally has a flat bed supporting both the printing plate and the impression bed, and the offset blanket is mounted on a cylinder that passes over the plate and then over the flat sheet of paper. The damping and inking reservoirs are usually at the ends of the machine.

The printing sequence is as follows (Fig. 115): (1) the inking system takes up ink; (2) the offset cylinder passes above the sheet of paper on the

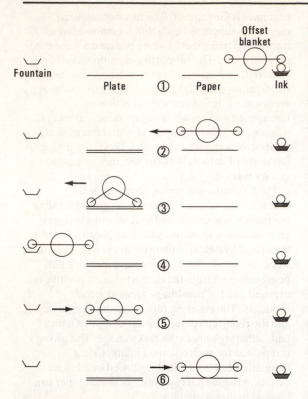

Offset blanket

Fountain

Plate  ①  Paper  Ink

②

③

④

⑤

⑥

**115.** Printing sequence of lithographic proof press

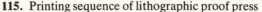

impression bed; (3) the plate is inked and damped; (4) the damping system is replenished; (5) the offset blanket takes an impression from the plate; (6) the offset blanket makes an impression on the paper. On a semi-automatic press the paper feed is also automatic, but on the more simple press the sheets are placed and registered by hand, the operator controlling the printing action.

## COMPARISON OF REPROGRAPHIC AND PRESS PROOFING

For large-format proofs, it is unlikely that the large production presses will be used, as these are only economic if employed on suitable volumes of work. For smaller formats, at which standing costs are not so high, the difference in cost between proofing on a production press, and on a proofing press, is not so great as might be imagined. Shearer and Weinreich (1973) give a careful comparison of relative production costs, both between rotary press and flat-bed proofing press, and between offset and Cromalin proofing methods. The small difference between proofing on a flat-bed press and a standard rotary press, using a single colour machine, was attributed to the fact that the main costs lay in plate making, labour and overheads. Under the specific conditions obtaining at that time, the experimental investigation also showed that the break-even point between a Cromalin proof and a press proof was about seventeen copies. For any larger number the additional proofs produced by the press were very inexpensive, whereas proofs made reprographically bear the same cost for every proof.

In practice, other factors would complicate the issue. For those organisations with their own printing department, the use of a rotary press for final proofing would seem to be realistic, but in a commercial printing firm it is unlikely that any production press would be available at the required time, assuming that the work load was programmed over a relatively long period. Even so, if providing press proofs to customers was important, as it often is, it might be worth while maintaining a high-quality proofing press. Whether or not this had any advantages over the use of an existing rotary press, which would also be used for production work, would depend on local conditions and requirements. In addition, the need for a simple check proof system remains.

Despite the development of special machines and systems that produce better quality proofs of colour, many cartographic organisations maintain dichromated colloid proofing in colour because it is simple, can be used for large formats, is not limited to process colours, can be used to build an image from different originals (for example by exposing a mask through a tint screen), and does not normally require extra investment in special equipment.

**PART FOUR**
# APPLIED CARTOGRAPHY

# 15 TOPOGRAPHIC MAPS: ORGANISATION, PRODUCTION AND REVISION

## SCALE, CONTENT AND CHARACTERISTICS

Because of differences in terrain, density of human occupation and scale, topographic maps exhibit an enormous diversity. They range from simple large-scale plans to very small-scale generalisations of the whole of the Earth. So far as the land masses are concerned, they provide the basic information on which all other maps are constructed, and therefore are of primary importance. They also function as the most widely used map type, because they serve the needs of many different types of map user.

Topographic maps fall into two main groups. The major group consists of the standard map series, usually produced by a national surveying organisation, which serve as the 'official' map series of a country. They are based on a common specification for each series, and the series themselves at different scales are usually inter-related. The other group consists of topographic maps devoted to a particular area, which may be produced for some special purpose, such as the construction of a large engineering project, or to assist recreational activities in a national park or nature reserve. Many of these are essentially variations on otherwise 'standard' map series, but some are produced quite separately. Because they are so widely used, the official map series tend to exert an influence on both users and map makers, for familiarity with the style of a map series tends to lead users to have particular expectations about maps, or to accept certain maps as being the norm.

Specialised topographic maps are dealt with separately, and therefore standard topographic map series form the basic subject matter of this chapter.

## SCALE

The choice of scale for topographic map series is related to the amount of detail required, the characteristics of the terrain, and the existence of map series at other scales. Because of great differences in the density of human population, reflected in the development of the built environment, many different scales of treatment are necessary to satisfy the need to give sufficiently detailed information. In most national mapping organisations, there is a distinction between those map series that cover the whole country, and those that are limited to particular areas. For example, very large scales may cover only urban and densely populated regions; large scales may cover the whole area actively occupied and developed; and the sparsely inhabited or uninhabited regions may only be covered at a smaller scale. In some cases these divisions are linked with differences in organisation and responsibility. The national mapping organisation may concentrate on medium and small-scale series of the whole country, leaving state, provincial or urban authorities to carry out larger scale mapping to their own requirements.

Choice of scale is also connected to systems of measurement. The early European series adopted scale ratios that made possible simple calculations in terms of scale and distance in the measurement units employed at that time. This convenient way of stating scale – such as one inch to one mile – generally resulted in an odd figure for the actual scale ratio; in this example, 1 : 63 360. The introduction of the metric system using decimals made possible a much more straightforward connection between scale ratio and calculation. At a scale of 1 : 100 000, for example, one kilometre is represented by one centimetre. The convenience lies not in the adoption of the metre as the measurement standard, but in the employment of the decimal system. The numerical scale, such as 1 : 25 000, is a ratio, and exists quite independently of the measurement unit. It is therefore quite incorrect to refer to 'metric scales'.

Present-day topographic map series exhibit a much greater conformity of scales than they did in the past. Because the choice of decimal ratios in round figures simplifies calculation and use, there is a concentration on scales such as 1 : 10 000,

1 : 25 000, 1 : 50 000, 1 : 100 000, 1 : 250 000 and so on. Although this gives apparently simple scale ratios, such numbers have little to do with the actual detail or density of information required about the Earth's surface. For example, Imhof (1982) shows that 1 : 50 000 is an inconvenient scale in terms of cartographic representation, particularly with regard to choice of vertical interval for contours.

Choice of scale is also directly affected by the relationship between scale ratio, topographic detail and map use. At 1 : 20 000 or 1 : 25 000 scale it is normally possible to show all the significant landscape features, down to individual land parcels and buildings, outside the built-up area. Consequently, this scale is often adopted as the 'basic' largest scale for the whole country. The scale of 1 : 250 000 is generally the smallest at which direct observation in the field, where the map is compared with what can be perceived in the landscape, is possible, and in countries with large land areas, overall coverage may be limited to this scale.

The distinction between topographic maps that can be used directly in the field, and those that summarise larger areas at smaller scales, is sometimes reinforced by a difference in terminology. Chorography is the description of spatial location in a broader sense, and therefore the smaller scale topographic map series (usually smaller than 1 : 250 000) are sometimes referred to as chorographic. The same distinction is made in some countries by describing them as 'geographic'.

## CONTENT AND CHARACTERISTICS

The content of a particular topographic map will obviously reflect the nature of the terrain that it represents, and this accounts for much of the diversity among topographic map series at the same scale. Despite this, what may be termed the major visible features of an area tend to be included in all map series: hydrographic features and drainage; relief; habitations, all types of buildings, and communications. Although some major boundaries may be marked in the landscape, most internal administrative divisions are not, but these important lines are a standard part of the map content. Names are the chief 'intangible' element, and the inclusion of place names is an important part of the total map information.

Major differences between map series occur, usually in relation to the description of the terrain surface and its vegetal cover. Although woodland and forest, which provide a prominent type of landmark, and in many cases an obstacle to movement, are normally included, there are considerable differences in the amount and type of information given about other landscape characteristics. It can range from minimal to very detailed. For example, maps of sub-tropical and tropical regions, in which agriculture and natural resources are often highly important, will often give a detailed sub-division of both natural and cultivated vegetal types. On the other hand, most North European map series describe only the relatively permanent vegetation, such as trees and bushes, and ignore the areas devoted to annual crops.

The degree of detail and the level of generalisation are influenced by several factors. If larger scale maps of the same area exist, then it is possible to reduce the amount of detail given, often in order to provide a more 'open' or legible map. On the other hand, if a certain map series is the basic (largest scale) coverage of an area, there is a natural inclination to maximise the amount of detail. But the policies of some mapping organisations, and the habits of map users, also affect these decisions. Some national map series are noticeably more detailed than others, and yet map users often regard their maps as 'normal'.

The less obvious contrast in topographic map content lies in the way in which particular features are classified. In a revealing account of this problem, Piket (1972) draws attention to the fact that even within a few European countries, their 1 : 25 000 scale map series show remarkable differences in the treatment of relief, built-up areas, roads, ground cover and hydrography. In this respect, the standard specification for the whole series may result in an inadequate treatment of either some areas (such as rural areas compared with urban), or some categories (such as relief compared with roads). Indeed he makes the point that it would be much more appropriate to devise the map specification according to landscape type, rather than by national territory.

### Content and the map user

Although the topographic map is sometimes regarded as simply an 'inventory' of the landscape, suggesting that the information it contains is both obvious and uniform, in practice it is strongly influenced by what is perceived as being of interest to a range of prospective users. For example, the distinct increase in emphasis on road classifications on medium-scale series of developed areas indicates the importance of one type of use, in addition to the frequency of roads in the landscape. In many cases the level of content desired by the users – which may be expressed through consultations between the users and the mapping organisation – may have considerable consequences in terms of both data collection and representation.

In those countries, now in the minority, where large- and medium-scale topographic maps are available to the public, there has been a perceivable increase in the attention given to map user groups. Although in a general sense this has always been the case – in Britain the history of the Ordnance Survey has involved numerous public and parliamentary inquiries – the formal consultation between map makers and map users has tended to increase (McGrath 1980). The difficulty is to distinguish between the map users' requirements for information (which has surveying and financial connotations), and the individual map user's ability to use the map for particular tasks. The distinction between content and design is not always obvious, and map users' complaints may result from either inadequate information or from inadequate representation.

## GENERAL FACTORS IN DESIGN

Most standard topographic map series make use of a specific number of printed colours, the main exceptions being achromatic large-scale maps of urban and developed areas, and medium- and small-scale maps of sparsely populated areas. In both these cases the range of information to be presented tends to be limited, and therefore a less elaborate graphic treatment suffices.

### Colour and printing

In the recent past, many such medium- and small-scale series used a comparatively larger number of printing colours, generally between four and eight. To some extent this was influenced by the gradual evolution of the topographic map from its initial engraved version, through the stage of a basic line image with colour additions, to the point where some general colour coding was introduced. It also

reflects the need for particular printing colours, partly because of colour association, but also because hues of a particular lightness value are desirable. The use of brown for contours is an obvious case.

Colour association, both natural and conventional, plays its part in determining the range of 'appropriate' colours for topographic features, although this is rarely consistently applied. The need for control of contrast is over-riding, and takes precedence over colour association. Even so, there is a general acceptance in most cases of a design basis which includes most of the fundamental cultural information in black, or black with other colour reinforcements; blue for hydrographic features; some variety of brown for contours; and various greens for vegetation. Despite the fact that in northern Europe water areas are rarely seen as blue, and much of the natural vegetation is only 'green' for a very limited season, these colour associations are strong. Although contours, as lines of equal value, are really abstractions with no physical existence, nevertheless the inference that these 'represent' the land surface in some way means that colour association is also applied to them. It is this interesting mixture of physical appearance and abstract meaning that makes much map symbolisation so difficult to define in simple terms.

The basic selection of printing colours is dependent on the need for a high level of contrast for fine line and small point symbols, and also many names. Against a white or light background, this means that only hues which are dark in the lightness value scale are appropriate. Therefore, black, dark blue, dark brown, dark grey, red and purple are most widely used. In contrast, colours that are to be used primarily for areas need to be relatively light. This means that either a light hue must be used, or a tint of one of the darker hues. As the primary hues are chosen for the line images, background area colour is frequently composed of tints or combinations of tints, rather than solids. At a time when large-format and fine-line tint screens were not generally available, it was often necessary to use many printing colours.

Although it may appear that the standard process colours are obvious choices if colour combinations are to be used, they can be inappropriate. Standard cyan is too light; standard yellow is frequently unnecessary as a specific hue, and where used in combination with other hues a lighter yellow is often desirable: and the magenta is too dominant to blend well with other elements in the composition. The fact that orange can be produced by combining suitable proportions (tints) of yellow and magenta should not be allowed to obscure the fact that a dark brown fine line is virtually impossible to produce consistently over large images by combinations of four tints, and that the extra complication and processing work is pointless. It is perfectly possible to design a topographic map in four colours, but in such a case the colours are individually chosen for the benefit of the map composition (see Christ 1966).

Care in choice of printing colours is important. There are many colours under the general description 'brown' ranging from a dull orange to a sepia. Similarly, there is a range of blues and greens. Although it is technically possible to produce shading as a half-tone of the black, better colour control and balance can generally be achieved if a separate grey can be independently selected, especially as this neutral grey can also be modified slightly, for example by being slightly blue or slightly green. The ideal arrangement from a design point of view would be to print a separate colour for every desired colour difference, as this would give maximum control over the design. In practice this would be uneconomic, and a compromise is reached between what is desirable and what is feasible.

For those maps that include hypsometric colouring, or extensive symbolisation of different types of vegetation and land use, yellow may be a useful component, even though it is of little or no value for a line image by itself. It is often suitable on its own for small areas, and in combination with blue and red it makes possible a range of greens and oranges. If brown is also available, a full range of hypsometric colours can be achieved. In this sense it is a very useful supporting colour, and often gives greater opportunities for colour mixtures than can be achieved by including a separate green.

# SHEET DIVISION AND ORGANISATION

Topographic maps usually have to be divided into a series of separate sheets, and this division is affected by a number of factors and requirements. It is closely connected with the coordinate system or systems used for the map, and in some cases the relationship of the map with other map series of the same area. In addition, the sheet size is affected by the known or assumed interests of the map users, and it also influences cartographic production.

## Coordinate systems

Most national surveys in developed countries make use of a national or local grid reference system as well as the conformal projection on which the map is constructed. Therefore, position can be determined either in absolute terms, by means of the geographical graticule, or in relative terms, by reference to a local origin, the grid. Those countries employing the Universal Transverse Mercator projection will normally also use the equivalent grid system. Although both reference systems may be included, it is usual for one to predominate by being shown in full on the map sheet, whereas the other is represented in a subordinate manner.

If the geographical graticule is used as the basis for the sheet, then the proportions of individual sheets will vary according to position in the projection of the whole country or region. The normal basis for the dimensions of each sheet will be to adopt fixed units of latitude and longitude – for example 1 ° × 1 ° – to determine the area covered by an individual sheet. In high latitudes the distance in longitude represented by a standard unit will be less than in low latitudes, and therefore the sheet size will differ with latitude. For countries with large areas in high latitudes, a common practice is to increase the dimension of the sheet in longitude to compensate for the decrease in sheet size. In addition, at small scales, the lines of latitude and longitude may be noticeably curved, due to the projection (Fig. 116).

If the grid is adopted as the sheet basis, then the size of each sheet can be determined in terms of a number of grid units, and of course this will give a standard format throughout the series. Because the grid consists of rectangular coordinates, all the grid lines are straight, which makes for easier sheet construction.

In practice, the choice of grid or graticule as the basic reference system is largely a function of scale. At large scales, where each sheet covers a relatively small area, the use of the graticule is cumbersome, and graticule references will involve relatively complex numbers in degrees, minutes and seconds. At small scales, the difference between grid north and true north becomes significant, and therefore it is more appropriate to use the graticule as the sheet basis. Most variation occurs with medium-scale series, in which either the grid or the graticule may function as the basis of the sheet division. For example, for the 1 : 50 000 scale series, the national map series in Switzerland and Britain use grid lines, while the equivalent series in France and Canada use graticule lines.

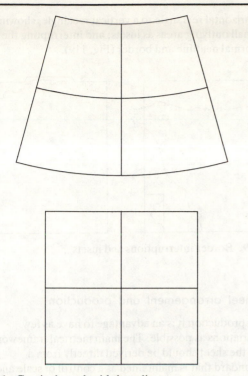

**116.** Graticule and grid sheet lines

## Sheet format

Whatever the basis of the sheet division, the overall dimensions of the sheets have to be determined. This involves decisions about both shape and size. Rectangular sheets are common, but so are square, or nearly square, sheets, especially at large and medium scales. A brief examination of existing map series will show that there is little uniformity or agreement about desirable sheet dimensions. For those series likely to be used in the field, there may be a benefit in adopting a relatively small size, in order to avoid handling large pieces of paper. On the other hand, the smaller the sheet, the larger the number of individual maps in the series, and the more separate units that have to be produced, marketed and bought by individuals. From a cartographic point of view, fewer large sheets are easier to control than larger numbers of small sheets.

## Common sheet line arrangements

In many countries, all or some of the national map series may be organised on a common sheet and reference basis, in order to facilitate the use of the

reference system, and also to enable the map user to determine the coverage of sheets at different scales. This is normally achieved by taking a standard sheet format at a small scale, and progressively sub-dividing this for larger scales. In such a case, the sheet limits for all the map series will have a common basis. If the first series is the 1 : 1 million, then the following arrangement is possible (Fig. 117).

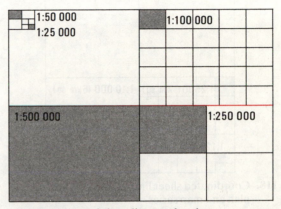

**117.** Coordinated sheet lines and scales

| 1 : 1 000 000 | ........... | 4 ° × 6 ° |
| 1 : 500 000 | ........... | 2 ° × 3 ° |
| 1 : 250 000 | ........... | 1 ° × 1 ° (or 1 ° × 1½ °) |
| 1 : 100 000 | ........... | 30′ × 30′ |
| 1 : 50 000 | ........... | 15′ × 15′ |
| 1 : 25 000 | ........... | 7½′ × 7½′ |

If the scale for each succeeding series is twice that of the previous one, then each smaller scale sheet will cover the same area as four larger scale sheets. Depending on the choice of scale and sheet coverage, it may not be possible to continue this symmetrical relationship throughout all the map series. This cannot be followed right through the set of scales listed above, because 1 : 100 000 is not twice as large in scale as 1 : 250 000. Attempts to follow a logical succession throughout the whole range usually run into difficulties with either the largest or smallest scales. For example, the United States 1 : 62 500 is four times the scale of 1 : 250 000. But to continue this progression would result in increasingly odd scale ratios, e.g. 1 : 31 250, etc.

In Britain, the larger Ordnance Survey scales are organised on a common grid basis. The single sheet unit at 1 : 25 000 consists of a 10 km square. Each 1 : 10 000 sheet represents an area one-quarter of this, that is a 5 km square. Each 1 : 2500 scale sheet covers a 1 km square, and each 1 : 1250 sheet covers a 500 m square. It is therefore possible to look at any individual sheet and identify the location of the corresponding sheets in the other series (Fig. 118). Common sheet line arrangements also assist in the production of derived map series, as a smaller scale sheet covers the same area as the group of larger scale sheets from which it is derived.

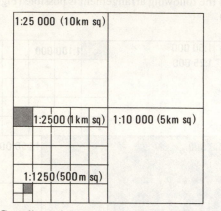

**118.** Coordinated sheet lines, Ordnance Survey plan and map series

## Sheet arrangement

The symmetrical arrangements referred to above can only operate if the sheet lines are followed consistently. This is sensible for large scales, but it poses difficulties for medium and small-scale series in developed areas or in small countries. Regular sheet lines are quite arbitrary in relation to geographical features and areas. It is a considerable nuisance for the map user if one small marginal strip or area appears on a different sheet, or if a small geographical feature is divided between two adjacent sheets. At large scales, with large numbers of sheet divisions, this problem cannot be avoided. But for medium and small scales, many national surveys attempt to organise the sheet orientation and position so that it causes minimum interference with complete geographical areas.

Irregular arrangements of sheets are certainly beneficial to map users, but of course they involve overlaps between adjacent sheets. Large overlaps are wasteful, because they duplicate the same information, but they also pose difficulties in cartographic production.

Once a free arrangement is adopted, then other variations on the regular layout become possible, and can be exploited to deal effectively with variations in geographical shape. These include changing the orientation of the sheet from a horizontal rectangle to a vertical rectangle; showing small outlying areas as insets; and interrupting the normal neatline and border (Fig. 119).

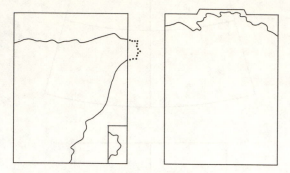

**119.** Border interruptions and insets

## Sheet arrangement and production

In production it is an advantage to have as few variations as possible. The main metrical framework of the sheet should be derived directly from a standard that is maintained as a control of scale and measurement. If a grid is used, then this grid can be duplicated exactly as the basis for each sheet, and the layout of border and margin can also remain constant. If many variations in orientation, sheet size, border interruption and insets are employed, then all these must also be incorporated in the sheet arrangement and layout.

If the graticule is used as the sheet basis, then a set of master graticules may be necessary, depending on the type of projection. For example, for a conformal conic projection, a different graticule will be needed for each latitudinal band. If the series extends into high latitudes, then the sheet dimension in longitude may be changed, and a sheet layout provided for each variation.

# SHEET DESIGNATIONS

Once a map is divided into a series of sheets, it is necessary to be able to identify and refer to the individual sheets in the series. If there is a common organisation between two or more map series at different scales, then it may be possible to introduce a sheet reference system that will also help to indicate the sheet relationships between different scales.

The series designation usually includes the map title or reference, indicating the area covered; the scale; and the name of the producing organisation or agency. If there is more than one edition, this is also included.

For reference to individual sheets, most users prefer a title, that refers to the most obvious or outstanding physical or cultural feature. At large scales this is usually impossible, and identification by little-known names is of no value. On the other hand, larger scale sheets may make use of reference to the title of the smaller scale sheet covering the same area, usually by indicating the compass quadrant; for example Glasgow NW. The alternative method is to give the sheet some numerical or alphanumerical code. Such a code may consist of no more than a number for each sheet in the series, or may be a complex code which indicates also the corresponding sheets at smaller scales.

Although a series of numbers may appear to be a simple arrangement, many variations on the sequence are possible. The numbers may run from left to right in rows, or from top to bottom in columns, and indeed several other possibilities exist. Especially for small countries, coordination of rows and columns can assist the user to identify adjacent sheets in the series, as in the following example:

05 06 07

14 15 16 17

23 24 25 26 27 28

series, in which coordinate references are important, often prefer a close mesh of grid lines across the face of the map. Grid lines at 2, 4 or 5 cm spacing are most common. Because a close mesh of lines causes an interruption in the map information, various methods may be employed to reduce the overall weight of the grid lines, without losing necessary information. Additional grid divisions are normally given in the map border; they can also be indicated by adding ticks along grid lines on the map, or showing other intersections by crosses (Fig. 120). Fine lines may be used, or the line weight reduced by screening.

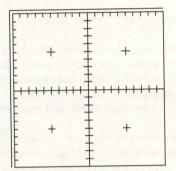

**120.** Methods of grid representation

If the graticule predominates, then some graticule lines will be shown in full across the map, and other divisions indicated in the border. If a grid is also included, then this may be shown only in the map border.

## GRAPHIC DESIGN

The coordinate reference system or systems must also be indicated on the map, and this in turn poses problems in design and arrangement. There is a great deal to be said for maintaining the dominant coordinate reference system on the black plate, along with most of the planimetry, in order to avoid the introduction of locational errors. But the amount of detail given is related to the use of the map, and the degree to which map users are likely to want to give grid references within scale limits.

Assuming that in many series both grid and graticule will be indicated, then if the grid is dominant in terms of sheet lines and arrangement, it is likely to be shown in full across the map. The frequency of grid lines has to be decided. The minimum is to show them only in the map border, so that a map user can construct them. Military map

## BORDER AND MARGINAL LAYOUT

The normal framework of a topographic map sheet contains a border, lying outside the neatline (Fig. 121), together with marginal information related to the sheet and the series. Because of the need to show the coordinate reference system or systems, and also to compensate for the interruptions caused by sheet lines, the map border usually consists of a band in which certain information in included. The necessary figures for grid and graticule values are placed adjacent to the neatline, frequently distinguished by a difference in typographic style or even colour. In addition, names of truncated areas may be completed and road destinations given.

The sheet margin must contain all the information necessary for identification and use of the map

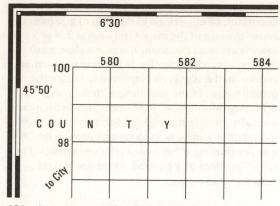

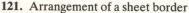

**121.** Arrangement of a sheet border

sheet. Frequently, this is divided into series and sheet identification information, usually placed at the top or head of the sheet, and explanatory information at the foot or in a side margin, depending on the sheet format and orientation (Fig. 122).

A large number of items may be included in the supporting information, and this material must also be laid out in a consistent manner. Because it may take a good deal of space, it has to be regarded as part of the overall layout and design. The major difference between different series is whether the explanation of symbols is given in full on each sheet, or issued separately as a booklet. The latter is more economical in terms of space on the map sheet, but many users find it inconvenient in practice.

Essential marginal information consists of the date of publication and subsequent editions, dates of survey and other sources of information, producing agency, publishing authority and copyright. The projection, spheroid and grid system are normally identified. For medium- and small-scale sheets, both geographic and magnetic north are shown. Details of datum, measurement systems, vertical interval of contours, and graphic scales are usually included. In addition, information may be given about a large number of items that will assist the map user, such as a compilation diagram describing sources of information and their dates and relative reliability. Adjacent sheets in the series can also be shown conveniently by a diagram. Some map series may include explanations of abbreviations, or even glossaries of terms in other languages.

If a full symbol explanation is given, this may be arranged simply as a series of items, divided into groups; or it may make use of a graphic display, showing the various symbols in their geographical relationships; or even a combination of both. For a map covering a large area with many differences in terrain, the full legend may be unnecessary for an individual sheet. In theory, it would be possible to modify the legend so that it contained only the information needed for a particular sheet; but in practice this would involve so much additional work

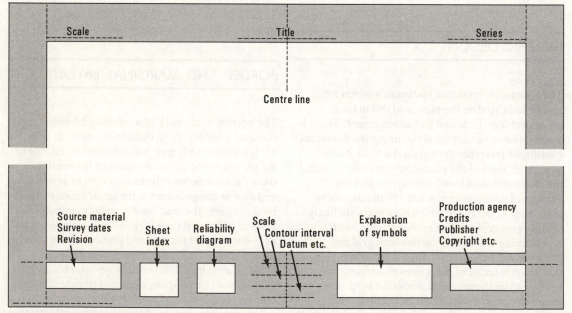

Series, title, scale, repeated

**122.** Margin arrangement for a standard sheet

for individual sheets that it is never done. However, it is another reason for using a separate legend for the whole series.

Like any other aspect of design, marginal layouts may be either good or bad. Attention has to be given to the order and logical grouping of items of information within a limited space, and the marginal layout is normally defined by a series of position lines and measurements, as part of the overall sheet design. In production, the standard items are repeated. Individual sheets will need some specific changes to be made. For example, the adjacent sheet diagram will consist of an outline, which is repeated, but the sheet reference figures have to be changed for each sheet.

On occasions, too much attention is given to symmetry rather than sense. In any legend, the symbol and its description should be close together, so that the map user can identify the symbol correctly. Separating the symbol from the description, in order to occupy an arbitrary space, is a common fault in marginal layouts (Fig. 123).

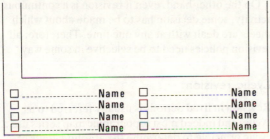

**123.** The wrong way of arranging a legend

## PRODUCTION METHODS

Most topographic mapping agencies use much the same range of methods for cartographic production. The differences lie more in the treatment of particular problems that arise from differences in terrain and style. The general factors that control the choice of production methods are the need for accuracy, the expected life of the series, the complexity of the image, the size of the production operation and therefore the involvement of many people, and the levels of skill and technical facilities available. In order to meet and maintain a specification and produce all sheets to the same standard, operations have to be carefully organised, and this normally results in a strict control of production methods and practices. In many cases a

method that will consistently give satisfactory results may be preferred to one that is potentially quicker or cheaper but is more difficult to standardise.

For basic maps the normal sequence is to produce the images defining the major physical components first (hydrography, heights and contours) and then the planimetry that is largely 'fitted' to them. The exceptions are the large-scale plans or maps of urban areas, especially if they are achromatic, as here the planimetric image of cultural features is dominant. The solid colours, tints and patterns, which occupy areas defined by the line images, are produced subsequently, and are sometimes referred to as 'secondary' images. For relatively simple maps, the correct fitting of the images may be obtained solely by tracing in superimposition, but generally it is more satisfactory if the line images are built up successively by using a series of guide images or keys derived from the preceding images.

Production of basic maps may be speeded up by direct output at the photogrammetric plotting stage, in which the photogrammetric operator constructs a final image to specification. Normally this is limited to contours and drainage, as the straight-line sections of the planimetry cannot be plotted freehand. Even so it is unlikely that the contour image will properly fit the drainage in fine detail. For basic map series produced photogrammetrically, the normal method is to produce the photogrammetric plots and then use these as the basis for cartographic production, but there are exceptions. In the Netherlands, the large areas of comparatively level terrain mean that a copy of the rectified aerial photography is used directly as a guide image by the cartographers, who plot and scribe the topographic detail in one operation. This is aided by notes about detail produced in the field by the topographers.

For derived maps using homogeneous material derived essentially from one larger scale standard series, the normal practice is to carry out a photographic scale reduction first, and then construct the generalisation and new specification on this. If the scale change is not too great, production and generalisation for the new map may be done simultaneously. Very often, if the graticule, grid and planimetry are dominantly on the black plate, this is produced first and the other colour plates for drainage, contours, etc., produced subsequently. If hill shading is to be included in the representation, the normal basis for this is a copy of the contours and drainage.

If there is a large difference in scale between the source map and the derived map, it is usually more efficient to re-compile first and then carry out a separate production stage on the new compilation. The advantage of this is that the relationships between physical and cultural features, which may be strongly affected by generalisation, can be properly examined and maintained, and due allowance made for the comparatively larger symbols used on the derived map.

## MAP REVISION

The revision and maintenance of a map series has always been an intractable problem. As many topographic map series contain thousands of separate sheets, the value of the map as a whole is largely a function of the relative up-to-dateness of the individual sheets in the series. Out-of-date maps are the chief complaint of topographic map users, and in most organisations the problems of revision tend to occupy an important part of the total work.

Theoretically, the aim of revision is to keep all published sheets up to date with changes. It is, of course, an impossible task, even if far greater resources were available for it than is usually the case. It is most critical in those areas where change is likely to be extensive and rapid – that is urban and highly developed areas – and yet it is these which often have the greatest requirement for detailed large-scale maps. Not surprisingly, it is the sheer size of this task that has led to the continued interest in faster production methods, and even the decrease in cartographic 'sophistication', where this is perceived as an obstacle to speed of production. Although errors on a topographic map rarely fall into the category of 'dangerous' to human life, nevertheless they are constant irritation to map users, and often lie behind a general dissatisfaction with the work and organisation of national surveys.

Revision of a topographic map requires a revision policy, the collection of information about changes, and the actual execution of these changes to the original material. Normally, some changes will be introduced before any reprinting of a sheet, but this partial revision is usually limited to the most significant items – for example changes to roads on medium- and small-scale maps. Revision normally implies a specific operation to bring the sheet up to date.

### Revision policy

The revision policy has to take into account the overall objective for the map series in terms of its general currency; the nature of changes and the need to obtain information about them; and financial problems of justifying the cost of issuing new sheets. All policies are complicated by the fact that some sheets in a series will change more rapidly than others, and some parts of some sheets will show more changes than others. Because of this, any simple policy is generally inadequate.

The major alternatives are between treating revision as a single operation to be carried out for a whole series at one time, and treating revision as a continuous operation. For the former, all sheets would be allowed to go unchanged over a number of years, and then a specific operation would be mounted to renew them all. This not only fails to take into account the varying degrees and rates of obsolescence, but also demands a great concentration of staff time and resources. And, even under these conditions, it would be impossible to produce revised versions of hundreds or thousands of sheets simultaneously.

On the other hand, even if revision is a continuous activity, some decision has to be made about which sheets are dealt with at any one time. Therefore, all revision policies need to be selective in some way.

### Cyclic revision

Revision in fixed periods is usually referred to as cyclic, and indeed most topographic map series use a cyclic principle to determine the absolute limits of revision. For example, it may be decided that sheets in a certain category will not go unrevised for a period of more than ten years. In practice, a combination of cyclic and selective approaches is common, by which a whole series or part of it is given a maximum period of obsolescence, but the sheets within the series are dealt with selectively, depending on the rate of change.

Much depends on the nature of the original survey material, whether or not the map is basic or derived, and the surveying methods used to obtain the new information. In urban areas, for example, it is possible to consider revision on a short-cycle basis, and indeed this will be necessary if the revision is to be based on photogrammetry, with new aerial photography being taken every year. In this case the aerial photography will be used as the main basis for identifying changes, as well as plotting new detail.

For derived maps, either the revision must wait on the revision of the larger scale source material – which will produce its own set of difficulties and constraints – or a separate revision operation must be carried out.

Apart from full photogrammetric plotting, aerial photographs can be used as a separate source of information, if only to identify points of change by visual comparison with the existing map sheet.

## Continuous revision

The expression 'continuous revision' normally refers on the one hand to the activity of obtaining information about changes and recording them, and on the other to a constant procedure of revising and issuing new sheets. As its operation usually depends on fixed working strength, it has to be selective. The advantage is that as changes are continuously recorded, the rate and amount of change can be monitored, and a decision taken as to the need to issue a revised sheet.

## Revision information

Information about changes may be obtained directly by the survey organisation, and may be provided by other government departments, organisations, or the comments of individuals. For continuous revision in particular, it is an advantage if regional and local offices are maintained, so that they can deal with local sources of information directly, and where necessary check them in the field. In Britain, the regional and local Ordnance Survey offices are largely maintained for this purpose. In some countries, all those government or regional departments concerned with the environment are obliged to report any intended changes to the regional or national survey organisation. Despite such a flow of information, planned or intended changes are not necessarily identical with what is eventually carried out, but the advance warning is of great importance. Obviously, the planning of a new industrial or residential estate, or the complete rejuvenation of part of a city, has major consequences for all the topographic maps of the area, and inevitably it is in urban areas that the greatest amount of work is involved.

A common procedure for all types of revision is to keep a file copy of the latest issue of a sheet, and mark on this the changes that have occurred. These are normally colour coded to distinguish deletions from additions. The changes may be marked directly on the printed copy, or on a registered overlay, but the need to show the change in proper detail often means that it is more satisfactory to identify the point or area of change by number, and to maintain separate sheets or notes containing the details. Even so, the record map is an index to all changes notified, and can be used to asses the overall amount

of change at any time. In Britain, the Ordnance Survey makes available the information about such changes to the large-scale plans to authorised users through the SUSI policy (supply of unpublished survey information).

If the sheets in the series overlap, it is important to record the information about changes on all the overlaps.

## The execution of revision

The chief cartographic task lies in carrying out the changes to the current materials. The methods used will depend on the nature and characteristics of the changes to be made; the nature of the original material; and the need to maintain standards of accuracy, completeness and correct fit.

The principal revision operations are deletion and addition. Addition is comparatively simple if a new feature is to be added in an 'unoccupied' area; and likewise, complete deletions of detail provide few problems. However, many changes are corrections that involve deleting the existing detail and replacing it with new information; that is, the revision involves both operations in the same area. The normal procedure is to carry out the changes to the line images, which are dominant in locational terms, and to amend the secondary images subsequently. Clearly if a new road alignment is added, any coloured infill is produced after the line change has been made. Apart from introducing the new information correctly, its connections with existing detail have to be preserved, and therefore it must be possible to refer to the existing information at all stages. It is this requirement that frequently makes it preferable to employ relatively complex methods.

Although the actual methods vary a good deal from one organisation to another, most of them are essentially variations on the same basic procedure.

## Deletion

Deletions are most easily carried out on a negative: frequently a duplicate negative of the image is produced first, and deletions made on this. In some cases, it may even be necessary to produce a negative of a final positive image for this purpose. The deleted items are opaqued out on the negative, working from the revision list that indicates all the deletions to be made on the sheet. Deletions can be made on positive images by scraping or dissolving away the ink or photographic image, but this tends to be less satisfactory than opaquing a negative.

## Addition

The addition of new material is more difficult, as it must be constructed and fitted to the detail of the original image, following the symbol specification. Various alternatives exist. The new information can be provided on a separate piece of material, produced as or converted into a negative, and subsequently combined with the unchanged information left on the deleted negative. Or the unchanged information can be copied on to a new piece of material, a key for the additions added, and then the new information produced to specification on the same image. The advantage of the latter method is that the new detail can be seen in position alongside the old, and thus any junctions between the two carried out directly.

Accuracy is assisted if a compilation of the new information is first printed down on a piece of material, and a copy of the old information is also produced in position on the same surface (Fig. 124).

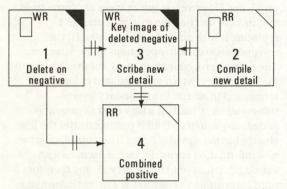

**124.** Deletion and addition using a combined guide image

Normally these are distinguished by colour. So if the old, unchanged information is in red, and the new detail in blue, the cartographer can see clearly the connections between the two.

If the original image is a scribed line, or a negative derived from scribed originals, it is also possible to repeat the deleted line image in a new piece of scribe-coated material by controlled etching; print down a guide image of the new detail; and then complete the corrected image by scribing the additions in place. This is most effective, provided

that the etching can be done satisfactorily, without affected symbol dimensions.

This also follows the general rule in revision execution, which is that unchanged information should never be constructed a second time, but is always copied reprographically from the existing material. This not only avoids unnecessary work, but prevents the introduction of subsequent errors.

Changes to smaller scale maps from the revision source material will also involve proper generalisation. Although in theory such changes should not be made until after the larger scales have been corrected, the need to issue a new small-scale sheet may result in carrying out all corrections known at the time.

For making changes to names and figures, it may be possible to go back to the original stick-up, assuming that these items will have been produced as a stick-up from typeset characters in the first place. In this case, names which are to be deleted, or moved in position, can be peeled away, and new ones added. As with all stick-up material, which is easily disturbed or damaged, it pays to make a new negative copy as soon as possible. The lettering stick-up corrections are made after changes to the line images have been completed. The same applies to any area patterns that have been produced as part of the stick-up images.

Changes to the secondary images are often more difficult to organise. In most cases, area colours will have been produced as either positive or negative masks. On a negative mask, small areas can be opaqued out, and new areas cut into the mask. But small changes, such as the road infill in a section that has been slightly changed, can be difficult to deal with. In some cases, it may be preferable to delete the section of road infill on the negative, produce a positive by contact copying, and then draw the new section in ink, producing a complete new negative subsequently.

Throughout the revision execution, it is imperative that any changes made should be seen in the context of all the map information, and are not dealt with in isolation. Names that interfere with line detail, patterns and tints that do not fit their enclosing lines, spaces in contour lines where figures have been removed: all these deficiencies are normally the consequence of not checking all the map information together.

# 16 REPRESENTATION OF TOPOGRAPHIC FEATURES

Although it is sometimes suggested that a topographic map is some sort of 'picture' of the Earth's surface, in fact it comprises a highly selective treatment of both tangible and intangible phenomena, and these are judged on the basis of their known or assumed significance to human activity. Although large- and medium-scale topographic maps need to show what is there, in the sense that the characteristics of the topography have to be represented, the degree to which detailed classifications of features are given reflects an assessment of relative importance. It would be possible to classify hydrographic features in the same degree of detail as road networks, but the latter are regarded as being of more significance than the former. It is this use of judgement that makes the topographic map useful, and exploits the advantages of different scales of treatment.

## HYDROGRAPHIC FEATURES

The distinction between land and water is an important topographic characteristic, and is of great consequence to human activity. Even in the most rudimentary maps, this distinction is always represented. Although generally classified as a 'natural' feature, in many cases water courses or areas of water have been modified or constructed by Man, and so combine both natural and cultural elements.

The representation of hydrographic features in blue is a strong cartographic convention in multi-colour maps. In areas of extensive surface drainage, where many small channels are likely to be present, it is desirable to use a blue that is dark enough to have sufficient contrast in fine lines. For regions where underground water reserves are important, the small point symbols used to indicate springs, water holes and wells also need a high level of contrast.

## Point features

Springs and wells are often of major importance as a source of water supply for both human and animal populations, especially in some tropical regions. Unlike many temperate regions, the availability and constancy of water supply is a significant factor in daily existence. If the information is available, symbol form or even colour can be used to indicate the difference between permanent and seasonal water sources.

A spring may be represented by a circle or a loop with an addition, and a water hole by a plain circle. A permanent water source is often shown by a solid symbol, giving it greater prominence, and a seasonal one by an outline or semi-circle.

## Lines and outlines

### Coastline and shoreline

The treatment of coast and shorelines depends mainly on scale and topography. In a tidal region, the division between land and water is a fluctuating zone (Fig. 125). If the coast consists of a vertical cliff the position of the coastline remains unchanged at any stage of the tide, but with other coastal types

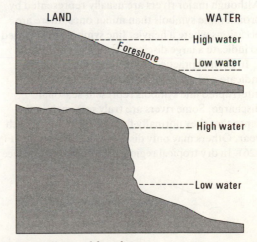

**125.** Coastline and foreshore

there is a horizontal difference between high and low water marks. The normal convention with topographic maps is to show the high water mark as the limit of permanent land, and if necessary to show the low water mark along the foreshore. At large and medium scales, the foreshore, which is exposed at low tide, will also be described by classified topographic features, such as sand, rock or mud flats.

As the coastline is of major importance, it is normally shown prominently, by using a heavy continuous line representing the high water mark. Quite often this has to be determined by a special surveying operation. If in some areas the exact coastline is not easily located, as happens in some swampy coastal regions, then it may be shown by an interrupted line.

### Surface drainage

Line symbols are used to show drainage and other channels along which water flows. If the area is sufficiently extensive at the scale of the map, so that both banks can be shown, then the water body is represented as an area. For channels that are too narrow at map scale to include both banks, single line symbols are used, although these may be varied to approximate the comparative widths of different channel sizes.

The major characteristics represented by the line symbols include channel width, the continuity of water flow, and where necessary the direction of flow. At small scales, where symbols are inevitably exaggerated, it is not easy to decide whether the gauge of the line symbol should indicate simply the distance between banks, often regarded as the nominal 'size' of the river, or the volume of water. Although major rivers are usually represented by broader line symbols than minor ones, there are occasions where a heavier line symbol may be used to indicate a large discharge.

In normal practice, continuous line symbols indicate a continuous discharge, whereas interrupted line symbols represent seasonal discharge. Some rivers are truly seasonal, in the sense that they normally flow for only part of each year. Others may only discharge intermittently (Fig. 126). In dry tropical regions, the numerous surface

channels may only be occupied by flowing water on infrequent occasions, and in these circumstances the channels may be represented as part of the surface topography, rather than as hydrographic features.

The characteristics of drainage systems also have to be considered in the general representation. In regions with high precipitation, extensive drainage networks are created, and normally the surface waters eventually discharge into lakes and seas. The overall pattern and structure of such a drainage basin must be maintained through generalisation, including the relative density of surface channels. Because in such climatic regions rivers tend to increase in volume as they converge, the increase in discharge and river size is often approximated by increasing line thicknesses downstream. The difference between these drainage systems and those characteristic of many drier regions must be appreciated. Many rivers or streams in arid regions will only discharge after heavy precipitation, without reaching any outlet, and the volume of water may diminish with distance from source (Fig. 127).

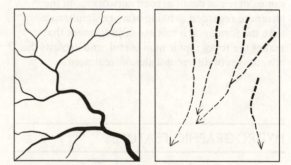

**127.** Characteristics of drainage systems

In this and other respects the generalisation of the hydrographic features needs careful attention. In many cases, not only does the representation of relief need to be adjusted to the drainage, but many significant landscape features, such as settlements, bridges and mills, depend upon the configuration of the drainage channels. As in map compilation the hydrographic features are normally dealt with first, correct generalisation and representation is an important aspect of the map information and structure.

Artificial water channels include canals, both for navigation and irrigation, and aqueducts. Some of these may be entirely constructed, others are modified natural features. They may be distinguished from natural water channels by different symbols, but in some large- and medium-scale maps they are only apparent by their more

Continuous

Seasonal

Intermittent

**126.** Continuous and interrupted drainage lines

regular form. This is particularly the case where the channels are primarily for drainage and irrigation. Navigable canals are often distinguished from non-navigable by differences in line addition (Fig. 128). Abandoned or disused canals (which remain as topographic features) are usually indicated by interrupted lines.

Navigable

Non-navigable

Abandoned

**128.** Line symbols for canals

## Area features

Areas of water are normally distinguished on a topographic map by being given an area colour; blue on multi-colour maps, or grey on achromatic maps. In some cases different blues may be used to distinguish fresh from salt water features.

The blue chosen is normally light, and formed by a tint of the dark blue used for the line features. Although this must be sufficiently light to allow names to appear clearly over it, care must be taken if small areas are present, as these require a stronger contrast than large areas.

Areas of water which vary seasonally are difficult to represent. In some cases, if the information is available, the permanent area of the lake is shown by a continuous blue, and the additional seasonally flooded area by a pattern of blue lines or dots (Fig. 129). Lakes which only exist intermittently may be shown only by a line pattern, to distinguish them clearly from permanent bodies of water. If the water is salt, rather than fresh, then a different line or dot pattern may be used to represent this distinction.

Lake with seasonal extension

Seasonal or intermittent lake

Salt lake

**129.** Permanent and seasonal lakes

## SURFACE CHARACTERISTICS

The main divisions in terms of surface characteristics are between areas of rock, or rock deposits, and those covered with vegetation, natural or cultivated. The distinction is not absolute, as rocky areas may have some plants, and even deserts may have short periods of extensive vegetation after rains. Although most types of symbolisation involve area colours and patterns, line symbols can also be used to effect.

## Rock forms

In mountainous regions, extensive areas may consist of exposed rock, and the degree to which these are represented will be affected by scale, extent and relative importance in the landscape. At large scales, the rocky areas may be large in relation to a single map, and form an important part of the overall character of the landscape. At very small scales it is unlikely that such areas could be separately symbolised.

Basically three approaches are possible. The rock surfaces may be directly represented by rock drawing; areas of rock may be indicated by standard symbols; or the presence of rock may be denoted solely by contour colour. Combinations of these methods are possible.

In relatively large-scale topographic maps, which are plotted photogrammetrically, it is possible to include a direct representation of rock forms (Fig. 130). These show the actual ridges, gullies and slabs, normally by shaded lines, and are often combined with hill shading. In the 'Swiss' manner, the index contours (in black) are maintained and built into the rock representation, which is also in black (see Knopfli 1970). In regions where extensive rock areas are unusual (as for example in much of the British Isles), it is more common to indicate steep rocky areas by 'cliff' symbols, emphasising them as a source of danger rather than attempting to represent them in detail graphically. Where outcrops of rock are small and scattered, it is much more difficult to devise a suitable representation, although they can be shown by a coarse tint or pattern in black or the

**130.** Representation of rock forms (enlarged)

contour colour. If minor surface forms, such as screes and moraines, are included, it is more logical to represent the major rock surfaces as well. It must be emphasised that proper rock representation is virtually impossible unless the rock forms are plotted while the stereo models are in the photogrammetric plotting machine, because the detail cannot be correctly located in the map by other means.

### Area colour and patterns

#### Superficial deposits

In some regions, large areas may be occupied by a dominant surface type, such as a sand or stony desert. On the other hand, there are often particular features composed of surface deposits, such as eskers, scree, and moraine. Depending on scale and the dimensions of the feature, these may be represented directly by contours and/or shading, but at small scale may be shown symbolically by a pattern that is intended to represent or give an impression of surface appearance. Because of the nature of the irregular natural surface, it is normal practice to employ irregular dot and line patterns for such features (Fig. 131). In some cases these need to be constructed by hand. With scree, for example, the grading of deposits, from large boulders to small stones, can best be suggested by a hand-drawn pattern. In a sense these types of area symbol are deliberately iconic in that they imitate the appearance of the features concerned.

This also applies to the use of colour. On a topographic map it is undesirable to show large areas with dark or strong symbols, yet at the same time the contrast with vegetated areas needs to be

made. Therefore, patterns of black or brown dots and lines are the most common, occasionally accompanied by a pale brown or grey tint to reinforce the visual impression.

#### Vegetation and land use

Although it is rarely possible to distinguish completely between vegetation which is 'natural' and that which has been modified by human intervention, there is an accepted contrast between vegetation which continues in its natural state, and crops (either plants or trees) which are deliberately planted and cropped.

The representation of plants faces the perennial problem of topographic maps – the difficulty of dealing with features that change or fluctuate in regular cycles. The general tendency is to concentrate on those types of plant that exist at least for several years, and to ignore planted crops that change seasonally, unless the planted crop occupies the same area on a perennial basis.

The other aspect that reinforces this approach is that tall vegetation can be an obstacle to human movement and other activity, and also provides landmarks. In this sense it is more important than low-growing vegetation. This is reflected by the fact that virtually all topographic maps will show areas occupied by trees, natural or cultivated, with the exception of those at very small scales.

Both natural and cultivated plants are predominantly shown by patterns of repeated point or line symbols, natural vegetation tending to employ irregular pattern arrangements, whereas cultivated plants are usually shown by regular symbol patterns (Fig. 132). Both plan and profile representations are used, and for forest or woodland, either the typical profile or the canopy may be the basis of representation. At large scales the relative density of woodland or bush areas may be indicated by the density of the pattern. Trees are usually shown by larger point symbols than low-growing vegetation. Areas of mixed vegetation may incorporate a combination of two pattern types.

Although major topographic series may be provided with sheets of prepared patterns of the

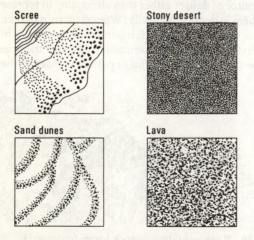

**131.** Types of superficial deposit

**132.** Natural and cultivated vegetation

type required, there is often considerable difficulty in generating suitable patterns for the different classes of natural and introduced vegetation. For single maps, there is little doubt that the most satisfactory solution is to draw or scribe patterns to the detail required, rather than depend wholly on those that are commercially available. It is perhaps significant that this is an area of cartographic representation that has to some extent deteriorated with the increasing use of prepared symbols.

Although in temperate regions only standing trees and bushes, both natural and planted, tend to be included, in many tropical regions areas of low-growing crops are frequently represented. This depends mainly on the degree of permanence of land use, and its relative importance. For example, rice paddies are frequently shown, especially where these are dependent on irrigation, as they are significant in the landscape. The presence of natural forests (jungle), cultivated trees (rubber, palms, etc.) and rice means that topographic maps of such regions frequently have a detailed classification and representation of vegetation and agricultural land use.

Where the treatment of vegetation and land use is detailed and extensive, then the combination of patterns and colours becomes a major part of the map information and design. Whereas in regions where only woodlands are shown a light green solid or tint may be adequate; numerous vegetation classes mean that light green alone is insufficient. Therefore, the choice of patterns and colours becomes more difficult. Several alternatives exist. If cultivated land is shown in a light green (either solid or tint), then patterns to identify individual crops or vegetation types may be added in black, or in green lines over a green, yellow-green or yellow tint. If small and irregular areas are present, then it is essential to keep the pattern elements small and close, in order to retain sufficient contrast between adjacent areas.

## Vegetation and other physical features

Where crops depend on irrigation or annual flooding, the importance of the water element may be indicated by introducing blue into the pattern. So rice areas may be represented by broken blue lines over green, or a combination of pale blue and a green pattern. A similar approach is also used in those areas of natural vegetation associated with water – that is marsh and swamp. The distinction between the two is essentially that marsh is shallower and can normally be traversed on foot, whereas swamp is deeper and can only be traversed

with boats. Therefore, to represent marsh the symbol for the associated vegetation (frequently grass tussocks) may simply be added to the 'land' in blue (to indicate the water element), whereas with swamp it is more usual to show the areas as occupied by water (in light blue) and add green vegetation symbols as required (Fig. 133).

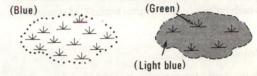

**133.** Marsh and swamp symbols

Apart from vegetation and water, there are many other landscape types that involve both surface composition or relief forms and vegetation. Stony steppe, salt steppe and grass steppe are all variations including both ground and vegetation components (Fig. 134). They are also a function of small-scale generalisation, where the individual features can no longer be represented, but the character of the landscape needs to be described.

**134.** Grass, stony and salt steppe patterns

### Ice and permanent snow

Those regions characterised by ice and permanent snow cannot be treated as similar to the rest of the topography, for indeed in many cases the topographic 'surface' cannot be mapped. Both line and area symbols may be used to represent this element. Contours across snowfields and glaciers can be shown in blue, along with minor relief features such as crevasses. Some small-scale maps, which include hypsometric colours, leave the areas of permanent snow white. In both cases the cold conditions and presence of ice and snow are suggested by white or pale blue.

## CULTURAL FEATURES

Many cultural features, such as buildings and roads, are relatively small in dimension at medium and small scale, yet they are highly important in the total topographic information. Other elements, such as boundaries, can only be indicated in relation to other tangible features. Consequently, it is with much of the cultural information that exaggeration of symbol dimensions has its greatest effect. In this respect, there is an intimate connection between symbol, legibility, generalisation and scale, all of which influence the symbol forms and dimensions.

## Point symbols

Point symbols are used extensively on topographic maps, mainly to identify different types of building. Most of these are exaggerated in size at medium and small scales.

Variations in type of symbol reflect either the physical appearance or character of the building, or its use. Because of the large number of point symbols, all the devices of addition and extension have to be used, although the basic forms of square, rectangle, circle and triangle tend to be retained for small symbols. Few small point symbols are directly iconic, but many make as much use as possible of some associated idea or concept. For example, the medium-scale topographic maps of at least eight European countries employ a point symbol with the addition of a cross, or simply a cross, to represent a church (Fig. 135). The same is true for a windmill, as

**135.** Point symbols for a church

even though the basic form varies considerably, the diagonal cross representing sails is widely used (Fig. 136). For other types of feature, such as monuments, there is much less uniformity. Either the plan or the profile may be the basis of representation.

**136.** Point symbols for a windmill

Because most of the symbols need to be small, there are fewer possibilities with addition. The chief variation is between solid and outline symbols. For important features, the greatest emphasis can be achieved by using a heavy outline with a high contrast coloured infill.

Small-scale maps in particular often have to bridge the transition from point symbol to area symbol for features of the same type but different degrees of size and importance. The classification of

inhabited places is the obvious example. In such a graded series, all the devices of point symbol addition are commonly used, but larger settlements will still have to be represented as highly simplified area symbols (Fig. 137). Once again, symbolisation and generalisation are connected.

**137.** Graded symbols, point to area

## Line symbols

This symbolic exaggeration is continued with the treatment of linear features. These comprise the most numerous symbol type on medium- and small-scale topographic maps, in complete contrast with large-scale plans, which normally employ little line symbol classification. The degree of exaggeration is considerable. A feature such as a minor road, some 5 m wide in actual dimensions, could be correctly represented at 1 : 10 000 scale by a line 0.5 mm wide, which would be sufficient for a double-line symbol. If the same symbol dimensions were maintained, for legibility and emphasis, then the actual dimensions would be exaggerated five times at 1 : 50 000 scale, and twenty-five times at 1 : 2500 00 scale (Fig. 138).

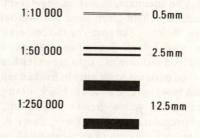

**138.** Line symbol dimensions and symbol exaggeration

All the devices of continuity, addition and extension need to be employed, together with colour, to produce a sufficient range of contrasting line symbols. Although in some ways linearity is obvious from the line symbol form, additions and extensions frequently attempt some degree of iconic representation as well. For other types of linear feature, the line between connecting points, or between supporting points, like power or telegraph lines, is usually shown by a line symbol with point additions or extensions (Fig. 139). The cross-line addition is typical of many representations of

| Form | Dimension | Addition and extension | Form and dimension |
|---|---|---|---|

**139.** Line symbol development

railway line. Although many features are represented by a single class, transport routes, and especially roads, are shown in hierarchies of related symbols, making full use of all graphic variations.

Series of symbols for road classes may make use of line continuity as part of the relative emphasis, but also to show whether or not a road is fenced. The extent of these sub-classifications is related to scale and the total variation in the feature category. The most detailed classifications are usually at medium scale (such as 1 : 50 000). At small scale the range is reduced, partly by the combination of classes, and partly by the elimination of smaller features.

Because of the importance of roads in medium-scale maps, colour addition is frequently used for emphasis, particularly for the major road categories. The usual order of red, green or brown, yellow and white provides a logical series of gradations, combined with symbol width and line thickness. Contrast and emphasis can be modified both by using tints for the road colour infill, and also interrupting the infill itself. Attempts at more uniform national and international road classifications have been influenced in turn by colours associated with road signs – blue in Britain, green in many European countries. Here the desire for consistency has provided problems, for both blue and green have a lower contrast than red, and therefore symbol dimensions have had to be further exaggerated.

The opposite starting point is employed in some topographic maps, where roads are left white against coloured backgrounds, the degree of emphasis depending entirely on the line symbol.

From a technical point of view there is little doubt that the use of double- and treble-line road symbols has been assisted by the introduction of scribing. Even so, the need to keep the road colour fill in correct position relative to the road outlines or casings also puts a premium on good technical construction and correct register. With small-scale maps it is more common to use single line roads in red or brown in order to avoid excessively broad road symbols, but these are also easier to construct and reproduce.

## Boundaries

The distinction between areas can be shown by either outlines or area symbols. Except for a few small-scale maps, which may represent different countries or regions by area colour, boundary lines on topographic maps are normally shown by line symbols.

At large scales the actual position of the boundary monuments or markers may be surveyed and shown by suitable point symbols. Few boundaries are marked on the ground by continuous boundary features, except in special circumstances. As with roads, boundaries exist in a hierarchy of importance, and the symbols chosen reflect this.

### International boundaries

International boundaries are normally based on treaties between the countries concerned, many of which have evolved historically over a long period. Those boundaries that are accepted internationally on this basis are *de jure* boundaries. Many other boundaries, or sections of boundaries, exist in practice, often as a result of war or conflict. Where these are enforced or tacitly agreed they are *de facto* boundaries. Many armistice lines are eventually accepted as *de facto*, although the countries concerned refuse to recognise them legally. The representation of international boundaries therefore has political implications, and in general national mapping agencies add disclaimer notes on topographic map sheets, stating that the representation of an international boundary must not be taken as a sign of political recognition.

Because of their importance, international boundary lines are invariably represented by a high contrast line symbol. Usually this is a multiple symbol, the black line indicating the boundary position, and the colour riband denoting its importance (Fig. 140). Although different types of international boundary could be represented systematically, there is insufficient international agreement on the status of boundaries to make this possible. Logically speaking, the fully demarcated *de jure* boundary should be shown by a continuous line symbol and a continuous riband. A *de facto* boundary could be shown with an interrupted riband

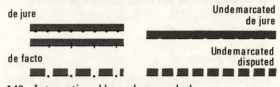

de jure    Undemarcated de jure

de facto    Undemarcated disputed

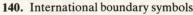

**140.** International boundary symbols

to denote its lesser status. An undemarcated boundary could be shown by an interrupted line symbol, although with a continuous colour riband if it was a *de jure* boundary. Provisional or disputed boundaries could have both the line symbol and the colour riband interrupted.

In order to indicate its importance, and provide a sufficiently strong symbol, either red or purple is commonly used for colour emphasis.

If an international boundary line is continued into offshore waters, or an inland sea, in order to delimit adjacent territorial waters, it should not include the black-line symbol unless the line is actually monumented. Only the colour riband showing the boundary status should be used.

On small-scale maps, groups of islands belonging to one state should not be enclosed by an international boundary line if this only serves to show their political affiliation.

### Internal administrative boundaries

These are an important element in the topographic map, and again they normally fall into hierarchies of administrative importance. This is reflected by the weight and continuity of the line symbol (Fig. 141).

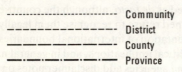

Community
District
County
Province

**141.** Administrative boundary symbols

### Boundaries in rivers

Many boundaries, both international and national, are located in relation to major physical features. Where a river is large enough at map scale to show both banks, then the boundary line may be placed along the centre line of the river, if this is in fact its location. There are cases where the boundary legally follows one river bank. With single line rivers the problem is more difficult. An interrupted line adjacent to the river is both visually disturbing and graphically difficult. The usual practice is to show it by short sections only, these being sufficiently close to enable the map user to locate it if necessary, with the boundary position being made clear at river junctions (Fig. 142).

### Area features

The predominant area symbol on topographic maps is that for built-up areas; relatively few other

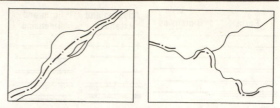

**142.** Boundaries and rivers

elements in the cultural environment are shown by continuous area colour or pattern.

The chief problems are the inter-relations between the built-up area and the road symbols, and the need to use relatively light colours or patterns in order to avoid interference with both topographic detail and names.

At large scales, where the building frontages are shown to scale, the roads are not separately represented by symbols, and their presence is essentially a matter of interpretation. At medium scale, the built-up area is generalised, and the major road pattern (spaces) is maintained by exaggeration. At very small scales, where the major road network is shown primarily as a series of routes, the built-up area is more likely to be shown as a continuous area (Fig. 143).

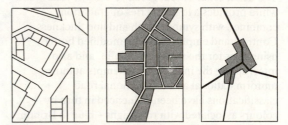

**143.** Roads and built-up areas

The chief problems arise with medium-scale maps. In order to bring out the built-up area, it is normally indicated by the use of a tint or light colour; frequently grey, pink or yellow. Where the map covers extensive built-up areas, the choice of the lightness value of this colour is critical. If the major road classes are represented with coloured infills, then these may be continued through the built-up area, or may be stopped at the edge. At small scales, either the major road line is simply overprinted on the built-up area tint or colour, or again the roads may be stopped at the edge of the built-up area symbol.

At medium scale the whole of the urban area may be represented by a single colour symbol, or there may be some classification within it. For example, public buildings may be separately distinguished, usually by black or a darker colour.

# 17 REPRESENTATION OF RELIEF

The representation of the relief of the Earth's surface poses many problems both in surveying and cartography. Because it includes the entire three-dimensional surface, it is normally described initially by measurement from sea level, the datum that separates the land and the water areas of the Earth's surface. The land relief is expressed in terms of elevation above datum, and the relief of the submarine surface as depth below the datum. Theoretically, this should be a single datum, approximating the surface of the spheroid, but in practice it has to be measured and defined locally. The variation in relief affects the measurement of all other topographic features, and of course the surveying of the submarine surface still remains as an enormous uncompleted task in world mapping. Historically speaking, the production of good topographic maps and nautical charts has depended to a large extent on solving the problems of measuring relief.

It is not possible to represent fully the third dimension on a two-dimensional map. It can only be indicated selectively, as otherwise the relief information itself would occupy the entire map. But a map, as opposed to a plan, has to show the third dimension in some way, as well as the two dimensions in plan. Consequently, the cartographic problems of representing relief so that sufficient information is provided without interfering unduly with other map detail have received a great deal of attention.

Relief comprises two main elements, elevation and slope. It is difficult to represent slope without first obtaining height information, except in an approximate way, because slope is determined by the relationship of change in height to plan distance. Whereas to some extent slope can be interpreted from elevation, the converse is not the case, and therefore inevitably elevation takes precedence on most topographic maps.

Whereas some elevation information can be represented directly on the map by point and line symbols, the relief features must be interpreted from the elevation information, or represented graphically by suggesting a continuous surface. The full representation of relief therefore requires a combination of methods, as no single method will accomplish both.

## Elevation

### Point values

The height or depth of any point is measured from a datum. There is no absolute zero that can be determined universally, so national mapping agencies use a specific point that approximates mean sea level, usually obtained by taking observations over a long period. This can also be used for measuring depths, either of land below sea level, or of the submarine surface. The chart datum used for establishing depth of water for navigational purposes is different, as in tidal areas it has to represent a low water level.

The elevation values, mainly heights, shown on a map are selected from those measured during the topographic survey, many of which are used as survey control points. The density of and distribution of elevation figures on a map varies with terrain and scale. Imhof (1982) shows that there are usually twice as many values given in mountainous areas compared with level areas, and that in general the frequency increases with decrease in scale. The increases in density that accompany scale seem to fall into two bands. For scales between 1 : 1000 and 1 : 50 000 the mean number of heights per 100 cm$^2$ ranges from 15 at 1 : 10 000 to over 40 at 1 : 50 000. Between 1 : 100 000 and 1 :1 000 000 the average number increases from 30 to 40. At smaller scales the larger area of terrain covered by a map sheet is likely to include a greater range in elevation and therefore a larger number of values needs to be shown. At scales most likely to be used in the field, such as 1 : 25 000 and 1 : 50 000, the selected values not only indicate high and low points in any area, but also assist the correct determination of location on the ground.

The height or depth values need to be of sufficient size for legibility. Like names relating to points or small areas, they should be adjacent to the point symbol marking the location, preferably placed to the right, and parallel to the base line wherever possible.

## Selection of point values

Although elevation figures are normally given for all the high and low points in an area, their judicious selection should also reflect the needs of the map user, particularly at large and medium scales. Mountain peaks, passes, hilltops and lowest points in depressions are obvious choices, but if possible figures should be included at breaks of slope, river junctions, and landmarks such as isolated buildings, churches, refuge huts, etc. These are most useful in helping the map user in the field to identify correctly the elevation of the places where he is most likely to be, or along routes which he is likely to use. Therefore, a large proportion should be adjacent to roads and footpaths. They are least useful in the middle of a long slope, where there are no clear landmarks to aid correct location of the point.

Point values should be strictly limited in areas of ice or permanent snow, or any other areas that are likely to short-term changes.

## Elevation values and lakes

For lakes the elevation information can be shown in two ways: either by depth to the lake bed from an assumed average water level (which acts as a local datum) or by the height of the lake bed above the datum used for the land elevations, that is, as part of the land surface (Fig. 144). For a dammed lake, the normal high water level is used. Where information is available, a common practice is to show contours for the elevation of the lake bed (being continuous with the land surface), and a selection of depth figures from the lake surface. If both sets of data are given, they are best distinguished by a difference in colour.

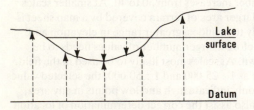

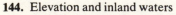

**144.** Elevation and inland waters

# CONTOURS

Whereas elevation figures give specific values for selected points, the overall variation in elevation of the three-dimensional surface is normally represented by contours, which are lines of equal height or depth, measured from a datum. Only a limited number of contours can be given, and therefore they are selective. The height or depth of any intermediate point can be judged approximately from the adjacent contours, as any point lying between two contours can be estimated as being within half the contour vertical interval. Given a fixed vertical interval between contours, slope can be interpreted visually by relating the change in height as shown by the contours to the plan distance between them. In level or near-level areas, a small difference in height will mean a large difference in plan position; in areas with steep slopes, a small difference in plan will equal a large difference in height (Fig. 145).

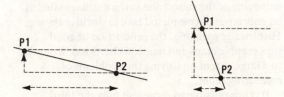

**145.** Slope angle and plan distance of contours

Contours do not represent either slope or land forms. They make factual statements about elevation, from which the map user can attempt to interpret slope, and therefore relief features.

Contours introduce three major problems: the choice of vertical interval between standard contours, and for both index and supplementary contours, if used; their graphic representation on the map; and the effect that both have on the rest of the map content and design.

## Types of contour

Contours may be either measured or interpolated. Contours plotted in a photogrammetric stereo-plotting machine are continuously measured lines. Contours constructed from an array of point values are interpolated, and have a lower order of accuracy. Specific contour lines can be measured by levelling in field survey, but this is a difficult and time-consuming operation. The quality of interpolated contours (like any other isolines) depends upon the density and distribution of the measured points, and any assumptions made about the gradients between them. If the point values are measured at regularly spaced small intervals (as occurs in modern hydrographic surveys) then the interpolated contours can be regarded as consistent with the point values. If the measured points are irregularly and widely distributed, then the modern practice is to compute a grid of values calculated from the known points, and then to 'thread' the interpolated contour lines.

Note that in many European languages, the equivalent of 'contour' is 'height-line' or 'height-curve', and the equivalent of the term 'contour' is used to refer to the planimetric outline.

## Vertical interval

The choice of the standard vertical interval is affected by map scale, the type of terrain, and the graphic factor, that is the number of contour lines of a given gauge which can be legibly represented close together.

Most topographic maps at large and medium scales also have index contours, which are visually emphasised, to help the map user to focus on particular lines and establish height more rapidly. The vertical interval for index contours is always a multiple of the standard vertical interval, usually four or five times. Supplementary contours may be added in areas of gentle slopes, where the standard contours are too far apart to provide a useful representation of elevation. They are usually at half the vertical interval of the standard contours, or even one-quarter (Fig. 146).

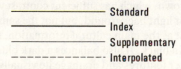

- —————————— Standard
- —————————— Index
- ———————— Supplementary
- – – – – – – – – – – Interpolated

**146.** Types of contour line

In order to decide upon a vertical interval, some assumption has to be made about the maximum slope that will occur in the area covered by the map, as it is here that the contours will be closest together. This is usually taken to be 45 °, as slopes steeper than this are more likely to be represented by rock drawing or cliff symbols. The problem is that the minimum vertical interval possible on gentle slopes would be too small for steep slopes, where the contours would run together. The choice is most difficult for standard map series covering large

areas, which may contain pronounced differences in elevation and relief. Some national mapping agencies attempt to accommodate this by using different vertical intervals for different parts of the country, but the vertical interval is always uniform for any individual sheet. If the area mapped contains both level areas and steep slopes, then the minimum vertical interval possible will be determined by the steepest slopes, and supplementary contours added in the flatter regions.

Although it is possible to calculate a theoretically ideal minimum vertical interval, in practice the actual choice is affected by the user's convenience, and for derived maps, the contours present on the source map. Finding the height of a particular contour from the nearest index contour, or the height difference between different contours, requires mental arithmetic, which is simple if the vertical is a convenient round figure, such as ten or fifty.

Imhof (1982) describes a simple formula for calculating the vertical interval, taking into account the graphic factor of contour thickness and separation. This is

$$VI = \frac{M \tan \alpha}{1000k} \text{ (metres)}$$

where $M$ = scale denominator

$\alpha$ = maximum slope angle

$k$ = the number of contour lines that can be legibly represented within one millimetre.

So if the scale is 1 : 100 000, $\alpha$ is 45 ° and $k$ is 2 (0.1 mm contours with 0.4 mm separation), then

$$VI = \frac{100\ 000}{2000} \times 1 \text{ m} = 50 \text{ m}$$

In less steep areas, where the maximum slope angle may be judged to be 30 °, then the vertical interval would be 28.5 m.

The theoretically ideal vertical interval would be given by

$$VI = n.\log n \tan \alpha$$

where $n$ = square root $\dfrac{M}{100}$

and $M$ = scale denominator.

For a scale of 1 : 100 000, in areas of steep slopes ($\alpha$ = 45 °) this would give a vertical interval of 47.4 m.

There are various formulae for determining suitable vertical intervals, but they all have to make assumptions about maximum slope angle for a given type of terrain. As the resultant figures may be inconvenient in use, they have to be modified to round figures, and these in turn should be chosen so that both the index contour interval and any supplementary contour intervals are also convenient in use.

Imhof points out that the actual choice of vertical interval for particular scales often departs from the theoretically ideal for a number of reasons. Maps of very flat areas at large scales tend to use a smaller vertical interval than the one calculated for areas of low relief, because the maximum slope angle may be less than 5 °. Derived maps are directly influenced by the vertical intervals used on the source maps. Generally, a multiple of the source map vertical interval is adopted, to avoid interpolation of a new set of contour lines. At very small scales (such as 1 : 1 000 000 and smaller), the vertical interval is usually smaller than the theoretical ideal because it is rare for more than two contours to be adjacent, and they are likely to be highly simplified lines at such scales.

It is of interest that topographic map scales, which appear to be convenient in terms of linear scale, pose considerable problems in the choice of vertical interval. The scale of 1 : 50 000, which seems to be a convenient round figure, is particularly difficult, a point demonstrated by the variety of vertical intervals used in practice for this scale. The theoretical ideal is 29 m (assuming that α = 45 °), which could be easily rounded to 30 m, but this value is inconvenient for index contours and for the map user. As is shown by other factors, scale ratios in neat decimal figures do not necessarily have any property that makes them effective representations of the topography in terms of map content.

## Contour interruption

Even when a vertical interval is chosen which is suitable for the steeper slopes within the map, it is always possible that in a few areas there will be very steep slopes on which the contours are crowded together to the point where they begin to be illegible. To compensate for this without increasing the vertical interval for the whole map, contour interruption can be introduced (Fig. 147). In such steep slope areas, the index contours are kept as

**147.** Contour interruption

continuous lines, but the intermediate standard contours interrupted for a short distance so that they do not run together.

## Contours and graphic design

The representation of contours has to take into account the quality of the information on which the contours are based, the nature of the information they give, the need of the map user to identify them individually, and their effect on the other elements in the map content. These different functions and effects are dependent on the contour specifications, using the graphic variables of colour, form and dimension, and the fact that the contours may appear as isolated lines or grouped closely together.

The effect of the contour pattern on other symbols varies considerably in different parts of the map. Where there are few contours, the fine irregular lines may be difficult to distinguish because of other information. On the other hand, closely grouped contours exert a strong influence on the total map colour, and may make it difficult to include other symbols legibly. Therefore, a line that is suitable for contours on a steep slope will tend to be too fine when it is isolated.

## Graphic specification

For the symbol specification of contour lines, hue, line gauge and form have to be considered together, both for the different types of contour lines, and the relationship to the rest of the map design. Contours that are too thick or high in contrast will have a serious effect where they are close together; therefore, a neutral hue is usually chosen, such as a dark brown, which has sufficient contrast with a white or light background, but not the same level of contrast as the main cultural information, frequently shown in black. It also balances a dark blue, so that the water features and drainage lie visually on the same surface as the contours representing elevation. A dark brown line can be a fine line, possibly as fine as 0.15 mm, and still retain sufficient contrast to be legible on its own. In this case the index contours need to be in the ratio of 5 : 3 for the best comparative effect, to enable them to stand out from the standard contours without being too intrusive. If the contour colour is lighter, such as a light brown or orange, then the line gauge needs to be increased accordingly.

A difference in line form can be used to differentiate the supplementary contours, if used, by employing an interrupted line; this gives a more perceptible contrast than a finer line, which could be difficult to distinguish from the standard contour symbol. If two sets of supplementary contours are used, then different interruptions, using longer and shorter segments, can represent the two types.

In some more sophisticated topographic map designs, where the representation of relief is important in the overall map content, three different colours are used for the contour line image: brown for contours over soil and vegetated or cultivated land; blue for contours over regions of ice and permanent snow; and black for contours over exposed rock. The contour image thus helps to classify the principal types of terrain. In practice, to balance the design, the black contour lines should be made slightly finer than the others, in order to complement them visually. This treatment is one element in the style characterised as the 'Swiss manner', due to its use on the official topographic map series in Switzerland.

## Contour numbering

It must be possible to determine the actual elevations represented by the contours, and therefore numerical information has to be included. It is neither necessary nor desirable to number all the contours simultaneously in any one area, but the numbers must be presented with a frequency and positioning which makes correct identification possible without prolonged searching. The general rule for a topographic map which may be used in the field is to number all the index contours, a set of figures being included on every major slope, and in between where necessary, and to number standard contours in any region where they cannot be identified easily from adjacent index contours. On a large sheet map, it is clear that the numbers must be repeated in different parts of the map. On a long slope, where many contours are present, then if possible it is convenient to number the index contours in a sequence. The row of figures will itself draw the attention of the user. Individual contours isolated from others will need to be numbered at intervals, and of course the figures are entered in places that are clear of other map information (Fig. 148).

The other main problem is the decision about the positioning of the figures relative to slope and the 'reading direction' of the map. A topographic map used in the field will be oriented in the direction of view of the map user. In such cases there is no point in adding the contour figures parallel to the base line of the map, as they will still be 'upside down' in

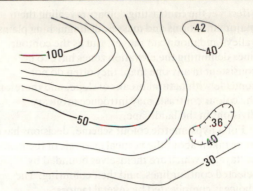

**148.** Contour numbering

other orientations. Therefore, the figures are placed consistently so that they read 'uphill', that is, following the direction of the major slope in the area. At smaller scales, where the map is unlikely to be used directly in the field, it will normally be read in its conventional position, with north at the top, and therefore the figures will be placed in the best position for reading from this direction.

Contour numbering is particularly important where there are closed contours, and where the topographic feature might be either a hill or a depression. Where there is a single closed contour, it is essential to add a spot height inside the contour line. Where the closed contour indicates a hollow the contour line itself may be given an additional symbol to show that it represents the top of a slope, and that the area so enclosed is lower than the contour itself.

# HYPSOMETRIC COLOURS

Hypsometric colour schemes, often referred to as 'layer colouring', are another means of representing elevation on a map. They divide the entire land surface of the map into a series of elevation zones, and they can be extended to include the submarine surface as well. Because they add colour to the entire map they have a much stronger visual effect than any other method of relief representation. They are unsuitable for large-scale topographic map series, as the elevation range on any one sheet is likely to be limited, and the entire map sheet might fall into only one or two coloured zones, which would give little useful information. At medium scales they can be combined with regular contours if desired. At very small scales they can be very effective, the contrasting colours revealing the major land forms and masses, distinguishing plains, valleys and mountain ranges. But as the contour lines delimiting the elevation zones have to be consistent in any one map, they often do not coincide with natural breaks of slope, and therefore the colour contrasts may introduce arbitrary divisions in the landscape.

For a hypsometric colour scheme, decisions have to be taken about the vertical intervals of the bands or 'layers', which are themselves bounded by selected contour lines, and their colouring. The choice is complicated by several factors. Graphically, the colours must have sufficient contrast to be perceptibly different, even where they occur over very narrow bands, but excessive contrast will have a disturbing effect on the overall appearance of the map. If other information is to be shown, then the layer colouring cannot be so strong in contrast that it totally overpowers the rest of the map information. If a long series of many colours is required, then there are technical consequences in the complexity of the map production processes and the number of printing colours needed. And finally there is the problem of colour association – the subconscious habit of the map user to associate particular colours with certain characteristics.

Like all design solutions which seek to encompass conflicting requirements, hypsometric colour schemes are inevitably compromises. Although the choice of vertical intervals and the choice of colours are interdependent – a large number of vertical divisions means that an equally large number of colours must be composed – it is necessary to consider these elements individually at first.

## Vertical intervals

The vertical intervals in a hypsometric colour scheme depend upon the total number of vertical divisions and coloured bands required, in relation to the total range in elevation covered by the map or map series. At relatively large scales, where the mapped area may comprise a limited elevation range, the division into coloured layers may be adjusted to the particular terrain. At very small scales, and particularly for maps that deal with the Earth or continental regions, the scheme must take into account the full elevation range of the land surface, from sea level to approximately 8000 metres, and if necessary the equivalent range of submarine depths, from sea level to some 11 000 metres. In practice, the total number of vertical intervals for the land surface tends to lie between a

minimum of four and a maximum of thirteen. But if more than ten are used, the consequent colour scheme becomes both graphically and technically complex.

If the overall elevation of the Earth's surface is plotted by showing the total area at particular heights, then the resultant graph shows the relative distribution of elevation against area, known as the hypsometric curve (see Imhof 1982). Over 50% of the land area is under 600 m, and 25% over 1000 m. If a constant vertical interval is used, the areas covered by the different bands will be disproportionate: and if a constant area is used, the vertical interval will increase rapidly on steep slopes (Fig. 149). Even so, unless the territory being mapped is limited to high elevation areas, it is clear that the vertical intervals used in a scheme need to be smaller at lower elevations than at higher. Even with maps of small areas, the relief differences at low elevations are of more importance for most human activity than at high elevations, so for this reason also it is necessary to ensure that the vertical intervals of the coloured bands are arranged in proportion to the areas at different elevations.

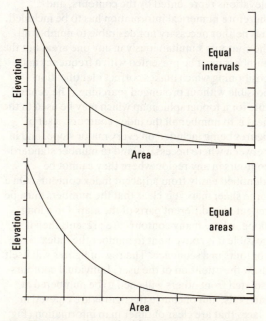

**149.** Hypsometric vertical interval, area and slope

There are several different approaches to the choice of vertical intervals (Fig. 150), the main alternatives being as follows:

1. *Equidistant*. The use of a constant vertical interval is similar to that of standard contours,

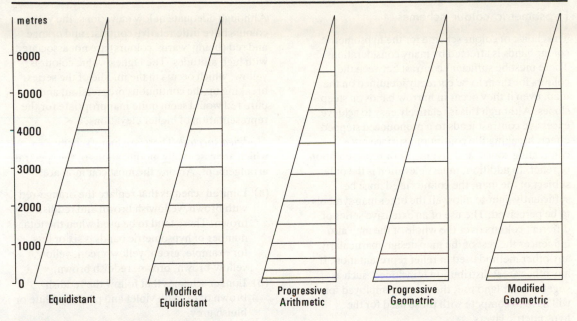

**150.** Hypsometric vertical intervals

but fewer are included. The method is only suitable for a small area with a limited elevation range, such as an island, and even then it will tend to give too large an interval for low areas and too small an interval for high areas.

2. *Modified equidistant.* The inadequacy of the equidistant method in dealing with low areas can be compensated by sub-dividing the lowest intervals. Sometimes referred to as 'double interval', it tends to give too many steps, and leaves a discontinuity in the middle.

3. *Progressive.* To accommodate the decrease in area with elevation, various schemes use the idea of increasing the vertical interval with increase in height in a regular fashion. This may be based on (a) an arithmetic progression, increasing by a constant amount, or (b) a geometric progression, increasing by a constant factor. For a large elevation range, the arithmetic method increases too slowly (e.g. sea level – 100 m, 100–300 m, 300–600 m, 600–1000 m) and gives too many steps if properly adjusted to the correct representation of lowland areas. The geometric method tends to generate inconvenient numbers at one end of the scale: for example, if a factor of 2 is used, then a scheme could be sea level – 100 m, 100–300 m, 300–700 m, 700–1500 m, 1500–3100 m, 3100–6300 m, etc.

4. *Irregular.* Although it is possible to consider adapting the vertical intervals to the topography of a region, generally speaking it is difficult to

devise any scheme that will do this satisfactorily. Ideally, the changes in the vertical intervals, and therefore the coloured bands, should coincide closely with actual breaks of slope. But these do not occur consistently at particular elevations.

The most used irregular scheme is a modified geometric, in which the uneven vertical intervals are changed to round figures, and very often the highest elevations are grouped into one band, simply described as 'over …'. In the example given above for the geometric progression, this method would change to 1500–3000 m, 3000–5000 m, and over 5000 m.

### Bathymetric intervals

The bathymetric curve of the submarine surface is quite different to that of the land surface. There are relatively small areas of very shallow and very deep water, and 50% of the oceanic depths lie between 3000 and 6000 m. For maps covering major oceanic areas, an irregular geometric progression is most useful, as the major deeps tend to be very small in extent. For maps that include only limited areas of coastal waters, the modified equidistant is preferable, using a smaller interval in the shallow coastal zone. This also concentrates information about the configuration of the seabed in the immediate offshore area, showing its physiographic connections with the adjacent land areas.

## Hypsometric colour schemes

The choice of colour scheme for the hypsometric colour bands is affected by many considerations. There must be sufficient contrast between the colours for them to be correctly identified on the map, even if they occur in narrow bands on steep slopes. Although this is relatively easy to achieve, excessive contrast leads to a pronounced stepped effect, breaking the continuous surface into contrasting zones, and giving a poor representation of relief. In addition, unless elevation is the only subject of the map, the colours used must be sufficiently light to allow all the other map symbols to be perceived. The use of an extensive series of different colours over the whole of the map also influences the rest of the map design, particularly any other method used in relief representation. If any other areal distribution is included, such as vegetation or land use, the colours employed for this will have to compete with those used for the hypsometric layers.

This complex design problem is made more difficult because of colour association. Blue is associated with water, orange and red with heat and drought, green with vegetation, and pale blue and white with snow and ice. These topographic connotations may be inconvenient, but they occur. Although hypsometric colour schemes may be devised using colours according to a cartographic convention, it is difficult to prevent map users from responding as though the colours were meant to be naturalistic, especially as some of them are. Naturalistic colour schemes, which attempt to reflect factors of elevation (increasing coldness), vegetation and land use, and climate, also become arbitrary because of seasonal changes. In practice the most widely used colour schemes are conventional, but are modified to take naturalistic factors into account.

## Types of colour schemes

Hypsometric colour schemes can be divided into two main groups: those depending primarily on contrast in hue, and those depending primarily on contrast in value. As all colours have both hue and lightness attributes, these cannot be separated completely. Schemes based primarily on hue include the following:

1. *Spectral*. Here the order of the spectrum is followed, using blue for water, green for lowlands, and then a sequence of yellowish green, yellow, light orange, orange, red.

Although adequate at low elevations, the yellow colours have little relative contrast, and orange and red (being 'warm' colours) are not associated with high altitudes. The highest value colour is yellow, which occurs in the middle of the series, breaking up the continuous progression, and a pure red would seem quite inappropriate for the representation of higher elevations.

2. *Modified spectral*. There are many schemes which are essentially modifications of the spectral arrangement. Among the most common are:

   (a) Limited schemes that replace the orange-red with brown, yellowish brown and reddish brown. These tend to be used when the total number of hypsometric bands is six or less; for example, green, yellow-green, yellow, yellow-brown, brown, reddish brown.
   (b) Longer schemes that follow the reddish brown of (a) with violet and possibly white or bluish grey.
   (c) Schemes that omit yellow. The sequence can be green, olive green, olive brown, light brown, brown, dark brown. Although these colours have a greater equivalence in value, they tend to give a comparatively dull appearance to the map.

Colour schemes based primarily on lightness make use of the lightness characteristics of both single and combined hues to obtain a graded series that increases or decreases from one end of the scale to the other. The principal alternatives are as follows:

1. *The higher the darker*. In this sequence the lighter hues are used at low elevations, and the colours become darker with increasing elevation. The advantages are that the lowland areas, which usually contain the greatest density of other information, remain light, providing good contrast. The colours normally follow a green-brown sequence, and can be extended by using white for the lowest elevation band, and purple for the highest band. This system is most effectively employed in medium-scale maps with a limited elevation range, and which therefore do not require a long series of steps in the total scheme. The scheme does not combine well with other relief representation methods. If additional contours are included (common with medium-scale maps) they are likely to have little contrast with the darker hues at higher elevations. Similarly, hill shading is made ineffective because of the reduction of contrast at high elevations.

2. *The higher the lighter*. In the converse arrangement, the darker colours are used at the lower end of the scale, and the colours increase in lightness with elevation. The required contrast may mean that relatively dull colours may be used at low elevations. The scheme is most advantageous in dealing with mountainous regions with relatively small lowland areas. If combined with hill shading it will provide a high contrast image at the highest areas and in regions of steep slopes. If a large number of bands is required, contrast between adjacent bands is likely to be poor, as the darkest hue at the 'bottom' of the scale must still be light enough for the map detail to be legible.

If the hypsometric information is used on its own as the method of relief representation, the stronger colouring of the 'higher the darker' scheme is usually more easily perceived by the map user. If the hypsometric colouring is combined with shading to give an overall impression of both elevation and slope, the higher the lighter can be very effective, but it is difficult to devise correctly from both graphic and technical points of view. A smooth and progressive change in both hue and lightness, allowing for the different extent of colour areas in different parts of the map and in regions of different relief characteristics, is one of the most difficult things to do satisfactorily in cartographic design.

## Limiting contours

The contours used to separate the vertical colour bands are usually included. They help in edge differentiation, particularly between adjacent narrow bands in light colours, but if too pronounced they also tend to reinforce the stepped effect visually. As they have to combine with a variety of hues they should be in a neutral colour, such as grey, and they should be as fine as possible.

## Bathymetric colour schemes

The oceanic areas, which are marginal to maps of land areas, differ in several ways from the land surface in terms of representation. They do not carry the same quantity of information as land areas, and colour convention (the use of blue) applies more strongly. Therefore, the usual colour scheme is to use tints of one or more blues on the 'deeper the darker' principle, leaving the lightest blue in the shallow water areas adjacent to the land. If a single dark blue is available, four to six bathymetric layers

can be produced. Two blues can be used to extend this range further to nine steps. If a second blue is not available, the deepest areas can be reinforced by the addition of a small percentage of red, magenta or grey.

## Legend

As hypsometric colour schemes are conventional, the vertical intervals and colours have to be identified in a legend. The simple form is a vertical series of 'boxes', usually with sea level at the base. If bathymetric colours are included, then a double scale is used. If the intervals change progressively, then the vertical scale of the legend should reflect the actual vertical distances covered by the layers, in order to give a correct impression of the nature of the scheme.

# SLOPE

Although it is possible to represent slope categories directly by classifying the land surface into zones within particular slope angles, on topographic maps slope is only represented by creating a visual impression, not based directly on an actual measurement. In this sense the form of the ground is portrayed by the use of light and shade, in the same way as an artist uses shading to give an impression of three-dimensional surfaces in a drawing or painting.

## Shading and shadow

Any three-dimensional object or surface, which is illuminated by a light source, will show tonal variations depending on the angle of each surface facet to the source of illumination. If the light source is strong and obliquely angled, the object will also cast a shadow (Fig. 151). The cast shadow is dependent primarily on the angle of the incident light, and varies accordingly. In the natural landscape, although a strong low sunlight will also result in cast shadows, the normal lighting is often diffuse. In such conditions, different tones will still exist depending on their relationship to the main light source. Although the terms 'shade' and 'shadow' obviously have a common root, it is wrong to confuse them in relationship to the use of shading to give an impression of a three-dimensional

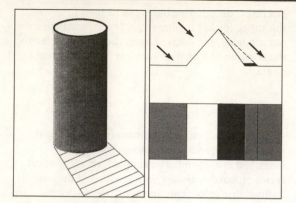

**151.** Shading and shadow

surface. Shadows frequently conceal variations in slope, rather than reveal them.

## Methods of shading

There are basically three methods of shading that can give an impression of a three-dimensional surface, although they differ in completeness of description. They are classified primarily on the basis of the assumed angle of illumination (Fig. 152). This may be either vertical or oblique, but in practice the most widely used method is a combination of the two. With vertical illumination, only slopes are shaded, and therefore only those parts of the surface that are perceptible slopes are distinguished. With oblique shading, the normal assumption is that the light source is from the upper left. With this method, level areas have an

intermediate tone (because they are at an angle to the light source); light-facing slopes are lighter than this, and slopes away from the light source are darker than this. This therefore separates all the main aspects of the relief by tonal contrast. In the combined method, the light-facing slopes and the dark-side slopes are tonally differentiated, but the level areas are left unshaded, thus combining elements of both the basic methods.

Each of these methods has advantages and disadvantages for both the representation of slope and the convenience of the map user. With vertical illumination, the steeper the slope the darker the shading, and this is constant for any angle of orientation of the map. However, it does not distinguish slope direction, and therefore whether a slope is up or down must be determined by the map user from other information. If used with good contours, the method can be very successful, particularly in areas characterised by many small local features.

Oblique illumination gives a complete visual impression of the surface, and the strongest relief effect. It has to be viewed from the normal map-reading direction, as otherwise its effect is lost or distorted to the point where the visual impression is inverted. Graphically it is the most complex, and the most difficult to combine satisfactorily with other map elements, as it is necessary to have a continuous light tone on the level areas (Fig. 153). Consequently, it is most successfully employed in regions with steep and mountainous relief, and relatively small areas of level ground.

Degree of illumination (%)

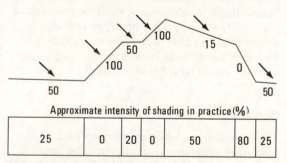

Approximate intensity of shading in practice (%)

| 25 | 0 | 20 | 0 | 50 | 80 | 25 |
|----|---|----|---|----|----|----|

**153.** Oblique illumination and shading

The combined method has the advantage that it does not require this continuous ground tone, and therefore interferes less with other map information, but the balance between the light and dark sides has to be finely judged if the three-dimensional effect is to be achieved.

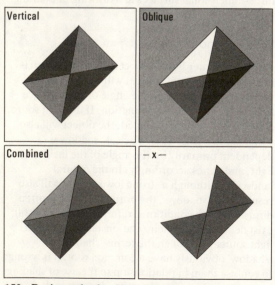

**152.** Basic methods of hill shading

Although the methods of relief shading are classified on the basis of the assumed source of illumination, they frequently depart from this in practice. With oblique and combined illumination, the assumed angle may be anywhere between the top of the map and the left-hand edge, depending upon the orientation of the relief forms in any area, and it may be varied slightly from one part of the map to another, for the same reason. Again, although theoretically in the oblique system the intermediate tone for level areas should be 50% (half-way between light and dark), it is more satisfactory in practice if it is made lighter than this, to avoid darkening the overall appearance of the map.

## Shading and scale

The relationship between the shading and the actual relief forms is affected by scale, and therefore the purposes the map is intended to serve. At large scales, where the detailed information will almost certainly be carried by contours with a small vertical interval, the main function of shading is to emphasise the major relief forms, which aids the interpretation of the relief of the area. Such shading is often lightly applied to give a broad effect. At medium scales, there is more likely to be a close correspondence between the shapes shown by the contours and the shading itself, but in mountainous areas the detail of the surface variation will be indicated more by the contours and other symbolisation than the shading. At small scales, where the contours, if used at all, are highly generalised and with a large vertical interval, the shading is likely to be more detailed than any contours shown on the map.

This scale difference also affects the methods of working and sources of information. At large and medium scales the contour image itself will be interpreted by the cartographer in order to represent the slopes. Even if not incorporated in the map, detailed contours and drainage are essential for the relief interpretation. For large scales the shading is then a form of generalisation, the land forms being simplified in terms of their major features. For medium scales in particular, a good and detailed contour image is an essential basis for the shading, for a degree of generalisation is present in both contours and shading. An understanding of the characteristics of the topography is just as important in generalisation in relief representation as it is for any other map element.

At very small scales, even detailed contours, if available, frequently give little information about complete topographic features. A single contour line crossing an extensive area is almost accidental in its relationship with the actual relief (Fig. 154). Either contour a or b might be the only one present. Shading along a single contour line is unlikely to produce any correspondence between slopes and the actual terrain. In such cases the limited contour information must be supported by a knowledge of the major forms of the topography, gained if necessary from the study of larger scale maps.

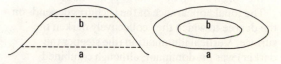

154. Vertical interval and shading at very small scales

## Shading and colour

The shaded image must not interfere with other map information. Consequently, a neutral hue is required, which has sufficient contrast to reveal the shaded forms, but is not so strong that it overpowers other map information. A neutral grey is most commonly used, although it may be given a slightly bluish or purplish bias, depending on the rest of the map design. When shading is combined with other methods of relief representation, then it must be sufficiently distinct in colour to avoid confusion or interference with the other methods, and again it must not interfere with the perceptibility of other features. In practice, if the contours are in brown, the shading is most likely to be in grey, and correspondingly, if the contours are in grey then the shading is likely to be in brown. It is clear that the maximum intensity of shading is limited not by any theory of illumination, but by the relationship between the shading and the legibility of the map.

With oblique illumination, the major classification of the surface into level areas, light-facing slopes and dark-side slopes can be reinforced by colour contrast. If the level or near-level areas are shown in a dull pale green, the light sides can be given a pale yellow or buff colour, and the dark sides left with the shading colour only, such as a grey. This makes use of the lightness value of different hues. In practice, the shading is normally completed in the neutral colour, and the level area and highlight colour added to reinforce the major slope contrasts.

## Hachures

Slope shading can also be carried out using patterns

of fine lines, called hachures. Although now little used, they dominated the early European topographic maps, during the period when such maps were produced by engraving. They preceded contours, as in the early period of modern topographic surveying it was possible to estimate slope categories by observation in the field, at a time when the huge amount of detailed measurement necessary for contouring seemed neither necessary nor desirable. Engraving was eminently well suited to the construction of patterns of fine lines.

The tonal appearance of the hachures depends on the degree to which they collectively darken the surface. Vertical illumination (the steeper the darker) was predominant, although combined illumination was also used in the most sophisticated form of hachuring. In addition, they show slope orientation and slope length, the three-dimensional surface being divided into slope facets which are approximated as slope angles. Therefore, hachures have four characteristics: direction, length, thickness and spacing (Fig. 155).

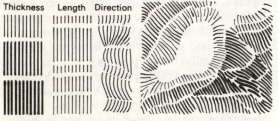

**155.**  Hachures (enlarged)

The hachure lines run parallel to the direction of slope, providing a useful visual clue to the map user. Their length varies according to the length of the slope facet they represent. They are nominally constructed on the basis of a given number of hachure lines per unit of distance (usually described as a number per centimetre) and therefore are equidistant in theory. The degree of shading of the surface is controlled by the thickness of the hachure lines, steeper slopes being represented by thicker lines. As the lines are fine and close together they give a visual impression of a shaded surface, but the various hachure densities can be matched against a legend description which may identify the degree of slope.

Although hachure systems with numerous slope categories have been described, in practice it is difficult to find examples on which more than three different slope classes can be clearly identified. With vertical or combined illumination the level areas are left unshaded, and in most schemes slopes steeper than 45 ° are combined into one group. They therefore distinguish gentle, medium and steep slopes.

Like other shading methods, hachures have advantages and disadvantages. If carefully produced, with a proper observation of the terrain, they correctly identify the actual breaks of slope, which helps to distinguish relief features clearly. Therefore, they are most effective in areas where there are abrupt and definite changes in slope, characteristic of many mountainous areas. They are least effective in regions of gently undulating topography, where many slopes are transitional, because the hachure bands break up the surface into discrete elements. In addition, on many engraved maps, where the entire map image was produced by engraved lines and patterns and printed in black, the hachures had to be interrupted to include other map information, and they tended to give a very heavy appearance to regions with accented relief.

Hachures based on the combined system were in many ways better suited to giving a visual impression of major as well as minor features, but the fact that the same slope class would have a different hachure density depending on whether it faced the light source or was on the dark side, meant that any direct interpretation of the hachures in terms of actual slope was no longer possible. In this form they more closely resembled the tonal shading which was introduced subsequently, giving a visual impression but not corresponding systematically to defined slope categories.

After contouring was introduced, it was possible to construct hachures by interpreting slope from the contours. This made for more consistent treatment, but in many cases meant that the hachures no longer corresponded to actual breaks of slope in the terrain, unless these happened to be revealed by the contours. Combinations of contours and hachures were sometimes used, in which the hachures operated as a simple slope shading method. At a later stage, mainly after the introduction of lithographic printing, hachures printed in colour were employed in very subtle combinations with other relief representation methods. Multi-colour methods used light coloured hachures on the light-facing slopes and darker-coloured hachures on the dark-side slopes. When well produced, the fine lines in relatively light hues gave a general surface colour, and at the same time did not interfere with other map detail.

## Individual slope features

The representation of a small steep slope (such as an

embankment) by a row of short parallel vertical lines is sometimes referred to as 'hachuring', but the term is inappropriate. Hachure systems classify all slopes in the terrain systematically, and are graphically more complex than simple line shading.

## COMBINED METHODS

Although some methods of relief representation are used individually, as any particular method can only represent either elevation or slope, but not both, maps in which the relief representation is a major component frequently use a combination of two or three methods. Common examples are as follows: heights and contours; heights, contours and shading; heights, contours and hypsometric colours; heights and hypsometric colours; heights, hypsometric colours and shading; heights and shading. As line and area symbols in different colours are graphically complementary (the line image being legible against a coloured background), the combination of

contours and shading, or contours and hypsometric colours, is not difficult to achieve. The combination of hypsometric colours and shading involves two sets of contrast in terms of area colour and tone, and can only be achieved successfully if the two elements are designed together, so that one does not destroy the effect of the other.

Hypsometric colour schemes on the 'higher the darker' principle inevitably conflict with the contrast requirements of shading, as their relatively dark colours subdue the light-facing slopes. If shading is to be combined with hypsometric colouring, then the coloured layers should be on the 'higher the lighter' principle in order to retain the tonal contrasts. The balance between colour bands and tonal variations has to be finely adjusted if both methods are to be employed in combination. Shading which is too dark will subdue or overpower the contrast in hue between different layer colours: shading which is too light will not have any visible effect if superimposed over strong hues. It is a classic example of the nature of cartographic design, in that choice of method is only a starting point; how it is actually applied in a particular map is the most critical factor.

# 18 ORTHOPHOTOMAPS

## Historical background

The idea of incorporating aerial photographs in a map, or using them in some way as a map substitute, did not arise in response to any one problem, but resulted from a number of requirements and developments. To some extent it was largely a technical problem – how to modify the structure of the photographic image so that it would have the same scale characteristics as a map – but the desire to solve this problem increased in relation to the perceived advantages of using aerial photography directly. Comparisons between aerial photography and maps stressed the following points:

1. There was a need to produce up-to-date maps in areas where it was impossible to carry out a ground survey, which was especially important for military surveying.
2. It was highly desirable to make more effective use of the large quantity of available photography, which could be employed relatively quickly.
3. It should be possible to reduce the time gap between taking the aerial photographs and producing something usable for the consumer, especially as conventional maps took a long time to produce, and were relatively expensive.
4. Specialists could carry out their own interpretations on aerial photographs, but it would be an advantage if they could be related directly to existing maps.
5. Some types of topography were poorly represented by maps, especially those with complex vegetation and land use.

It was also frequently suggested that the aerial photograph is inherently 'more informative' than the map, or even 'more accurate' because it does not operate through subjective generalisation.

## Aerial photographs and orthophotographs

Like most pictorial images, the aerial photograph has a central perspective, and scale changes outwards from the centre. In addition, if the camera is tilted when the photographs are taken the resulting images will be distorted. In areas of high relief, the variation in elevation, and therefore distance between the camera and the ground surface, also affects the scale and relative position of features (Fig. 156). Although this distortion may not be noticeable within a small area near the centre of the photograph, it is impossible to 'fit' a series of aerial photographs to a map of the same nominal scale. For normal photogrammetric plotting, overlapping pairs of photographs are used, so that a stereoscopic image is viewed by the photogrammetrist. This has the great advantage that the ground surface can be seen in three dimensions, and therefore the relief of the surface can also be measured.

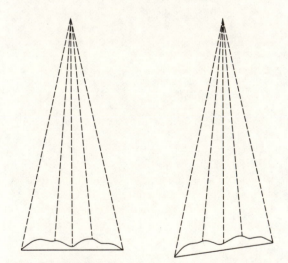

**156.** Scale differences in vertical and tilted aerial photographs

## Orthophotography

Although the consistent scale differences in the aerial photograph can be removed by optical processing known as rectification, the major difficulty is to compensate for the scale variations and displacements introduced by relief. In effect this can be done by using the height information measured in the stereo-model to reconstruct the aerial photograph. To do this it is scanned in parallel narrow strips, and thus re-composed with the scale differences eliminated (van Zuylen 1969). At the same time it is also possible to use this elevation model to generate a set of contours at a fixed vertical interval.

For areas of flat or level terrain, rectified aerial photography can be used to produce a photomap. In such areas contours are not usually included, and height information may be limited to a network of height points. From the point of view of subsequent reproduction and processing, these photomaps are treated in the same way as orthophotomaps.

The first stage is the production of the orthophoto image, the fully corrected image that is used as a basis for the orthophotomap. The orthophoto does not contain any coordinate system information, or names, and therefore usually the minimum requirement to produce an orthophotomap is to add a grid or graticule and whatever names are needed. To produce a map, height information must also be included, normally consisting of point height values and contours. Therefore, an orthophotomap requires first the geometrical processing of the aerial photographic image, and then some additional cartographic input to provide the locational framework and other necessary information.

Orthophotomaps have been produced in a great variety of forms, largely as individual topographic maps, but also as a basis for cadastral and other information. Therefore, they have to be considered as a varied group, and not as a single uniform type.

## Photograph and map

Although aerial photographs and maps can often be used for ostensibly similar purposes, the comparison between the two as sources of topographic information is not as straightforward as may first appear. The claimed advantages of the aerial photograph include absolute surface scale, complete information, no subjective generalisation, and therefore consistency. In fact, the informational detail depends entirely on the scale and resolution of the photography, and the quality of the image. The essential difference is that the photographic image can be interpreted directly by the user, and therefore a skilled interpreter can concentrate on the information relevant to his enquiry.

In contrast, a map classifies and completes the selected information. Its surface scale is relative, in the sense that the dimensions of symbols are controlled independently of overall scale. It includes information that cannot be present on aerial photography, such as names, coordinate systems, boundary lines and so on. It deals effectively with well-defined features, but less effectively with irregular and complex distributions.

Consequently, the object of producing orthophotomaps is to combine the advantages of both types of information. Although in theory this could be attempted either by adding cartographic elements to a photographic base, or by adding photographic elements to a cartographic base, current practice is dominated by the former.

# CARTOGRAPHIC AND REPROGRAPHIC PROCEDURES

The cartographic aspect of orthophotomap production can be said to have three main objectives. First, to make the photographic image, which of course is a continuous-tone image, reproducible by standard printing and other reprographic processes. Second, to make the image more legible, especially at those points where it may lack contrast or be inconsistent in visual appearance. And third, to make it more informative by adding additional information, and completing and classifying features where necessary.

In practice many of these possible modifications overlap. For example, one procedure used to make the continuous-tone photographic image reproducible by lithographic printing can also be used to modify the image to make it more legible (by increasing the contrast) and to introduce feature classification.

The degree of cartographic processing can extend from the addition of a small amount of information to a complex processing of the initial image in several colours. In this sense, orthophotomaps, like conventional maps, may be graphically simple or graphically sophisticated. This in turn creates its own problems. If the photographic image is extensively modified and a great deal of information added, the orthophotomap may be highly informative, but it then begins to lose its perceived advantage over the conventional map of being quicker and cheaper to produce. Indeed, the most subtle and complex orthophotomaps are probably among the most sophisticated of all cartographic products, both in graphic and technical terms.

## Reproducibility

The initial technical requirement was to make the orthophotomap reproducible by standard printing and reprographic processes. Although photographic reproduction could be used to generate individual copies, it is highly uneconomic for numbers of large images, especially in colour. The central problem was to make the continuous-tone photographic image reproducible by lithographic printing, diazo copying, or some other economical method. The obvious solution was to convert the continuous-tone image to half-tone, and then treat it like any other half-tone or tint image. But it was realised that this affected the resolution of fine detail, which was one of the perceived advantages of the orthophoto, and resulted in loss of contrast. Consequently, many methods were tried to find ways of making the image reproducible without direct conversion to half-tone in the normal manner.

The principal alternatives were between conventional photography; other reprographic processes such as diazo which have some tonal reproduction capacity; conventional lithography with half-tone; conventional lithography without half-tone; and other printing processes, such as collotype, which could reproduce tonal variations without a half-tone process.

# LITHOGRAPHIC REPRODUCTION OF CONTINUOUS TONE

Many avenues were explored in order to find a method of reproducing the continuous-tone image without loss of contrast or image detail. These included screenless printing, in which the continuous-tone image is exposed to a specially prepared lithographic printing plate, capable of reacting to the variation in density of the tonal image; and photographic processing of the continuous-tone image to convert it into discrete elements, which could then be reproduced by a normal lithographic plate. This could be attempted either by using special processes, or by using specialised photographic film emulsions.

## Screenless lithography

From the 1950s onwards there were several attempts to develop a method of forming a reproducible

image directly on a lithographic plate without using a half-tone screen (Keates 1978). This was regarded as of great commercial potential in general graphic arts colour reproduction, and the initial experiments were not directed at maps as such. Progress in this area reflected the development of the pre-sensitised lithographic plate, normally making use of polymers, both oxides and resins. The basis of the procedure is that as the developed continuous-tone image consists of minute 'clumps' of silver-halide particles, the perceived image density in terms of tone is dependent on the aggregation of these fine discrete units. Therefore, the development of the image on the printing plate reflects the tonal density of the original image, but also breaks it up into image and non-image elements.

Several companies successfully produced such a plate, but none was commercially successful. The Howson-Algraphy Contone Alympic plate was the latest in this line, producing an image with an extremely fine structure, but although highly effective as a means of continuous-tone reproduction, it failed to attract a large enough market to be maintained.

## Specialised photographic emulsions

The next alternative was to copy the original continuous-tone image on to a special photographic film, which would bring about a modification of the tonal characteristics. Kodak produced the Integral Random Dot (IRD) emulsion which makes it more 'grainy'. This high-contrast lith-type film produces an image consisting of randomly dispersed hard dots.

## Specialised photographic processing

Any change to the photographic image can only be brought about by modifying the densities of the original image. This can be carried out by photographic masking, by variations in exposure, or by making use of the properties of different photographic emulsions. Most methods involve all three.

## Photographic masking

The continuous-tone photographic image actually consists of tiny clumps or groups of dense points, resulting from development. The object of photographic processing is to emphasise this characteristic in order to introduce a 'texture' or graininess that can be reproduced directly on the lithographic plate. If this tonal range is exaggerated but maintained, then the resultant image can be called 'phototone'. The same basic procedure can be extended in order to bring out the edges marking the contrast between light and dark areas, and this is termed 'photoline'.

To produce phototone, the normal procedure is first to make a continuous-tone negative with a density range limited to 1.2, and from this to produce a positive copy with a density range limited

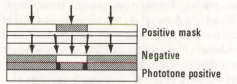

**157.** Positive masking for phototone

to about 80% of the negative. To produce a positive image, the negative is placed face down on lith film, and the positive mask registered on top of the negative, emulsion side up (Fig. 157). The two film images are thus separated by the thickness of the bases. The resultant positive image is a function of the combined densities of the negative and the positive mask. This retains the tonal contrast, but maximises the change in contrast along edges. Exposure is critical, and the whole process has to be carefully controlled. The effect is to produce a discontinuous structure of light and dark areas, which on the lithographic plate provide the printing and non-printing elements (Fig. 158). For the procedure to be successful, a high-quality negative is essential; the positive mask must have the correct density and be in perfect register, and exposure must be exact.

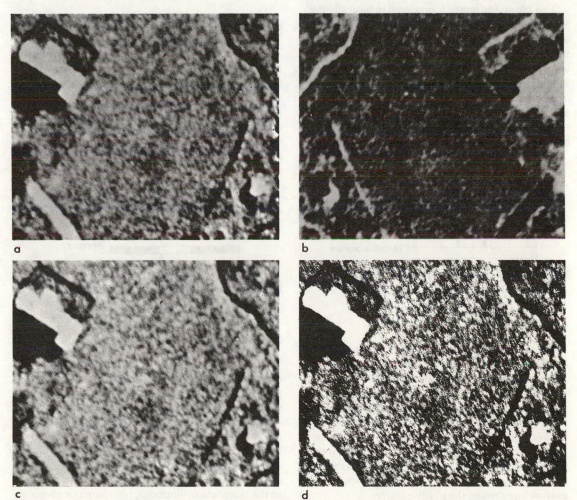

a  b  c  d

**158.** Phototone production by positive masking (reproduced in half-tone and enlarged)

## Unsharp masking

In all contact copying, a 'sharp' image can only be produced if the original image and the unexposed emulsion or coating are in close contact. If the images are separated, for example by the thickness of the film base, then contact is 'unsharp', and there will be some diffraction of light rays at edges. In unsharp masking this is deliberately exploited in order to emphasise the 'edges' between areas of different density, so that these print as lines or outlines. Therefore, they are then reproducible by any standard printing method. Theoretically, this could produce a complete 'line' image directly from the photography, but this extraction only works if there is a constant level of contrast along the edges of all the desired features. Because of differences in photographic tone, and of variations in light and shadow, this consistency is rare over a whole series of aerial photographs.

This photoline image is produced in basically the same manner as phototone, but both the negative and positive mask are prepared as high-contrast images (Fig. 159). The positive mask is also separated from the negative by a transparent spacer, to increase the light diffraction. The width of the 'outline' is affected both by the degree of separation and the duration of exposure. It can be increased if the light source is angled, and the combined image rotated.

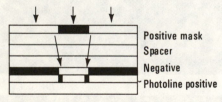

**159.** Unsharp positive masking for photoline

## Other photoline methods

Various other methods have been tried to produce line images directly from tonal photographs. The Klimsch *Variomat* can be used to construct a photoline image. The continuous-tone negative is placed in the copyholder, and the dense positive mask (the converse image) placed in front of the unexposed film in the camera back. A transparent or low density area on the negative transmits light that will be blocked by the positive mask, but as the light rays are slightly displaced they also produce a narrow line around the positive image.

Equidensity film is a special emulsion developed for the extraction of a particular density range from a tonal image. The film is given a double characteristic curve, that is, both light and dark areas develop equally in the same film. It requires normal exposure but special development. If used in contact copying of an aerial photographic image, the light areas of the original positive image appear on the negative as dark areas, but the dark areas of the original image also appear as dark areas on the negative. On development a narrow band is left between them. This band is the equidensity, lying about midway between opaque and transparent, or high and low density. At the edge of any tonal area there is a general change from light to dark: the equidensity film process extracts a narrow line from the middle of this band, where it is midway between light and dark tones. If this density happens to coincide with the density of a whole area on the photograph, then this will reproduce as a solid area.

## Shadowline

This is a variation on photoline, using high-contrast negatives and positives. The mask is deliberately displaced in one direction, normally to the north-west, creating a fine line around the south-east edges of areas (Fig. 160), giving an effect similar to the appearance of shadow. It is a fairly simple operation, but again its success depends on the degree of contrast between topographic features on the original image.

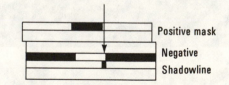

**160.** Positive masking for shadowline

## Supplementary photographic images

The continuous-tone image of the orthophoto, reproduced directly or in half-tone, together with phototone and photoline, provide the principal elements for representing both the tonal characteristics and the emphasis of edges of areas or linear features. Just as in the conventional map the line images can be complemented by the addition of solids and tints (area colours and patterns), separated into different colours for greater contrast, so the complete orthophoto, phototone or photoline can provide further supplementary images and colour combinations. Although these may be reproduced in colour, it is also possible to use tint screens to control the overall strength of the image.

## Density separation and colour addition

Although it is possible to make orthophotomaps from panchromatic or false-colour aerial photography, the density processing of orthophotomaps is normally operated achromatically. Even so, it is possible to print the tonal image in any chosen colour. Using further photographic processing of the original tonal images, or phototone or photoline, different parts of the image can be distinguished by different colours.

In many areas the high vegetation will often appear on the photographic image as darker in tone than low growing vegetation, such as grassland. If the orthophoto negative is copied on to lith film, with controlled exposure, the effect is to divide the whole image area into either dark or light components. On the normal negative, the 'dark' areas will be light in tone, and therefore on the lith positive copy, these will appear as solid dark areas (Fig. 161). The high-contrast lith film will not record the intermediate densities between light and dark. If this image is then used as though it were a negative, and copied to the printing plate, it will reproduce the light tones of the original image, that is the converse of the positive. From the same positive a contact negative is made, and when printed down on the lithographic plate this will represent all the dark tones of the original image. If

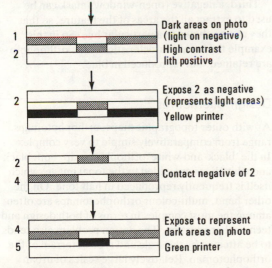

**161.** Separation of 'dark' and 'light' areas of continuous-tone image

the light tones are printed in a light green or yellow-green, or even yellow, and the dark tones printed in dark green, then their contrast is enhanced by a colour contrast.

Images derived from these high-density masks no longer contain the original photographic tonal detail of the orthophoto. Therefore, an alternative method is to use the masks to separate the orthophoto image into different areas, reproducing these in grey or in different hues.

The printing colours chosen for the orthophotomap may also introduce difficulties in controlling the background or land colours. For example, if a brown is used for contours, it may well be too dark to be used directly as the printed colour for any of the phototone elements. It is in these circumstances that the 'solid' phototone or high-density mask image may be modified by using a percentage tint screen to control the density. In this way, 'light' areas without tall vegetation could be shown as a pale brown in contrast to the vegetation green.

## CARTOGRAPHIC COMPONENTS

The images referred to so far are all derived from the original orthophoto by special photographic processing. In order to produce an orthophotomap, other information must be added which is not present on the orthophoto, and these additional components are produced cartographically, using the standard methods of scribing, masking, lettering and so on. The minimum requirement is to include additional essential information, such as the graticule or grid, names and contours. But cartographic modification can be extended to replace inconsistent or incomplete photographic images with map symbols, and to improve contrast and legibility by introducing additional contrast and colour.

This also raises the question of cartographic design, because the cartographic additions, along with the tonal images, have to be composed, in a design sense, like any other map.

## Cartographic design

Orthophotomap design is fundamentally different to that of other maps because there is an overall continuous tonal image, which may range from light to dark at any point. Normally the starting point in map design is the 'white' paper, and contrast is judged and controlled in relation to this. In an orthophotomap, the starting point is the tonal image, and this presents quite different problems. Like all other map design problems, it is essentially a question of the control and manipulation of contrast.

### Line images

In the most simple type of orthophotomap, which is normally achromatic, the orthophoto appears as a 'grey' of various densities between dark and light. The line images, including lettering, can therefore only show against this background by being either darker or lighter: that is, by being introduced as either black or white. Much depends on the overall tonal density and range of the orthophoto base. If this is generally dark, then, for example, contours may be added as white lines, by using the contour line image as a positive mask against the orthophoto negative. However, any contours that appear in 'light' areas will have poor legibility. This method is used in many achromatic orthophotomaps, occasional deficiencies being accepted against the overall economy and simplicity of production.

For orthophotomaps in colour, the possibilities of contrast are increased. Even light-coloured lines can be made to contrast with a grey or other background colour. Although dark hues can simply be superimposed on the orthophoto image, better contrast is normally obtained by first masking out the tonal image and then 'inserting' the coloured line. This procedure of employing a 'hold out' mask is also referred to as undercolour removal. As before, the process is quite simple, as the line positive used to reproduce the coloured line is also used as a positive mask against the orthophoto or other tonal negatives (Fig. 162). Where this undercolour removal is used to delete the image on more than one coloured plate, register during processing and register during printing are crucial.

In some cases, particularly with the addition of lettering, the background image removal may be deliberately exaggerated in order to increase the contrast further. In this case the positive masking of the negative is enlarged, usually by making a 'spread' version of the positive, so that the mask is slightly larger than the printed positive image. In this way, the names, for example, can appear with a slight 'halo' effect, outlined in white, which reinforces their contrast with the background.

### Symbolisation and colour coding

The line image components such as grid lines, contours and lettering are all additional information. Apart from this, cartographic symbolisation may be added to clarify or even complete images that are unsatisfactory on the orthophoto.

In many cases the visual appearance of a particular feature on the orthophoto varies from one part of the image to another, and thus is inconsistent. Water areas are a typical example, as these may be either dark or light, depending on their surface characteristics in relation to the incident light. Several possibilities exist.

A positive mask of all water areas can be constructed cartographically, and used to delete the tonal image. The negative of this mask can then be used to introduce a light-blue solid, or a blue tint, so classifying all water areas by colour symbolisation. This replaces the original tonal image completely, but makes interpretation easier.

A second possibility is to leave the original image as it is, but to add a pale blue tint to all water areas, in order to distinguish them more clearly.

Third, a negative (open-window) mask can be used to extract all the areas of the feature, so that they can be shown in a particular hue. So in this example the tonal characteristics of the orthophoto are retained, but reproduced in blue.

### Design alternatives

As with other topographic maps, orthophotomaps range from comparatively simple to very complex. In the 'black-and-white' orthophotomap, both black and white lines are added to the tonal image, which itself is frequently reproduced in half tone. On the other hand, multi-colour orthophotomaps are often among the most complex in terms of both design and technical production. The design problem also tends to be affected by the scale and type of terrain of the orthophotomap. Relatively large scales of urban areas, for example, are frequently left as simple tonal images, and other information added with

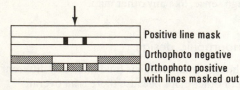

Positive line mask
Orthophoto negative
Orthophoto positive with lines masked out

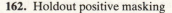

**162.** Holdout positive masking

black lines. On the other hand, medium-scale multi-colour orthophotomaps of terrain with complex vegetation/land use characteristics may involve extensive processing in order to improve the contrast and legibility of different features. Some possible lines of approach to orthophotomap design are as follows.

1. Contrast for the line image can be improved by weakening the tonal image, so that it is relatively light. This tends to reduce the legibility of the orthophoto, thereby decreasing the value of the photographic image for interpretation.
2. Contrast can be improved by increasing the weight of line image elements, so making them apparent even against relatively dark tonal backgrounds. This can be effective, but produces an unattractive map.
3. Contrast with the line image can be improved by reproducing the orthophoto image in colour, normally using light hues. This can be very effective, but always runs the risk that the increase in overall lightness value for the orthophoto will decrease its legibility. Therefore, the balance of the different hues is critical.
4. The deficiencies of merely lightening the overall image can be reduced by reproducing the orthophoto image in black or grey to maintain the legibility of detail, but using photographic and cartographic masks to add colour to particular features. As long as there is not too much colour superimposition, this can be very effective, as it combines both the tonal properties of the orthophoto with the symbolisation of the map. If holdout masking, photoline emphasis and colour tint addition are carefully used, then it is possible to make maximum use of the photographic imagery and yet compensate for some of its inherent deficiencies.

## Technical and cost limitations

As the design and informational potential of the multi-colour orthophotomap was explored, it became clear that the objectives could change from the initial impetus of a 'cheap' and rapidly produced orthophotomap, to an elaborate representation of topographic features and terrain characteristics. In addition, although specialised photographic and cartographic processing could greatly improve the image, this could only be done by accepting the additional technical work and costs. The difficulty of this is not only that complex processing takes resources, but much of the processing needed for orthophotomaps depends on careful control, especially if the same specification is to be applied to many different maps. Despite the many interesting examples of photographic processing, in practice many national agencies have relied extensively on the straightforward half-tone reproduction of the orthophoto image, with a minimum of cartographic addition.

Despite its limitations, the half-tone process is well known, and the equipment for it readily available. Many agencies have experimented with variations in half-tone reproduction, particularly with different types of screen and screen ruling (van Zuylen 1974). In order to reduce the interference with detail legibility, a very fine screen would seem to be advantageous. The problem with this is that such fine screens – up to 100 lines per centimetre – can in turn present difficulties in plate making and printing.

## Orthophotomap applications

Orthophotomaps have been produced in many different countries, and they range in scale from about 1 : 2000 to 1 : 100 000. In Belgium, the 1 : 10 000 series has the orthophoto and names in black, with a white grid. In Germany the topocadastral base map has a black half-tone orthophoto, black contours and white roads and names. In South Africa, the 1 : 50 000 uses a dark grey orthophoto, with black, blue and red lines, but not masked out. A 1 : 100 000 scale series of Saudi Arabia employed a brown half-tone over yellow, with blue line and tint (to make green), black names and a purple grid.

The pioneer of the multi-colour orthophotomap as part of the standard national topographic map series was the United States, and the 1 : 24 000 quadrangle sheet of the Okefenokee was the first of the sophisticated orthophotomaps for civilian use. This introduced combinations of land tone, photoline, half-tones of vegetation and water areas, brown contours and black names with holdout masking.

The inclusion of orthophotomaps as part of the standard national topographic series has been the exception rather than the rule. In many cases, the official mapping agencies have tended to employ them where coverage was needed quickly, either for reconnaissance, or because of the rapidity of change in urban areas; or for specialised base maps dealing with forestry, urban development, planning, cadastre, or the monitoring of agricultural change.

# 19 TRAVEL AND RECREATION

In addition to standard topographic map series, and individual topographic maps for particular areas or projects, there are also many kinds of topographic map produced for special purposes. Some of these, for example road maps, street guides and maps for tourism and recreation, are among the most familiar of maps used by the general public.

## ROAD MAPS

Road maps cover a great range of scales and individual types, from the comparatively small-scale 'outline', in which often the roads are shown diagrammatically, to elaborate representations of both detail and topography. The latter tend to be most fully developed in regions with high densities of population and a well-developed road network. Road maps of this kind overlap into the 'tourist' category, as many of them are designed for both local and tourist use.

The small-scale maps, usually at 1 : 500 000 or smaller, are essentially for route planning over long distances. They are necessarily restricted in the amount of detail that can be shown in densely populated areas, and concentrate mainly on major through routes. A map of this type is usually included in the commercially produced road atlas. Graphically simple, they contain little information apart from the major road network and names of places.

The detailed road map, map series, or road atlas encounters problems with scale. A uniform scale is needed to give a consistent representation of different areas and distances, but the scale that is suitable for this – usually between 1 : 100 000 and 1 : 300 000 – is too small for urban areas, which need separate treatment. This may be done by adding insets on sheets, or additional maps in an atlas.

Most road map series and road atlases are commercially produced, and in many European countries there is considerable competition for both national and international markets. In North America the position is rather different, as the provision of 'free' maps by petroleum companies has tended to distort this aspect. Many of these maps are cartographically simple, but the large areas involved, relatively low population densities and limited major route network of much of North America perhaps favours this approach. Even so, many state, provincial or recreational maps are largely road maps with additional information, similar to the tourist maps of Europe, but often smaller in scale.

### Sources of information

Road maps are essentially specialised versions of medium- and small-scale topographic maps, and the basic source of information is the topographic map series of any country. This raises problems of copyright – in which practice varies considerably – and also suitability of scale. It can also be affected by the maintenance and up to dateness of the official map series.

Using this information as a basic source, the road map content is selected and re-designed. The symbolisation of road classes and categories is a primary requirement, as is the selection of places to be named. The generalisation of road networks is critical, and poses interesting problems in simplification. The relative positions of road junctions may have to be displaced to demonstrate clearly their connections, allowing for the comparatively large symbol dimensions required (Fig. 163). In addition, there is considerable variation in the nature and amount of other information that may be needed.

Because road networks are constantly being renewed and developed, the review of detail is an important part of the map production requirement. As with topographic map series, to some extent revision information can be obtained in advance from government departments concerned with roads

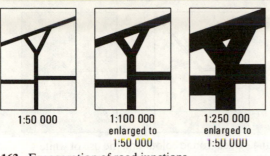

| 1:50 000 | 1:100 000 enlarged to 1:50 000 | 1:250 000 enlarged to 1:50 000 |
|---|---|---|

**163.** Exaggeration of road junctions

and the built environment. But plans and proposals do not always coincide with what conditions exist at a given date, and new editions of standard topographic maps are not likely to appear at the desired time interval. In small European countries, at least, it is possible to traverse every route at least once a year, checking ground detail against the map. Changes to roads, road junctions, and through-route systems are highly important to the road traveller. Therefore, the rate of revision has to be much higher than that usually given to small-scale topographic map series, and in most cases there has to be an annual re-issue of each map.

This need for constant revision also influences the overall design. The more sophisticated the map design, the more complex is the technical revision. Therefore, there is a struggle between devising the most attractive and informative map, which is likely to be cartographically sophisticated, or concentrating on a more simple map, which can be revised and produced economically. In this respect, the road map has some similarity with the chart. For example, double-line road casings with a different coloured infill are widely employed for major roads, as these can be suitably emphasised and distinguished for the user. But corrections to such a symbol are technically more difficult than changing a single line symbol, which will usually affect only one colour. Therefore, some producers use strong double-line symbols in a single colour, without a black casing, as these are easier to produce and correct.

Information about roads is supplemented by additional information about travel directions and facilities. It is in this respect that the basic road map overlaps into the general 'tourist' map category, including information of interest to visitors as well as travellers. Approaches and connections to major interchanges, emergency services, telephones in non-built-up areas, toilet and parking facilities, and in remote areas fuel supply and overnight accommodation, can all be important and useful

items. But these also change and develop, and the collection and checking of this information also has to be carried out. Motoring organisations usually concentrate on providing this information, and indeed in some countries such organisations are among the major suppliers of road maps.

## CONTENT AND DESIGN

### Roads

Like other topographic maps, the design objective for a detailed road map is to provide a series of visual levels or layers, separating the total information into grades of importance. Obviously, roads and named places are the first priority, usually reflected in the use of high-contrast hues and strong line symbols. In congested areas, the difficulty is to avoid a close pattern of inter-connected and confusing lines, which excessive use of high-contrast line symbols can easily induce. Multiple carriageways are easily represented by multiple line symbols, distinguishing dual carriageways from single, and of course these are necessarily exaggerated in width. Red, orange, green, brown and even bright blue are used to give prominence to these large line symbols. Less important roads may be represented in brown, pink or yellow, with the least important left with black or grey lines only.

Although this method of multi-colour symbolisation is well established, a lighter image can be obtained if the topographic 'outline' is reduced to grey, limiting the direct use of black to names and important point symbols. This also helps to avoid the interruption of black road casings for names and other detail on the black plate, and helps to 'lift' the place names to a higher visual level. In many cases, the symbolisation of a road by a coloured line only also helps to achieve this contrast.

Choice of road symbol dimensions and colours is affected by complex factors, including the habits of the road user and familiarity with existing maps; the perceived relative importance of different colours, especially red; and the type of road classification used nationally and therefore familiar to the map user.

An examination of a variety of road maps of this type demonstrates the importance of small differences in the specification of individual symbols, which have a cumulative effect on the overall weight and appearance of the map image, and the importance of not using high levels of contrast too freely.

## Topography

One of the major contrasts between different designs lies in the treatment of the 'background', which can range from simple to elaborate. Of course the simple solution is to leave it blank, indicating only some defined areas, such as built-up areas, national parks, woodlands, or nature reserves, by suitable patterns or solid colours. Such an approach fails to make use of white in contrast to the printed colours, and also tends to be unattractive aesthetically.

The choice often lies between using the surface as a contrast basis for the roads, or introducing some representation of relief. Although full layer colouring, once widely used on the 'tourist' versions of medium- and small-scale topographic map sheets, is no longer fashionable for road maps, hill shading of some kind is often introduced. Although this gives more information about the topography, there is little doubt that it also serves to add visual variety and form to the landscape, and indeed to provide contrast between regions. In most cases a simple combined shading is given in a neutral background colour, such as grey, although the exact specification of this grey can be important in the overall visual effect. The addition of a highlight yellow or buff can serve to enliven the impression, without reaching the point where it would interfere with the contrast level needed for the road colours. When unsupported by contours, at these scales the shading needs to give a complete representation of the surface, and for this a fully developed combined shading is necessary to give consistent form to the landscape.

The other approach is well represented by many road maps produced in the Netherlands, where obviously the representation of slope by shading would be pointless. By adding a very light grey background tint, and by using a very pale blue tint for water areas, even the minor roads indicated by 'white' with a grey line casing stand out very slightly from the background. The addition of such a light tint, which is barely noticeable without examination, also lends a solidity to the topographic surface (Fig. 164). The use of a blue-black for names and outlines of point symbols, also gives a different impression to the map as a whole, showing the value of reconsidering carefully the apparently 'obvious' design features of most maps. It is in the treatment of such low-contrast components of the map that the

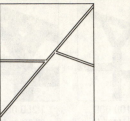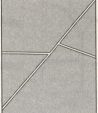

**164.** Background colour and the use of white

overall design quality is often revealed. Variation in design is not simply a matter of choosing different methods, but in the particular way any method is applied to a given problem.

# SPECIALISED ROAD MAPS

The normal road map, as a version of the topographic map, concentrates on giving what may be termed factual information about roads and attendant facilities. It describes what is there, and leaves the user to make decisions about appropriate journey routes. In many cases, what the motorist is most concerned with is total journey time and ease of travel, which of course is often a function of traffic density. Many attempts have been made to include this variable as part of the road information, and this approach has been thoroughly described by Morrison (1980, 1981). The difficulty is that though it is at least theoretically possible to obtain this information for a given period, it varies with time of day and season, and therefore it is very difficult to provide any general information that can be applied to all journey time considerations.

## Navigation

Road maps can be used in various ways, and in this sense are multi-purpose maps. Obviously, motoring to a distant destination, especially if it is unfamiliar, will usually involve pre-planning of the route, and possibly the memorising or noting of a sequence of road numbers or codes, or places to be passed *en route*. The objective is to anticipate the requirements of the journey, and may include deciding on stopping places for meals, etc. For the tourist, the planning of a route may well involve considerations of places to visit, or simply scenic views.

If there are many possible alternative routes, then the map may also be used during the journey for actual navigation, to identify landmarks in unfamiliar territory, or to make decisions about where to turn. As this cannot be done directly while driving, many users regard this as the least effective part of road map use, and it reaches maximum difficulty in built-up areas that have a high density of information and also traffic.

Many avenues have been explored to overcome this difficulty, based mainly on the principle that as the driver must use his vision for driving, some other means of obtaining or checking information is needed. Experimental developments include database systems which in theory can control the navigation of the vehicle by both planning and following a chosen route, thereby taking away the problems of visual navigation. Although such systems are relatively meaningless for long-distance travel on single routes, they are obviously intended to be most useful in urban and built-up areas. Whether their expense, response to minor road changes and navigational 'accuracy' can be justified, remains to be seen.

Other possible methods include moving-map displays (derived from aircraft navigation) which continuously project the immediate map area on to a screen. The expense of such systems is likely to be prohibitive for private car users. Even so, continued interest in this problem highlights the difficulty of interpreting a map for navigational purposes, as compared with route planning.

## TOWN PLANS AND STREET GUIDES

A number of different but related maps and plans can be grouped under this heading. The traditional street plan or street guide is essentially a road map of an urban area. At one extreme it may be no more than a basic outline of buildings with street names, often accompanied by an index of names and places, either on the reverse of the map sheet or as a separate index in an atlas. Frequently, such a plan tries to include the maximum detail at the smallest possible scale. At the other extreme, town plans designed specifically for tourists or visitors are not only intended to serve as guides to buildings and streets, but also to provide other information about interesting places, sights and services. This is the tourist equivalent of the recreational/road map of rural areas.

Given the density of information in an urban area, relief is unlikely to be represented, apart from very prominent features. But the treatment of urban areas is essentially a question of scale, generalisation and the use of contrast. Although tourist maps usually concentrate on city centres or particular city areas, the general town plan faces similar problems in composition and treatment.

Assuming that many visitors will need to find their way to particular locations either on foot or by public transport, a clear indication of the basic pattern of buildings at street level is essential. Although obviously this can be achieved most easily at a sufficiently large scale, for large towns and cities such a scale becomes too unwieldy in terms of sheer size. Consequently, the problem is to provide the information at the smallest possible scale for the user. This generalisation problem cannot be solved by omission; a visitor passing through the city centre needs to be able to identify individual streets and open spaces. Therefore, simplification and exaggeration, especially of street frontages, are essential. To include street names at legible sizes, street widths must be exaggerated wherever necessary, but the correct relative positions of buildings and the angles of street interconnections must be maintained.

The solution to this problem is invariably based on reducing the line image, and making use of area colour as far as possible. Unlike the smaller scale road map, where the linear road network is coloured against a plain or neutral background, in many town plans the buildings and other areas are coloured, leaving the roads principally 'white' – although some road colour may be added in particular cases. Exaggeration of street width in combination with showing buildings in a pale, bright colour (solid or tint) also means that street and other place names can be superimposed in black, giving high visual contrast (Fig. 165). These names (and indeed other point symbols) can overlap on to the coloured areas without introducing any confusion. Given that the shapes of urban areas are familiar to most people, the building colour may be comparatively light without losing the contrast needed for clear identification. Although a pure yellow may be too weak, a yellow-ochre, yellow-orange, pale brown or pale purple are usually sufficient. Of course different colours may be used to indicate classes of building or classified areas.

The other difficulty is one that applies to all town plans and maps. Buildings are three-dimensional, and roads and other means of movement and activity extend in three dimensions, not two. This is the weakest point of the two-dimensional map, as all

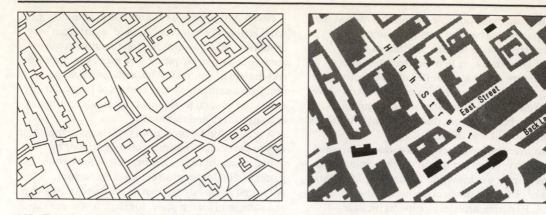

**165.** Exaggeration and displacement in the representation of built-up areas

levels of movement have to be represented on a single surface. Although underground railways may only be indicated in terms of their 'surface' stations or terminals, the vertical extent of buildings is a major factor in landscape character, and also the identification of individual buildings and places. Not surprisingly, attempts to include a three-dimensional 'view' of urban areas have often been favoured, despite the difficulties of representation.

## Perspective views

These urban landscape views, or 'prospects', which often reflect the oblique view of the terrain employed centuries ago, encounter problems with choice of viewpoint and angle of view. Tall buildings interfere with lines of sight. Judicious exaggeration, simplification and omission are essential. Even so, such representations no longer depend on symbolisation and interpretation in the normal cartographic manner, but involve direct representation of the appearance of the buildings from a given angle. The less obvious problem lies with the presentation of the terrain. Many of these representations are still essentially plans, in the sense that the ground surface is treated as a plane. If they attempt to include the differences in elevation of the surface, that is the land relief, then this further complicates both perspective and lines of sight.

This type of three-dimensional view of an urban area obviously demands a great deal of detailed work in construction, and is not easy to revise. In addition, the indication of other types of information, such as names and the location of services, becomes more difficult against the complex background of building shapes and details. Although many users find such 'maps' initially

appealing, their value is less conclusive in actual use in a built-up area.

## Names, symbols and identifications

Given a mass of detailed information, and the need to identify clearly many individual locations and places, it is easy for the urban area map or plan to become over-congested. Separation into visual levels is essential. As many town plan users may be inexpert, discrimination needs to be assisted by avoiding the absolute threshold of detection, especially with names.

Names of streets, important places and services need to be clearly seen without becoming unduly large. The only way to achieve this satisfactorily is to avoid very small type sizes (less than 1.5 mm) and to reduce the competition from other information. If black is used sparingly for other detail, its full contrast can be realised for names.

Similar considerations apply to point symbols. In many cases important locations and services, such as post offices, car parks, information offices or public buildings cannot be represented solely on the basis of their extent in plan by an area symbol, as they would be too small to have sufficient contrast to be apparent. Therefore, they are located and reinforced by high-contrast point symbols. Various approaches are possible. A classified symbol may be used to identify the location, and its description added along side as an annotation. So 'Post Office', or its abbreviation 'PO' may be included on the map. Or the symbol may be included where necessary, and indexed to a description in the legend. The latter method, although it may save 'space' on the map, involves more visual search, both of the legend and in the map. If the point symbol on the map can also include some abbreviated identification of the function (such as P for parking) it can be very effective, as it probably provides the map user with the most rapid identification of places and services. But it also poses problems in symbol size and design.

A common way of dealing with this is to use a 'negative' symbol, such as a black, dark blue or red solid square or circle, with the identifying abbreviation or symbol in white against the dark background. By first isolating the abbreviation from surrounding detail, and then giving it a high contrast, relatively small symbols can be easily distinguished. The method is most successful with simple alphanumeric characters that can be easily recognised by the map user. It is least successful with small iconic symbols, that have to be explained in a legend, found and remembered, and then found and identified on the map by the user.

Where an important feature or place occurs only once on the map, there is little point in giving it a classified symbol and then explaining this in the legend. Naming it directly on the map provides the most rapid identification for the map user.

# MAPS FOR OUTDOOR RECREATION

Although the road map is widely used for individual or private recreational activities, it also overlaps into another and different need, which is the representation of the topography for activities which take place in essentially rural areas. Such activities need not only information about road access, but also what lies 'between the roads'. For quite a long time this requirement was at least partly served by 'tourist' maps, normally modified versions of standard official topographic maps of particular areas. Very often these make use of a standard topographic sheet as a base, adding extra information mainly by special point symbols (see Forrest and Castner 1985). In the United States, the maps of national parks at various scales, using different sheet lines from the standard series, are a good example. In Britain, tourist or special sheets have been provided by the Ordnance Survey for more than half a century, and there are many special sheets of mountain areas in northern and western Europe.

With the increased emphasis on road travel, both as a means of access to an area, and for travelling within it, many 'tourist' requirements are increasingly served by the normal 'road' maps. But in recent times, the growth of more specialised outdoor activities has meant that in many cases other types of map are necessary or desirable. Therefore, there is a distinction between general recreational/tourist maps and those that are specially produced for sporting or other events. Maps for mountaineering, orienteering, etc., often have demands which cannot be properly met by slight modifications of existing topographic sheets, as they need both an appropriate scale and a particular specification.

## Topographic information

For some activities, the map user is principally interested in the detail of the topography, in terms of rivers, streams, steepness of slope, beaches, suitable camp sites, and so on. For such purposes, the standard topographic map, if at a suitable scale, should provide the appropriate information, assuming that the map is both available and can be interpreted by the user. Specialised activities such as climbing, canoeing and orienteering may need both a different level of topographic information, and also the inclusion of additional information as well.

## Maps for sporting activities

Although it is relatively easy for a national or provincial survey to provide special sheets of particular areas, and even add some additional information, the basic specification of map content and design cannot be easily changed. Carrying out a full topographic survey, simply to publish a few sheets of interest to a small number of potential users, is difficult to finance unless subsidised in some way. Therefore, in many cases the major discussion lies in whether or not the official survey can or should provide the desired maps. This problem is closely examined by McGrath (1984), who points out that many users encounter difficulties with out-of-date maps, and many potential users are unaware of the availability of suitable official maps that would serve their purpose. The difficulty for a national survey organisation is that satisfying one user group is often only possible at the expense of another user group, and a compromise between the two may satisfy neither.

Much depends on the level of organisation of those sports or recreations that demand a specific organisation for planned events. Because orienteering events are normally staged by local clubs, or national associations, and the competitors use a detailed topographic map of a small area, the provision of a suitable map is fundamental to the activity (see Palm 1972; Petrie 1977). Many such maps are produced voluntarily by enthusiasts. If they can be aided by making available suitable aerial photography and photogrammetric facilities, the quality of the map, which is always field-checked in great detail, can be very high. Similarly, long distance mountain races in Britain have special maps provided, and these need to be produced to a very high standard.

In many cases, scale is a critical factor. Sporting activities which are based on a small area, or a narrow zone, generally need a comparatively large scale – between 1 : 5000 and 1 : 20 000 – to give sufficient ground detail. The representation of small topographic features is usually important, and therefore the specification for the survey must take this into account. Such maps must be up to date if they are to be of any use.

Where the sport or field activity is practised by individuals or small groups in many different areas, the problems are much greater. For example, it is perfectly possible to make detailed topographic maps of rock areas for climbers, given suitable aerial photography and a large enough scale. But such a map-making operation would usually require investment out of all proportion to the number of maps that might be used.

## The representation of relief

Although for the general 'tourist' or motoring map the representation of relief may mean no more than making clear those areas in which roads have steep gradients, and giving a general indication of the overall topography, the outdoor recreation map depends in the first place on an effective representation of the topographic surface. If this is achieved fully by the national survey series, then many users may be satisfied. For example, the official map series in Switzerland, Austria and many German states, provide sufficient detailed information for all general activities, the scale of 1 : 25 000 or even 1 : 50 000 normally being regarded as adequate for navigation in mountainous areas. Much depends on the choice of vertical interval for contours, their correct generalisation, and the extent to which minor terrain features are identified and symbolised. The detailed representation of rock surfaces, moraine, scree and boulder fields, snow and ice fields and crevasses of the Swiss national series indicates both a need for and an interest in the representation of mountainous regions. In other countries, the representation of mountainous or wilderness areas may not have such a high priority, and in this case the specification may be inadequate.

For the serious map user, the quality of the topographic information is often more important than the map's appearance. The difficulty is that although additional information may be collected and inserted for the specialist user, the scale and the vertical interval of the contours cannot be changed. But the quality of the relief information is fundamentally dependent on these two factors.

# 20 NAUTICAL CHARTS

Using a map to plan and follow a route is one of the oldest and most important map functions. The term 'chart' has been reserved traditionally for maps used to find directions at sea, and it is now extended to route finding with aircraft as well. In the particular case of travelling by water, three requirements dominate: the need to find position at sea; the need to plan and follow the desired course; and the need to avoid danger.

With such a tradition, the modern nautical chart is the product of gradual evolution and refinement over a long period. Produced for a particular group of users, it is possible in most cases to define clearly what information is wanted and in what form. In addition, the nautical chart is vitally important in safe navigation, and largely because of this the relationship between quality of information and representation is considered more thoroughly than with any other type of map. In this sense, marine cartography often provides the best examples of cartographic practice.

Descriptions of coasts and harbours have been made for a long time, but the introduction of accurate charts, like topographic maps, depended on the development of instruments for measuring angles and distances. The additional problem was that of determining position when a ship was out of sight of land. Before this was achieved, charts were essentially small-scale geographical maps for locating ports and harbours, which could not be used directly in aiding navigation in coastal regions. The first great series of charts in Europe, the Portolan charts, were initially specialised descriptions of the Mediterranean and adjacent coasts; but actual navigation depended still on a knowledge of winds and approximate distances.

Lack of adequate charts was reflected in the importance of verbal descriptions of coasts and harbours. In the sixteenth and seventeenth centuries, the seaman's book or 'rutter' was a valuable source of information about landfalls, landmarks, dangerous areas, currents and so on, but these could not be plotted in position. The development of trans-oceanic sailing emphasised even further the problems of finding position. In place of estimates of latitude and approximations of longitude, it became possible to fix position by observation and measurement, and for this the development of an accurate chronometer was essential. The failure of the compass began to be understood, and indeed the needs of navigation were one important strand in the development of modern science.

The detailed mapping of shallow water coastal areas therefore took on a different nature from that of oceanic areas, and at an early stage the two main types of chart could be distinguished: the sailing or route chart, and the coastal or inshore chart. The production of coastal charts requires not only the accurate mapping of the coastal topography, but also the depth of water and the positions of rocks, reefs, wrecks and other hazards.

For a long period the only method of measuring depth of water was by lead line, by which a weighted line was dropped from a boat until it reached the bottom. Necessarily limited to relatively shallow water, and concentrated in areas most used by ships, the information was irregularly distributed, and therefore could not be used for interpolating depth lines or contours. The importance of depth values has continued to the present, even though it is now possible, at least over limited areas, to produce enough data to make the interpolation of good depth contours possible.

## The development of the modern chart

As with national topographic mapping, the end of the eighteenth century and beginning of the nineteenth saw the replacement of sporadic individual efforts by organised national surveys. Early harbour charts were often made by ships' captains and published privately (if at all), rather like the county surveys on land. The establishment of hydrographic departments saw the beginning of the regular and consistent surveying of coastal waters. One great contrast with land mapping is that from early times travel on the high seas has involved aid and cooperation internationally. With few exceptions, chart information is openly published and made available.

Although the most important function of the chart was to aid navigation, the development of improved methods made possible the collection of much more information about the bathymetry of the oceans – the parallel problem to that of surveying the terrestrial surface. In many respects this remains the last great unfinished task in acquiring scientific information about the Earth.

Modern charting has seen the introduction of faster and more accurate surveying methods. Electronic distance measurement on land is paralleled by electronic position fixing at sea. Sounding instruments can record water depth in continuous profiles. At the same time, the function of the coastal chart has been widened by the development of larger craft for transport, and of small-boat recreational sailing.

## GENERAL CHARACTERISTICS

Nautical charts are divided into different types on the basis of their principal functions. Although different countries have their individual classifications, four main groups can be distinguished.

Planning charts are used for general route planning, and as they need to deal with oceans and seas, the scales are small, usually less than 1 : 5 000 000.

Sailing charts are used by the navigator to plot courses in the open seas and are usually at scales 1 : 500 000 and smaller.

Coastal charts, used on approaching land, and navigating in areas that contain submarine dangers, range from about 1 : 50 000 to 1 : 300 000.

Harbour charts are at the largest scale, from about 1 : 10 000 to 1 : 25 000.

Apart from these main groups, in modern times many specialised types of chart have been introduced. Routeing charts are designed specifically for very large ships, which have particular navigational problems in shallow water and land approach. Latticed charts show the hyperbolic lines used for fixing position with radio aids. Fisheries charts are issued for fishing vessels in important fishing areas, and small-craft charts are provided for recreational small-boat seamen. The most noticeable contrast here is between very large ships and small boats.

## Chart format

Unlike many topographic map series, which divide the entire map area into a series of regular sheets, coastal charts must show particular areas on individual charts. As exact scale ratios in terms of representative fractions mean little in marine navigation, both scale and sheet size can be modified so that the chart covers the desired area. On the other hand, planning charts are frequently produced at a given scale, but even here the need to represent complete areas means that chart formats vary. For coastal charts, in particular, sufficient land must be shown for common landmarks to appear, and there should be enough space in the offshore area to ensure safe navigation. Overlaps with adjoining charts are arranged so that a conspicuous feature appears on both.

## Chart accuracy and reliability

The overall accuracy of any chart depends on two factors: the quality of the original information, and the relationship of this with any changes that have taken place since the chart was issued. Compared with topographic maps, this time factor is very important. The published chart information is affected in the first place by the quality and completeness of the original surveys, and the accuracy with which this information has been generalised. But because the printed chart cannot be kept up to date, new information has to be issued, and where necessary added to the chart by hand.

Where information about depth of water is sparse, taken from old surveys, or represents areas that change rapidly, this has to be made clear to the chart user, if necessary by adding warnings on the chart itself.

## CHART CONSTRUCTION

In navigation the fundamental requirements are to be able to determine position, and to plot and follow a course. This involves direction and distance. These operations are affected by the choice of projection and the way in which distance can be calculated in terms of chart scale.

## Projection

Although gnomonic projections are sometimes used

for plotting great circle routes at small scales, the Mercator projection is principally employed for the construction of sailing charts, and approach and coastal charts. Lines of constant compass direction are straight lines on the chart, and therefore the navigator can plot a course which can be followed by steering a compass bearing. For long routes, a great circle course can be approximated by a series of sections using constant compass bearings. The disadvantage is that the chart scale is not constant.

Although the normal (equatorial) case of the Mercator projection is widely known, it must be appreciated that for charts the line of correct scale is usually the mid-latitude of the chart, or group of charts. Therefore, scale is only correct in the central band of the chart. In order to measure a distance, the navigator can use a unit of latitude in the chart border. This is sufficiently accurate for practical navigational purposes. As great circles appear as curved lines on the Mercator projection, the Decca and Loran lattices have to be constructed as hyperbolic curves. Plotting position in latitude and longitude is comparatively simple, as the parallels and meridians are straight lines.

As courses are plotted as compass directions, all navigational charts also carry at least one compass rose. The magnetic variation is also displayed on the compass rose.

## Datums

The shoreline datum (the high water line) should coincide theoretically with that used on the official topographic map series, but this is not necessarily the case.

Chart datum is the plane of reference for depths on the chart, and therefore its establishment is of vital importance. Because of tides and meteorological conditions, this reference plane fluctuates. The normal practice is to make the chart datum equivalent to a low water level, so that the navigator finds the actual depth of water in a coastal region by adding the appropriate figure from the tide tables to the depth shown on the chart.

The datum should be chosen to give realistic information about both depths and drying heights (areas which are above water at low tide but under water at high tide). If the predicted heights of low water should fall below chart datum, then these would have to be shown in the tide tables by using negative values, which would be confusing. The chief alternatives are between some version of low water, mean low water, or lowest low water if there is a double tidal range; the lowest observed or

lowest possible low water; and lowest astronomical tide. Lowest astronomical tide (based on tidal predictions) has now been adopted for all charts of British coasts. It avoids the inclusion of very unusual tidal levels, which arise under freak conditions and which can affect the lowest possible or lowest recorded tide, but is less likely to exception than any mean value for low water.

## CONTENT AND REPRESENTATION

In order to aid navigation, the coastal chart has two essential elements: the description of the topography and hydrography; and the description of dangers to navigation and navigational aids. Because the legibility of a large amount of detail is so important, modern charts provide excellent examples of the development and elaboration of point and line symbols, many of them iconic in character.

### Topography

The representation of topography concentrates on the coastal zone. Aspects of land relief are only important in so far as they affect navigation and provide landmarks. Emphasis is always placed on how the topographic features appear from seaward.

On small-scale charts, detail of the coastline which would not be visible from a distance is omitted, and the shoreline is simplified. On large-scale charts used for navigation inshore, the configuration of the coastal zone is highly important.

The coastline, however defined, is a very significant line on the chart, and is normally represented as a continuous bold black line. If its position is in any way uncertain, the line is interrupted. No distinction is made between the natural coastline and any part that is built up or constructed.

On large- and medium-scale charts, the position of the low water line, which defines the limit of the area which is exposed at low tide, is taken to be the chart datum. The foreshore, the area between high and low water lines, is usually described in terms of its composition, using abbreviations such as $S$ for sand. Coastal features (such as cliffs or sand dunes) are symbolised, whether or not the land surface as a whole is contoured. Many different forms of relief representation may be used for the land area, ranging from simplified contours to hill shading.

Along the coastal zone, and particularly if the coast rises steeply, landmarks are selected and described. The most useful are those that can be observed from a distance at sea, and these may be either natural or artificial features that stand out in profile, or for which there is a sharp contrast in colour. Isolated trees, woods, prominent buildings, towers, chimneys and so on are indicated by both point symbols and abbreviations (Fig. 166). If necessary these are annotated to aid identification.

Chimney    Tank    Fort    Mosque    Windmill

**166.** Landmark point symbols (Crown Copyright. Reproduced from *Admiralty Chart No. 5011 Symbols and Abbreviations used on Admiralty Charts* with the permission of the Controller of Her Majesty's Stationery Office)

## HYDROGRAPHY

The detailed representation of the submarine topography is an important part of the chart information. Unlike most land areas, the submarine surface is often affected by quite rapid erosion and deposition in coastal regions.

### Depth of water

The navigator is primarily interested in whether there is sufficient water for the ship to pass safely. In shallow water areas the depth of the seabed is vitally important, but in deep water the actual depth is of little consequence to surface ships.

### Depth figures

On modern charts depth of water is shown by a combination of point values (depths or soundings) and submarine contours. Although traditionally British charts used fathoms and feet, modern charts use metres and decimetres. Depths are shown on Admiralty charts by italic figures, decimetres being given in subscript down to a depth of 21 metres. The position of the sounding is taken to be the centre of the figures. Values taken from older surveys or smaller scale charts are shown in light upright figures. Drying heights show the height to which a

point uncovers at low tide, and therefore these figures are heights above datum. They are also in italic, but underlined.

In some cases the actual depth is unknown, but the area has been swept by a wire drag to a given depth. These are shown by normal depth figures, but with the wire drag symbol (Fig. 167). Soundings to depths without reaching the bottom are placed beneath a bar, and soundings of doubtful value are annotated SD.

| $12_8$ | $0_6$ | $12_8$ | $\overline{220}$ |
| drying height | wire drag | no bottom |

**167.** Depth figures (Crown Copyright. Reproduced from *Admiralty Chart No. 5011 Symbols and Abbreviations used on Admiralty Charts* with the permission of the Controller of Her Majesty's Stationery Office)

### Depth contours

Where sufficient information exists, contours are added, and this in turn affects the density and selection of depth figures. The vertical interval of the contours is progressive, as most information is needed in shallow water areas. As with all other chart content, the purpose is to provide information relative to the navigator's need, not a systematic description of the submarine topography. Additional contours may be added for particular charts, but the principal ones are as follows: 2, 5, 10, 15, 20, 30, 50, 100, 200, 500, 1000, 2000, 3000 and 4000.

The contours are shown as continuous fine lines in black, with upright figures. Approximate depth contours are symbolised by an interrupted line.

### Colour emphasis

On older charts, important fathom lines (such as 2 fathoms and 3 fathoms) were reinforced by a vignetted dotted line. Modern charts use colour, and this practice is gaining acceptance internationally.

The land area (above high water) is usually given a buff or yellow colour, and the shallow water zone between chart datum (low water) and 5 metres a light blue. The two colours are combined in the foreshore zone, producing a dull green or olive (Fig. 168). These area colours emphasise the most important distinctions on coastal charts. Beyond 5 metres, other contour lines may be reinforced by a blue band. The most common one is 10 metres, but

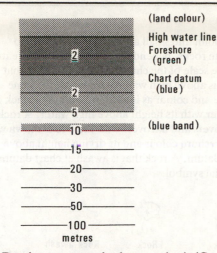

(land colour)

High water line

Foreshore
  (green)

Chart datum
  (blue)

(blue band)

metres

**168.** Depth contours and colour emphasis (Crown
Copyright. Reproduced from *Admiralty Chart
No. 5011 Symbols and Abbreviations used on
Admiralty Charts* with the permission of the
Controller of Her Majesty's Stationery Office)

others may be added, depending on the scale of the
chart and the submarine topography.

## Generalisation and the representation of depth of water

Except in those cases where there is so little
hydrographic survey data that all of it must be
included, depth information has to be generalised
on the published charts. Naturally the problems
increase with smaller derived scales. The
generalisation processes of selective omission and
simplification are most important.

Depth figures and contours are used in a
complementary manner to give the most useful
representation of depth of water without misleading
the navigator as to the amount and quality of
information available. As Russom and Halliwell
(1978) point out, 'Each detail should be assessed for
its usefulness to some important class of chart user
in the context of the surrounding details and the
scale of the chart.' In this respect it is impossible to
apply standardised rules or practices.

The navigator is chiefly interested in being able to
navigate the ship safely along the coast and into
harbour. Therefore, safe passages are just as
important as warnings of danger. In some areas,
only a careful selection of depth figures will reveal
the necessary detail: in others, contours reinforced
by selected soundings are adequate. In some cases,
very limited information is given so as to avoid
misinforming the chart user. Much depends also on

the range and quality of the surveyed data. A
modern survey of a harbour area and its approaches
will provide complete and consistent information,
and so both contours and depth figures can be used.
In other areas there may be only an irregular
distribution of depth values from older surveys.

Some of the most important principles are as
follows. The least depths (shallowest points) in
shallow water areas and along channels are always
included, as these represent the chief source of
danger, and determine whether or not a ship can
pass safely. The density of soundings tends to
emphasise visually the most important areas and
lines, both for channels and dangerous banks. The
gradient along the side of a channel is important, as
many ships following shallow channels will
continually monitor water depth with the echo
sounder. If the sounded depth does not agree with
that in the charted position, this provides an
immediate indication that the ship's position is not
the one intended, and is dangerous (Fig. 169).

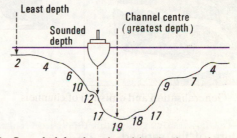

**169.** Sounded depth and position in channel

If contours can be used, the soundings should
indicate depths that are away from the contour lines,
not values that are close to them. In navigable
channels, depths need to be shown along the
deepest part of the channel as well as along the
edges. An even overall spacing of depth figures
suggests that the information is consistent, whereas
irregular spacing suggests variations in source
information. Where depth information is limited, no
attempt is made to conceal this fact, and the
junctions between surveys carried out at different
times are not smoothed out (Fig. 170).

In some cases, the amount of information is

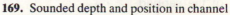

**170.** Junction of surveys

deliberately restricted. These include any areas that are subject to rapid change, and charts or parts of charts that are covered by larger scale and more detailed charts. In the same way, on areas cut off by the chart border all depth information is omitted, so that the navigator must change to the appropriate chart.

On small-scale charts, where much less information can be given, care has to be taken to ensure that shallow water areas are not maintained at the expense of passable channels. There is an inevitable tendency with contour simplification to remove small channels within a shallow area, thereby producing an overall reduction in depth of water (Fig. 171).

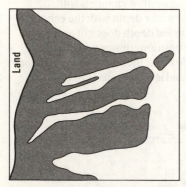

**171.** Generalisation and omission of channels

## Nature of the seabed

This is important not only for its effect on anchoring, but also for small craft which are beached. In many coastal areas, ships need to anchor offshore before entering harbour, and in some cases the lack of harbour facilities means that cargo has to be loaded at anchorage. Sand and shingle hold better than soft surfaces such as mud, and of course rock needs to be avoided.

The information is also of value to fishing vessels, as it can assist in locating fishing grounds, and also affects nets.

## DANGERS TO NAVIGATION

The second major part of the hydrographic information represents dangers to navigation, including those that are partly visible as well as those that are submerged. These are primarily rocks, reefs and wrecks.

## Rocks

Both point and small area symbols are used to indicate rocks, depending on the area of the feature and the scale of the chart (Fig. 172). A rock that remains above high water is shown by the same outline and colour as is used for land, or a black dot, together with its height above high water. A rock that covers and uncovers with the tide is shown with the foreshore colour and its drying height above chart datum. A rock that is awash at chart datum has a special symbol.

**172.** Representation of isolated rocks (Crown Copyright. Reproduced from *Admiralty Chart No. 5011 Symbols and Abbreviations used on Admiralty Charts* with the permission of the Controller of Her Majesty's Stationery Office)

For submerged rocks, the standard symbol is a cross, usually with its sounded depth. Rocky areas are outlined with a dotted line. Reefs are treated in the same way, with the appropriate abbreviation such as *Co* for coral.

## Wrecks

Wrecks in shipping lanes are a major source of danger. Basically two approaches are used in symbolisation. If any part of the hull is visible at chart datum, then either the plan of the hull is represented by the equivalent of the land symbol, or a profile is used (Fig. 173). If only the masts are visible, then plan symbols are used. Submerged wrecks are given the depth figure in the normal way, which the annotation *Wk*.

Wrecks that have been swept by a wire drag are given the standard wire drag symbol.

**173.** Representation of wrecks (Crown Copyright. Reproduced from *Admiralty Chart No. 5011 Symbols and Abbreviations used on Admiralty Charts* with the permission of the Controller of Her Majesty's Stationery Office)

## Other dangers

Various other natural features can be dangerous, including eddies, tidal rips, breakers and kelp. Most of these are given iconic symbols, such as that used for kelp (Fig. 174).

kelp       tidal eddies

**174.** Representation of obstructions and dangers (Crown Copyright. Reproduced from *Admiralty Chart No. 5011 Symbols and Abbreviations used on Admiralty Charts* with the permission of the Controller of Her Majesty's Stationery Office)

# AIDS TO NAVIGATION

It is difficult for ships to proceed safely in coastal waters without rules for navigation and fixed indicators to define dangers, channels and directions, both in daylight and dark conditions. The most important fixed aids include buoys and beacons, lights, fog signals and radio and radar stations.

## Buoys and beacons

In addition to the fixed lights provided by light houses and light ships, buoys are used to mark the position of obstructions, to indicate channels and fairways, and to show the presence of delimited areas such as spoil grounds and prohibited zones.

The representation of buoys on charts provides a classic example of the development of iconic symbols, and the full use of variations in form and dimension. The position of the buoy is indicated by a circle or dot in the base line, and six basic shapes are used: can, cone, sphere, spar, pillar and spindle (Fig. 175). Colours are shown by abbreviations, or by shading patterns in the buoy symbol. Extensions are made in the form of topmarks using the cross, rectangle (can), cone, circle or solid sphere.

With the adoption of a standard buoyage system in European waters, three main classes of buoys were introduced, further distinguished according to whether they are lit or unlit.

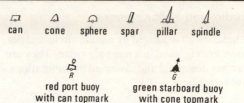

can    cone    sphere    spar    pillar    spindle

red port buoy
with can topmark

green starboard buoy
with cone topmark

**175.** Types of buoy (Crown Copyright. Reproduced from *Admiralty Chart No. 5011 Symbols and Abbreviations used on Admiralty Charts* with the permission of the Controller of Her Majesty's Stationery Office)

Lateral buoys show the port and starboard limits of defined channels. Port hand buoys are all red, with a can topmark, and starboard buoys are all green or black with a cone topmark.

Cardinal buoys show directions from the point of interest. These are all pillar or spar buoys in two colours, and with double topmarks (Fig. 176). Pillar and spar buoys are also used to indicate isolated dangers; the topmarks are two black spheres (solid).

If lights are added, these follow the standard practice on Admiralty charts of using a magenta flash to indicate the presence of a light. The lights are colour coded, and where necessary both the colours and the periods of flashing and occulting are shown by annotation (Fig. 177). For example, *V Qk Fl(3) 5s* means that three very quick flashes appear every five seconds. For full details of these and other symbols, reference should be made to *Symbols and Abbreviations* (1976).

north mark
black above yellow

isolated danger
black with red band

**176.** Cardinal direction buoy and isolated danger buoy (Crown Copyright. Reproduced from *Admiralty Chart No. 5011 Symbols and Abbreviations used on Admiralty Charts* with the permission of the Controller of Her Majesty's Stationery Office)

East
mark

*V Qk Fl(3)5S*

*(magenta flash)*

**177.** Light signal annotation (Crown Copyright. Reproduced from *Admiralty Chart No. 5011 Symbols and Abbreviations used on Admiralty Charts* with the permission of the Controller of Her Majesty's Stationery Office)

## Radio and radar stations

These are all indicated by a small circle and dot, and most are enclosed in a magenta circle. They are further described and classified by annotations and abbreviations.

## Fog signals

Because light signals may not be visible in fog, many buoys also carry fog signals. A great variety of devices is used to produce different sounds, including explosion, siren, horn, whistle and gong. These are also shown by annotation or abbreviation.

## Limiting lines and linear features

Most of the dangers and aids to navigation are represented by point symbols, often accompanied by abbreviations and annotations to provide further information. Line symbols, sometimes reinforced by coloured bands, are used to show leading lines (courses for ships to follow); sectors, usually for the limits of light visibility; and arrows for traffic separation schemes in congested areas. A variety of other features, such as submarine pipelines, also have line symbols (Fig. 178).

Outlines of areas include such items as spoil grounds and prohibited areas. Because of the large number of line symbols needed, both black and magenta are used on modern Admiralty charts.

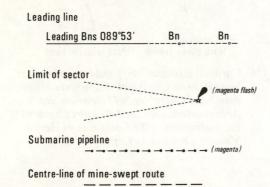

**178.** Line symbols: directions, sectors and submarine obstructions (Crown Copyright. Reproduced from *Admiralty Chart No. 5011 Symbols and Abbreviations used on Admiralty Charts* with the permission of the Controller of Her Majesty's Stationery Office)

# INTERNATIONAL STANDARDISATION

Navigational charts are produced by many countries throughout the world, but almost entirely by national charting agencies. Some countries, with an extensive interest in maritime trade, produce chart series covering all shipping areas. Because of the degree of similarity of requirements, many of these charts and chart series overlap, in the sense that charts of the same area, and at much the same scale, may be produced by several different countries.

With the increase in output of hydrographic data, and its rapid circulation, it has become even more difficult to keep published charts up to date with changes. Naturally this has led to attempts to improve the interchange of chart products, but this in turn has depended on reaching agreements in relation to chart design and specification.

Two possible approaches exist, and both are presently employed. Small-scale charts could be co-produced by a number of maritime countries, and then this material could be circulated to other interested countries to produce their own charts, essentially as facsimile copies. This can only operate if navigators are familiar with the symbolisation of the chart. Secondly, large- and medium-scale charts of coastal waters could be produced by different countries for international shipping, provided that sheet coverage and scales could be agreed internationally, together with a common symbol specification. In this case each maritime country would produce the charts for areas adjacent to its territory. In both cases, the aim is to decrease the amount of duplication, and to avoid the risks involved in the use of unfamiliar charts.

Possibly because of the long tradition of mutual aid at sea, the International Hydrographic Organisation is one of the few organisations that could attempt such a degree of international cooperation (see Ormeling 1978). Developments have been in progress since 1967, when the co-production of a small-scale chart series for world coverage was begun. In 1972 and then in 1982, further programmes of standardisation were agreed internationally. After 1967, some small-scale series at the scales of 1 : 10 000 000 and 1 : 3 500 000 were published, and the reprographic material made available to other countries to print their own charts. After 1972, a much more detailed attempt was made to standardise medium- and large-scale charts, and initially this was directed at the charting of the North Sea. This pilot study had to agree on questions of chart coverage in relation to harbours and ports as well as a standardised symbolisation and content. Eventually all the problems of both content and representation were solved, including the information to be given about any feature, annotations, the use of colour, the coding of lights and a standardised buoyage system. The eventual aim is to ensure that all charts, including large-scale charts produced by one country, and which are not part of any international scheme, should use a standard symbolisation.

After further study by other countries, international agreement was reached in 1982, with the publication of standard chart specifications by the IHO. As Newson (1983) points out, '... the world will ultimately be covered by an internationally conceived homogeneous series of three to four thousand sheets, of manageable size, and on adequate scales, to suit the needs of international shipping and to be accepted without question in all parts of the globe'.

## PRODUCTION, REVISION AND MAINTENANCE

Because of the need to maintain published charts and frequently revise them, chart production has always encountered difficulties in making radical changes in production methods. For a long period the achromatic engraved chart was dominant, and indeed engraving persisted long after it had been abandoned for topographic and other maps. In many cases a copy of the engraved map was subsequently produced by pen-and-ink drawing for lithographic printing, and this in turn was superseded by scribing and masking with photoset lettering and symbols. Given the density of point and fine-line symbols on coastal charts, engraving was an effective means of image composition, but it always provided difficulties in correction.

Modern charts make use of the standard production and reprographic processes.

Standardised symbols are mass produced. Because of the limited colour range, black, blue, magenta and buff or yellow are the only printing colours used on Admiralty charts, and the information is basically contained on the black and magenta plates. Computer-driven plotters are frequently used for the construction of chart projections, borders and lattices. The digitising of soundings, which is a necessary preliminary to any digital processing of chart information, has led to many experiments with suitable methods, including the voice recording of sounding values.

### Revision

All charts issued to the user should be up to date at the time of distribution. Therefore, printing runs are kept small, in order to avoid wastage. On the other hand, this means that revision cycles are usually very short, and revision is a major part of any chart organisation's total work.

Each chart has a file that records all the details of compilation, production and revision. A master copy of the latest printing is maintained, and on this are marked or indexed all changes. This information can also be circulated to other sections if it affects other charts that overlap or are produced at smaller scales.

With modern materials, the corrections are carried out by the usual methods of deletion and addition. Eventually it may be desirable to replace a chart entirely if extensive corrections have been carried out. Both modernisation and changes to international specifications have to be coordinated with the chart revision programme.

### Maintenance

Because the accuracy of the chart information is so important for navigation, out-of-date charts with incorrect information cannot be used. In order to provide information for the up-dating of charts already published, the Admiralty issues *Notices to Mariners*, normally every two weeks. Navigators then make manual corrections to the charts they have in use.

# 21 SPECIAL-SUBJECT MAPS

## General considerations

A large group of specialised maps deal with a particular subject or theme, and are often referred to as thematic maps. Unfortunately, the term 'thematic' tends to be applied also to special-purpose maps, but for these the 'theme' is the map user's requirement, not the subject matter itself. For this reason it seems preferable to distinguish between the two types, as they pose different cartographic problems. Special-subject maps may be, and frequently are, used for quite different purposes. They are concerned with the representation of a particular phenomenon, or some aspect of it, not the satisfaction of a given user requirement.

Initially, the apparently endless diversity of special-subject maps – in scale, content, and complexity – suggests that they are equally diverse in cartographic terms. In fact, this is not so. The graphic problems involved in representation can be divided into a few major groups. Therefore, although the maps differ individually, essentially the same cartographic problems are repeated in many cases.

Accounts of special-subject maps tend to approach them in terms of their content, or of the method of representation involved. Some types, such as 'statistical maps' are distinguished on the basis of the origin and nature of the source information. It is all too easy, in this respect, to confuse cartographic problems with those engendered by the nature of the available information. There are no 'cartographic techniques' that will make good maps from inadequate source material.

A proper analysis of special-subject maps must take into account three principal factors: first, the characteristics of the phenomena; second, the characteristics of the available data about the phenomena; and third, the problems of graphic representation, including the technical and economic factors that circumscribe the production.

## THE CHARACTERISTICS OF THE PHENOMENA

Considering the great diversity of the environment, it would seem to be an impossible task to make a complete classification of all phenomena. However, the point of interest cartographically lies in the spatial and temporal characteristics of the subjects for which maps may be made, and of these it is the spatial characteristic that usually takes precedence. The following major groups may be distinguished.

### 1. Continuous phenomena

#### 1A. Classified by type

Many natural phenomena are visible and relatively static. The major characteristics of the Earth's surface are of this type: geology, soil, vegetation and land use. They are described by dividing them into major classes, frequently with numerous related sub-classes. Although they are all three-dimensional, the relative importance of the three-dimensional structure varies with individual cases. With geological formations it is highly significant, and therefore information is needed about concealed deposits as well as surface types. All land surfaces can also be described as having a particular use. However, some of these may change seasonally or annually, as well as over longer periods, and indeed the nature of the classification is essentially a judgement. Many different classifications may be made, depending on the factors considered.

The common element is that phenomena occupy particular areas, and are sub-classified by type. The number of sub-classifications depends on the nature of the phenomenon and the detail in which it is examined. It is also characteristic that some classes will exist uninterruptedly over large areas, whereas others will occur over small areas. Therefore, there is a contrast between uniformity and diversity. It is also the case that some class boundaries will be sharply defined, and even marked by discontinuity, whereas other sub-classes may have transitional boundaries that are difficult to define.

### 1B. Distinguished by quantity or value

#### (i) Continuous variation

These consist of a single phenomenon, relatively fixed or permanent, which varies in amount. An obvious example is the elevation of the land surface. A value can be attributed to any point on the three-dimensional surface.

#### (ii) Spatial and temporal variation

These phenomena are marked by a relatively rapid rate of change, and some are intangible. This change in value or quantity may consist of accumulation over time (precipitation), or variation above and below a datum or zero (temperature). The rate of change may be rapid, as with air temperature, or relatively slow, as with surface water temperature. Many other natural phenomena, such as magnetism, are in this group.

Strictly speaking, precipitation is not continuously distributed, but as it can occur universally it is convenient to place it in this category.

## 2. Discontinuous phenomena

### 2A. Location of a single type

Maps of such phenomena often deal with an individual category within a larger group, for example a map of peat bogs, or a particular botanical type.

### 2B. Spatial and temporal variation

Although precipitation is of this type, it is usually treated as if it were continuously distributed.

### 2C. Variation in quantity at specific locations

The usual focus of interest is the absolute or relative amount, which may be treated as a cumulative quantity or a mean value. Many types of industrial production or transfer of goods fall into this category, such as volume of manufacturing, or exports from a port. In many cases the activity is one that occurs over an area, but such areas may be treated as point locations at small scales.

### 2D. Discrete units, spatially variable

These differ from (2C) above, as they exist as individuals, even though they may be qualified as aggregates. Generally they are not fixed in location, and therefore can only be recorded as being present at or within a location at a given time. The quantities are enumerated, not measured. Population, of human beings, livestock, or fauna, is the principal example. In some cases they are aggregated in small areas treated as points at map scale (for example population of cities), or may be represented as quantities in relation to area (density).

## 3. Change, rate of change, or physical movement

In the first two categories, the point of interest lies in the nature and distribution of the phenomenon itself. Other maps deal specifically with the way in which it changes. The change may be either temporal or spatial, or even both simultaneously.

### 3A. Change over time

Temporal changes are difficult to deal with cartographically. The alternatives are either to represent the degree of change within a specified time, such as population increase or decrease, or to provide a series of maps showing the distribution and/or quantity within different time periods, or at selected intervals. This leaves the interpretation of the location and degree of change to the map user, but can provide more information than one map limited to a single period.

### 3B. Change over space: physical movement or transfer

Most of these deal with tangible quantities, such as volumes of goods exported or imported, or the migration of populations. They can also include intangibles, such as the flow of capital. Those maps which purport to show actual route of movement, or direction from one place to another, face the cartographic difficulty that both quantity and movement require symbolisation.

## SOURCES OF INFORMATION

In dealing with special-subject maps, it is important

to understand the connections between the nature of the phenomenon and the nature of the available information. Sources of information can be classified in various ways in terms of their cartographic applications. Different forms of information can be described: first, by the mode in which the information is collected and presented, which may be graphic, alphanumerical or digital; second, by the geometrical relationship with the intended map; and, third, by the graphic relationship with the intended map in terms of scale and degree of generalisation. In this sense the perfect informational source should give complete and consistent locational information about a fully defined feature or phenomenon, expressed in the chosen coordinate system and at the desired compilation scale.

Such conditions can only be met, even partially, if a specific data collection operation is carried out for the map concerned, or if the map is matched to a set of locational data that satisfies these conditions. Even then it is unlikely to yield more than a certain proportion of the desirable data, for technical or economic reasons. For example, a good map of precipitation over a given region should be based on a regular network of recording stations, placed at a density that would be satisfactory for the scale of the intended map. Given such a network, it would be possible to plot isohyets in a consistent manner, and these could be said to give a reasonable approximation of the three-dimensional 'surface'. In fact, such perfect conditions are never realised, and interpolated isohyets range from consistent to highly approximate.

An attempt to meet these conditions is usually implied in a basic map, which results from a specific data collection operation, and is not derived from the processing of existing material. Although special-subject maps tend to be associated with secondary data, it must be realised that many important ones are indeed basic maps, and are the product of considerable investment of resources over long periods of time. Basic geological map series, or soil survey series, are the specialised equivalent of basic topographic maps, and indeed in some countries have taken precedence over topographic maps at particular times.

The information needed for specialised maps must satisfy the same fundamental conditions as for maps of any other kind. In order to be mappable, the data must include information about both location and type. The locational information may be specific, as in a large-scale topographic survey, in which the content is determined before the features are identified and measured; or very approximate,

as for example in the enumeration of population by census districts.

The data may be entirely numerical, as in the sounded depths of a hydrographic survey, or the coordinates of control points in a topographic survey. It may be a combination of alphanumerical and graphical, as in the record of a continuous barometric pressure gauge, or a geological survey field sheet with annotations. It may be entirely graphic, as in outlines of identified areas plotted photogrammetrically.

# CHARACTERISTICS OF INFORMATION SOURCES

In making a map of any phenomenon, the source of information is vitally important. It can range from the exact to the highly approximate. There is an important difference (rarely commented on) between making a map of a particular subject, and making a map of the available information about it. Many special-subject maps are very limited in their informational base, but unfortunately this is rarely made evident either by the graphic representation or the title.

The following are the main types of data collection from a cartographic point of view.

## 1. Specific location recorded

### 1A.

The simplest type of data record is to list the specific locations at which a certain type occurs, or to obtain or use quantities or values attributed to point locations. These may be expressed either by coordinate positions or by reference to existing topographic features which can be treated as points. Industrial plants, for example, are located in this way at small scales.

### 1B. Actual areas located

These can include direct description, using observation and identification in the field to determine the boundaries of the phenomenon or its various sub-classes. The degree of detail will depend on the complexity of the feature and the scale. Surface geology, geomorphological forms, types of land use, areas of a botanical type are all examples.

Description can also be done indirectly, either by using a photogrammetric interpretation and plot, or through an existing source of basic data, such as in the classification of slope zones based on a detailed contoured topographic map. Most of these involve static and visible phenomena. They can also include small areas treated as points at map scale, for which location can be determined from existing maps. In all such records, the information should be continuous and complete.

These subjects are in many ways the easiest to deal with cartographically, as location and extent in two dimensions are naturally suited to a two-dimensional map.

## 2. Measurement by sampling

### 2A. Measured or recorded at points

In many cases, it is impossible, even theoretically, to measure the variation over the entire area concerned. Therefore, it can only be sampled, either at points, or along lines. This applies to invisible phenomena, such as sub-surface geological forms, or to intangible and rapidly changing phenomena, such as temperature. The frequency and spatial distribution of such points of record or measurement is vitally important in determining what map, if any, can be made. Conclusions about the distribution and characteristics of the whole have to be interpolated from the sampled data. All weather and climatic records are based on point sampling for actual measurement, even though remotely sensed images may be used to indicate distribution of certain phenomena by type. A whole range of natural phenomena are included in this group, from measurements of river discharge to boreholes for geophysical investigation. Where the phenomena are in fact continuous, as with geological strata, the boundaries between different classes or levels have to be approximated. Clearly, this can be done satisfactorily only if the network of measured points is sufficiently dense to detect points or lines of change within narrow limits.

Where the natural phenomenon constantly changes, either by variation above or below a datum, or by accumulation, it is not possible to represent this directly in a fixed graphic. The data must therefore be summed over time periods, or converted into mean or average values, or minima and maxima.

If the phenomenon is actually continuous over areas, for example the depth of a water table, a sufficiently dense network of measured points can be used as a basis for the interpolation of lines of near-equal value to represent the 'surface', as a dense network of measured numerical values is difficult to represent or to interpret in a two-dimensional graphic form.

### 2B. Measured or recorded along lines

Sampling along lines can take place in either of two ways. A line of constant value (isarithm) can be measured or recorded, in which case the variation is described by using a series of such lines with equal intervals between them (Fig. 179). Continuously measured lines (contours) have a higher accuracy than other types of isarithm which are interpolated from points. Alternatively, a profile or cross-section can be followed and the variations along the line recorded. If these measurements are continuous, and in closely spaced parallel lines, then a set of such profiles will give full information about the variation over an area. Whereas the former is the familiar 'contour line', measured at a specific constant value, the latter is exemplified by either a hydrographic survey of the seabed depth, or by a forest inventory, where the presence or density of certain tree types within an area is interpolated from a regular series of traverses at selected intervals.

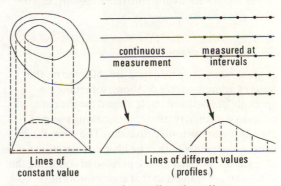

179. Measurement and sampling along lines

Continuous measurement can only be achieved if the 'surface' can be seen, or detected by some kind of measuring instrument.

## 3. Recorded within data collection areas

Data of this type can include both discrete units or individuals, and also other quantities, which may occur at particular locations, or within certain areas, but are only recorded as being present within an area of data collection.

### 3A. Constant areal units

Phenomena that vary in distribution may be recorded by using a network of fixed areal units, such as grid squares, to enumerate the total number present. This is one method of population mapping by census, and indeed is necessary if the data are to provide systematically for map output. It can bring out spatial variations successfully if the resolution of the grid units or cells is sufficiently fine, in relation to both the distribution of the phenomenon and the intended map. The difficulty is that a fine resolution involves much greater effort at the data collection stage, and much more work in data storage and processing. For population records, it has most often been applied to limited areas, such as large cities. Remotely sensed data also fall into this category, as they record the presence or absence of certain phenomena in terms of pixels, which are recording units, although they may have unequal dimensions.

The information so acquired may be used directly in the representation, or as the basis for the interpolation of isarithms, by which the value or quantity within one unit is treated as though it occurred at a point. It has the great advantage that the resolution is fixed, and therefore variation is represented consistently, so that aggregation into larger units, such as administrative regions, can be done systematically.

### 3B. Arbitrary areas

A great many statistical data are collected on the basis of administrative divisions, sometimes specially devised and implemented for the data collection authority, sometimes making using of standard administrative divisions. The normal census represents both. The basic enumeration districts are devised for the purpose of collecting the data, and are framed so that enumerators have approximately the same amount of work to do. Consequently, small street-based units are used in urban areas, whereas rural areas are often treated on the basis of existing parishes or other civil divisions. Location can only be ascertained within the limits of the minimum administrative division for which the data are published (which may be larger than the enumeration districts themselves). The units of data collection differ in shape and dimensions.

Apart from discrete phenomena, such as population, other quantities are often recorded and published in relation to administrative divisions, rather than their actual locations. Output from industries, for example, is often aggregated deliberately, so that the quantity or value connected with a particular plant or location cannot be extracted from the total.

Whereas with fixed grid units, the resolution at least is uniform, with arbitrary units the resolution is variable. Even when aggregated into larger units, these themselves are also likely to be arbitrary. They have no inherent correspondence with the characteristics of the phenomena and their distribution, and indeed the locational information available may conceal more than it reveals.

The distinction between fixed and arbitrary units is just as important in relation to derived values. In many cases, the map attempts to show the relationship between quantity or number and area. Thus mean values, such as densities, can be calculated consistently for data based on a fixed unit, but become virtually meaningless for data based on arbitrary units, especially if the distribution is discontinuous. Human population is not 'continuous', but it is often represented on statistical maps as though it was, essentially because the information comes in a form that relates total numbers to arbitrary areas. If no attempt is made to relate the data to the actual areas in which the phenomenon is located (which may be only a small part of the whole), the resultant map may be entirely misleading.

## CARTOGRAPHIC REPRESENTATION

This brief examination of the characteristics of different phenomena, and the information about them, brings out not only the great variety of subject matter of specialised maps, but also the importance of the different types of data from which maps have to be composed. Consequently, cartographic representation is confined to relatively few tasks, which constantly recur. The fundamental reason for this is that although the subjects themselves are very different in their characteristics, the information available largely controls the nature of the representation which is possible. In addition, the two-dimensional nature of the map, and the fact that it is a static image, means that it can only deal effectively with certain types of representation.

The principal types of representation can be distinguished as follows.

# 1. Continuous phenomena

In a few cases the area is divided into a set of different types, each of which is unique. Theoretically, each should be represented by a different symbol, although all symbols should be visually equal. The only obvious case is the division of the map area into independent states, or a state into separate regions.

## 1A. Division of the surface into a series of sub-classes

These are divided by type, and therefore of equal value (Fig. 180).

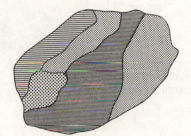

**180.** Sub-classification by type

The range may be from a few to a very large number of sub-classes. Geological or soils maps are typical examples.

## 1B. Continuous phenomena, varying spatially, but relatively static

The 'surface' can be properly represented by isolines (contours) if sufficient information is available, or it can be divided into a series of graded sub-classes, ranged above a datum or zero, or varying above or below a datum or zero (Fig. 181). The relief of the Earth's surface is the obvious example, in which elevation can be shown either by contours or by hypsometric and bathymetric layer colouring.

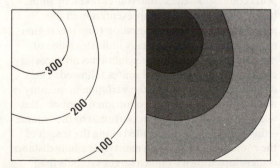

**181.** Isolines and classification by value or quantity

## 1C. Continuous phenomena, varying spatially and temporally

As the temporal variation cannot be represented directly, graphically the subject has to be treated in the same way as 1B, either by taking the quantity at a given time, or by obtaining mean, maxima or minima values. Maps of temperature, for example, may use isarithms interpolated from measured points, thereby producing a 'contoured' surface related to a given time period, or may use 'layer colouring' to further distinguish a series of sub-classes by quantity.

In both cases, the quality of the map is essentially a function of the density of the measured points or lines, and may range from complete and consistent information, such as measured contours with a fixed vertical interval, to isarithms interpolated from a few widely separated recording stations.

# 2. Discontinuous phenomena

## 2A. The representation of the distribution of a single phenomenon by type

This is graphically the simplest task, as it falls naturally within the scope of the two-dimensional image (Fig. 182). A map of a subject such as the distribution of woodland in a region represents the actual areas of occurrence, whereas a map of historical battlefields will indicate a series of locations by point symbols.

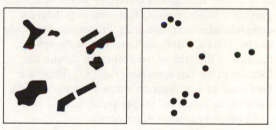

**182.** Specific areas and point symbol locations

## 2B. Distributions that vary both spatially and temporally

These have to be treated graphically in the same way as (1C) above. Although precipitation is spatially and temporally discontinuous, the records lead to aggregations over time. This quantitative 'surface' is then treated either by constructing isolines interpolated from recording stations, or again can be further emphasised by division into a series of classes, ranged above a zero.

### 2C.  Quantities related to points or small areas

These quantities or volumes may vary over a great range. The common method of representation is to use proportional symbols representing the quantitative differences at different locations (Fig. 183). Very often the total data range has to be broken down into a series of graded sub-classes in order to obtain sufficient contrast between the symbols. Such maps can pose extremely difficult problems in representation, as the two-dimensional graphic image of the map is a very limited medium for representing quantities. Both single and multiple variables can be represented, and either Cartesian or polar diagrams can also be employed.

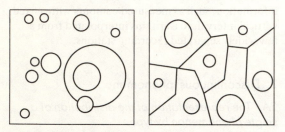

**183.**  Proportional symbols located at points and applied to areas

### 2D.  Quantities aggregated over areas

The difficulty of dealing with this type of subject matter is increased when the information is only available by aggregating quantities occurring at particular locations, usually by summing them for complete administrative units. Proportional symbols can be 'placed' within such areas, even though the location of the symbol has nothing to do with the location of the phenomenon (Fig. 183). These are best described as 'diagram' maps, for they combine some of the locational properties of the map with some of the properties of a diagram.

### 2E.  Discrete units

These are very difficult to deal with cartographically, essentially because of the nature of the source information, which tends to lack locational detail. Attempts to represent such a distribution either endeavour to obtain further information about the likely location of the data by reference to external sources, or treat the data solely on the basis of the areas for which they are provided. The first of these leads to the use of the 'dot' map, in which a symbol representing a unit quantity is distributed over the area in accordance

with what is believed to be the approximate location. The second depends on relating the total quantity to the area, leading to the density map typified by choropleth mapping (Fig. 184). In areas of great concentration, the discrete units may be aggregated into a single quantity and represented as occurring at a point location. Thus populations of cities or large towns are often represented by proportional symbols, treating them like any other quantity or volume located at a point.

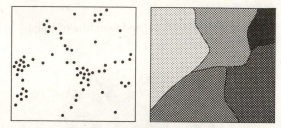

**184.**  Dot distribution (point symbols) and choropleth (graded area symbols)

### 3.  Representing change

The representation of change may take place in basically two ways. Changes in volume or quantity at given locations, or within specified areas, have to be treated like any other quantity applied to a point or area. For example, changes in urban population over a period of time may be shown by proportional symbols. Changes in total population within administrative areas can be shown as a series of graded sub-classes, in the same manner as other quantities applied to areas, whether the distribution is actually continuous or not.

The physical movement of goods, or people, poses quite different problems, as both the destination and the route have to be shown by the symbols. The route itself may be real or purely symbolic. For example, trading routes between ports could show the actual routes taken by ships, but this is unlikely. The representation of the movement of people in migration from one region or country to another actually indicates a line of movement or flow, but the symbol has no necessary connection with the actual routes followed. The difficulty is that although the variation in quantity is normally indicated by a proportional symbol – the width of the 'line' being proportional to the volume – this symbol is then extended along the length of the route (Fig. 185). Movement over a long distance is therefore shown by much larger symbols than movement over a short distance. Although in some

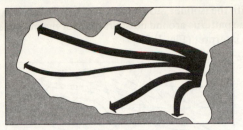

**185.** Proportional symbols and movement

cases this may be a significant part of the information, in other cases it may be irrelevant. However, the main problem is to represent variation in quantity by proportional symbols, and in this respect such maps are basically similar to others of this type.

## PROBLEMS IN REPRESENTATION

### Reference base and compilation base

The terms 'base' and 'basic' also tend to cause confusion with specialised maps, as they are used in several ways. Of course, there are basic special-subject maps, just as there are basic topographic maps. In this case the reference is to the fact that the map concerned has been produced from original information, and is the largest scale treatment of that topic for that area. Consequently, it can also act as a source for small-scale derived maps or map series.

All special-subject maps must include some locational information, which will identify the area of the map in geographical terms, and also provide information relevant to the use and interpretation of the map. This reference base will at least include the major land-water divisions, and major political boundaries. It may also show other administrative boundaries, inhabited places, or physical environmental factors such as relief, land use, vegetation, etc. Accordingly, there may be very few or a great many names, used to identify locations, inhabited places or areas. The reference base therefore may range from a simple topographic 'outline' showing only shorelines and international boundaries, to a quite complex representation of physical and cultural features. To some degree this is affected by scale, but it depends much more on the nature and intended use of the map. Large- and medium-scale special-subject maps, which are likely to be used in the field, necessarily must include

sufficient information to enable the user to establish location correctly. A simple choropleth map of a given region, at small scale, is little more than a diagrammatic representation of statistical values applied to areas, and as such is more likely to be used to obtain a general sense of spatial distribution of a limited phenomenon. So the degree of locational information will also vary considerably from one type of special-subject map to another.

Confusion occurs largely because many specialised maps are compiled on a 'base map', which may contain, for example, the boundaries of administrative or other units within which quantities occur, or a topographic base, which is used in the field to plot the locations of features. Specialised maps of any kind can only be produced if this 'base' information is available at the stages of collecting and preparing the specialised information, or is itself part of the whole map-making operation.

This compilation or working base, or 'base map', is not necessarily the same as the reference base used on the final map. Frequently, it is at a larger scale, and is more detailed in some respect. In a population map based on census enumeration districts, the boundaries of these districts need to be plotted to begin with in order to locate the census data, but this detailed outline is unlikely to be included in the final map. A geomorphological map may be produced by observation in the field, using a topographic map of suitable scale for locational detail. If this topographic base map is inadequate, or out of date, it may prove difficult or even impossible to establish ground locations correctly, thereby effectively preventing proper data collection. Again, the whole of the detail of the map used in the field is unlikely to be incorporated in the final map, though information taken from it may be used to construct the reference base itself.

In some cases the compilation base may be an existing map, containing all the necessary information, or with some information added to it. In others, especially for primary specialised maps, the compilation base itself may have to be produced by a ground or photogrammetric survey in order to plot the specialised information, and in turn will be used as a foundation for the reference base. At very small scales, the compilation of the base map in order to produce a specialised map may involve reference to many map sources, and the production of this itself may be a large task.

### Reference base

The reference base information for a specialised

map should be relevant in content, appropriate in generalisation, and complementary in graphic design. Therefore, it is strongly affected by the nature of the subject matter, the methods used to represent it, and the degree of generalisation. Failure to observe these related requirements is responsible for many of the deficiencies of specialised maps.

Graphic design must take into account the necessary contrast with the graphic elements used to represent the special subject, and also the requirement to keep the reference information at a secondary level, so that it does not intrude into the major subject matter. Legibility requires that the reference information has sufficient contrast to be distinct, but secondary visual level means that this degree of contrast needs to be less than that used for the main subject of the map. The difficulty of controlling this varies from one type of map to another, and it is most complex in specialised maps where the whole of the surface is affected by variations in coloured area symbols, or patterns of complex lines.

Achieving a secondary visual level is assisted by using neutral colours as far as possible (especially grey and brown); by using very fine lines if black is necessary; and by avoiding strong contrasts with any area symbols or patterns.

Complementary symbolisation is important (see Spiess 1978). If the main subject consists primarily of area symbols, then it is better to represent the reference information by line and point symbols (Fig. 186). Conversely, if the main subject is shown

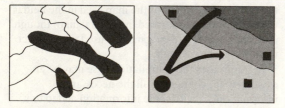

**186.** Complementary use of line and area symbols

by line arrangements, or consists of line and point symbols, the reference information should as far as possible make use of area symbol contrast. So on a geological map, for example, relief is best represented by contours, as these can be perceived against the geological area symbols. But if the map consists essentially of patterns of isolines, then areas of high elevation would be better shown by a contrast in area colour, possibly limiting the information to one or two hypsometric layers to show the elevation as 'background' information. A pattern of contour lines would compete for attention

with the isolines, and interwoven sets of different line patterns are the most difficult for the map user to interpret (Fig. 187).

**187.** Undifferentiated lines

In this respect, the common practice with large- and medium-scale specialised maps of using an existing topographic map, or certain elements of it, unmodified, can introduce many problems. Sometimes this is justified on the grounds of economy, as the cost of producing a different version of the detailed reference base would seem to be prohibitive, or even beyond the cartographic capacity of the organisation. But the effect is always likely to be inadequate, as the symbol design of the topographic map will have made use of the full range of contrast available, not to present the topographic information at a secondary level. Simply converting an existing map, or 'outline', to grey, ignores the fact that the line widths, point symbol sizes, and lettering weight, will have been chosen originally to be legible in high-contrast hues, and correct presentation in grey would require different symbol dimensions to be effective.

## STATISTICS, MAPPING AND CARTOGRAPHY

Although the term 'statistics' applied originally to data collected by the state, it is now used in a general sense to refer to any collection of numerical data.

When such statistical data describe spatial phenomena, then theoretically at least it can be used for 'mapping'. The relationships between mapping and cartography are rarely examined, despite the overlap between the two.

Scientists and researchers in many fields dealing with the human and physical environments are concerned with spatial location, in both two and three dimensions, and therefore the map is an important means of both representing and analysing spatial phenomena.

The objective may be either to present spatial

phenomena in map form, or to construct maps that can be used for spatial analysis. For this purpose, the map is only one means of two-dimensional representation. Pictures, diagrams, block diagrams and aerial photographs may also be used. In some cases the scientist is little interested in the true geographical location of the phenomena, but only in their internal spatial relations. Therefore, the dividing line between the map and other types of two-dimensional representation is relatively unimportant.

The point is well illustrated by publications dealing with spatial data and its analysis. Unwin (1981) gives a cogent and logical view of the map as perceived by the geographer concerned with spatial data analysis, and firmly avoids what he calls 'explicitly cartographic questions such as the choice of paper, pens or symbolism'. On the other hand Davis (1986), having noted the importance of maps to Earth scientists, states that 'Maps, in this general definition, include traditional geologic and topographic maps and also aerial photographs, mine plans, peel prints, photomicrographs and electron micrographs. In fact, any sort of two-dimensional spatial representation is included.' Unfortunately, because these specialists are equally interested in using diagrams and other two-dimensional graphics, there is a tendency for some of them to regard all graphic representation as 'cartography'.

The expression 'mapping' also has related but different connotations in other fields, and these in turn have clearly influenced the use of the term in spatial statistics. In mathematics, especially geometry, mapping can be described as 'a translation from one vector space to another' (Unwin 1981), and Hagen (1986) observes that a function 'maps one set of values on to another'.

The difficulty for the cartographer is that the treatment of such statistical data frequently requires elaborate methods of the data processing, and similarly the analysis of the map involves complex problems in spatial analysis. What should be evident is that the application of statistical methods to geology, geomorphology or social geography must depend primarily on the specialised knowledge and interests of the scientist. Indeed, although many geographers use statistics, they do not regard statistics as 'part of' geography. If they want to make maps in order to represent or analyse spatial data, then of course they are drawn into cartography, or at least 'mapping'.

In this respect four possibilities exist. The specialist map author may learn to be a competent cartographer, and then carry out, or at least control and supervise, the entire mapping operation. Or he may conceive the map and prepare the data, but depend on the 'cartographer' both for some contribution to graphic design, and execution (essentially the 'drawing' of the map). Third, a cartographer working in a specialist organisation, such as a geological survey, may learn enough about geology to understand properly the nature of the phenomena and the data being represented. But the ideal solution is a proper dialogue between the map author, who understands the purpose and informational content of the proposed map, and a cartographer, who can anticipate and solve the representational problems involved.

What should be clear is that no individual cartographer can possibly become an expert in every branch of environmental science, or be in a position to deal with the classification, data preparation and generalisation of specialised subjects. Unfortunately, elements of statistical 'mapping' are often described in textbooks on 'cartography', which suggests to the student that specialised knowledge, 'mapping' and cartography are all part of the same field.

It so happens that in many cases the statistical map is graphically simple, frequently in black and white and limited to a small format, even though the statistical data processing may be intellectually complex and demanding. The advent of simple graphics packages linked to small computers has meant that the difficulty of 'drawing', in the sense of depending on manual skill, can be avoided, thus making it possible for the specialist to produce his maps directly. As often they are essentially for personal study and research, serving a limited and temporary purpose, the graphic limitations of many such systems are no great hindrance.

## LAYOUT AND ARRANGEMENT

Special-subject maps dealing with particular areas often provide problems in presentation within a limited regular format. In many cases an extensive legend is also required.

Fitting an irregular shape into a rectangle means that the space between the map area and the sheet or page limits consists of several irregularly shaped areas (Fig. 188). The problem is to make use of these to include the descriptive and explanatory information needed for the map. A basic approach is to position the necessary information adjacent to the rectangular outline, so that continuity is

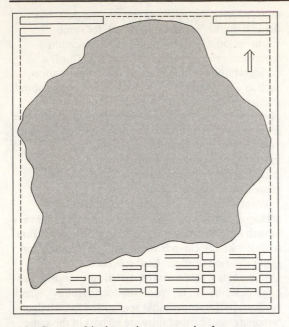

**188.** Geographical area in rectangular format

largest continuous space, with credits, sources, etc., at the foot. The descriptions of the symbols in the legend should be placed close to the symbols themselves, and not separated by spaces.

The inclusion and treatment of other information should be judged in relation to the nature of the subject, the informational basis of the map and potential map use. There is little point in adding a bar or graphic scale to a small map of the world. Indeed its inadequacy as a scale description may not be apparent to the inexpert map user. Whereas the nature and quality of source material may be highly significant (although in many cases it is not given at all), lengthy details of the personnel involved in making the map may be of little interest to the user.

If the legend includes a large number of items, these should be sorted out into suitable groups. With specialised maps, the principal legend normally describes the representation of the subject. A subordinate legend may give sufficient information about the topographic map base. If the legend has to be divided and placed in different marginal areas, the divisions should be made between complete legend sections. A complex legend requires considerable organisation. In order to arrive at a satisfactory solution it is essential that all the relevant information is assembled at compilation stage, so that it can be properly designed as part of the map layout.

preserved around the rectangular edges, allowing the uneven line lengths to face the irregular map area. In this example, if the map title can be placed top left, and the scale and bar scale top right, columns of legend symbols can be arranged in the

# 22 REPRESENTATION OF CONTINUOUS PHENOMENA

There are two main kinds of continuous phenomena, those that are divided into a series of classes by type (qualitative differences), and those that vary continuously in quantity, value or volume.

## CLASSIFICATION BY TYPE

Maps of this kind may include only a few sub-classes within the same subject category, or a large number. They depend fundamentally on area symbols and patterns, and graphically can be among the most complex maps produced.

### Compilation

A basic map of this kind, such as an original geological map, should ideally be composed from complete and consistent source material, which can be fitted properly to the topographic features identified for locational purposes on the reference base. This assumes that a number of factors have been taken into account. The field observations, or air photo interpretations, usually need to be at a scale larger than the final output map, but the scale difference should not be so large that considerable generalisation is needed at the cartographic production stage. If this occurs, it is essential that the generalisation should be supervised by the geologist. In many cases the same topographic compilation base can also be used to provide detail for the subsequent reference base. If, for any reason, another map is used (either directly or as a source of information) for the reference base, then the fitting of the geological survey plots to the new base may introduce further difficulties.

There are also many derived maps of this kind, in which the generalisation may pose special problems. Normal processes such as simplification may be carried out by the cartographer, but selective omission and combination need to be judged or agreed by a geological expert, as such changes require a proper knowledge of the subject matter and its classification.

The design specification of such a map is likely to be affected by established conventions, either nationally or internationally. Despite this, in many cases the specification has to take into account local characteristics. All area symbols have to be described in terms of their graphic components – solids, tints and patterns – and if the map is complex, the arrangement and combination of these images needs to be carefully worked out in terms of the production requirements.

In compilation terms, an outline is marked up to show the actual area symbols to be used (according to their proportions of solids and tints), and a production guide is made to identify the specific masks and patterns of which the symbols are composed. So that if a yellow at 50% is used in two or more different symbols, all the areas that contain the 50% yellow will appear on the same mask.

### Design problems, multi-colour maps

If the map is complex, and contains many different area symbols, then colour will be necessary for a clear representation. Although in theory all area colours can be composed of proportions of cyan, magenta, yellow and black (the four process colours) this may fail to take into account the needs of the reference base, and indeed may be inefficient in terms of the actual symbol specifications. If a dark blue line is preferred in order to show the drainage clearly, then cyan may be too light. If contours are to be included, they cannot be represented adequately at a secondary level by any of the single process colours. If grey is required, either for the reference base information, or for outlines of the geological or other classes, it is easier to construct and reproduce this as a grey rather than as a tint of the black. Therefore, the selection of printing colours should be made with the whole map in mind.

If the subject matter is not affected by any 'standardised' specification, then the choice of colours and patterns must be done in relation to the needs of the individual map. In fact this is easier, as the distribution, size and frequency of particular sub-classes will strongly affect the choice of suitable area symbols.

## Visual balance

The objectives of the map design can be stated clearly. Each area symbol should be readily identifiable wherever it occurs. None of the sub-classes should appear to be more important than the others; therefore, they should be visually in balance. And it should be possible to perceive the reference base at a secondary level. Achieving this will require full use of the graphic variables, particularly in the control of lightness and saturation.

In designing such a map, much will depend on the total number of sub-classes required; their relative areas on the map; and whether or not the total range contains major categories with minor sub-classes. In some cases, the map content will be divided into a limited number of major categories, with several sub-classes within each of these. In this case the necessary colour coding will tend to assign a dominant hue to each major category, and then make variations within each of these by modification to the principal colour, probably with the addition of patterns. This is the most difficult design problem. Using the contrast of dominant hues provides the most obvious design basis, and helps to identify the sub-classes within a common group; but the dominant hues differ markedly in lightness, and therefore their relative effect on the map will vary according to hue. For example, if one group is represented principally by red, and another major group by green, the red area symbols will be much more prominent visually than the green, especially if they cover large areas. This therefore opposes the objective of giving all classes equal visual emphasis. On the other hand, if the map is composed of a limited number of distinct classes, with no sub-divisions, contrast in hue can be more easily employed to produce a series of visually equal colours.

## Contrast by single colours

Because of the unequal lightness value of the primary hues, it is impossible to achieve a balanced map design by depending on single hue contrast as the only graphic variable. The strong hues such as red, which have maximum saturation at relatively

low lightness value, will dominate wherever they occur.

The reason for this can be demonstrated by reference to the Munsell system. The *Munsell Book of Color* (1976) displays some forty different hues, each with its full range of saturation and lightness variations (chroma and value in the Munsell notation). If a pure yellow is considered, such as 5Y, it is clear that the maximum saturation (chroma) occurs at the high lightness value of 8.5, and indeed there is only one lightness value (9) above this (Fig. 189). This colour (5Y 8.5/12) is at maximum saturation and can be regarded as the equivalent of the solid printed ink, assuming that a yellow ink could be matched with the Munsell hue. All the reductions in saturation within value 8.5 contain grey, and all the colours below value 8.5 also contain grey. Therefore, none of these can be matched in print without the addition of grey to this yellow, that is by adding a shade to either the solid or tint. The general rule is that only the maximum saturation colour at values above the solid ink can be matched by simply making tints. The comparison with a 'pure' red makes this clear. The 5R in the Munsell system shows that the greatest saturation (14) occurs at a lightness value of 6. Therefore, three other

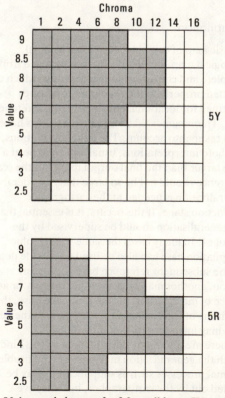

**189.** Value and chroma for Munsell hues 5Y and 5R

lightness values are possible as pure tints: 7/10, 8/4 and 9/2.

Theoretically, it would be possible to make a very small number of equal-appearing colours (those with equal lightness values) by using solid/tint variations of single hues. This can even be done with a yellow, but of course such a yellow in printing ink would be spectrally less pure than the Munsell 5Y, and therefore would be a darker colour. The range of tints would be controlled by the lightness value of yellow at maximum saturation, which lies at 8.5, or 8 on the shorter Munsell scale (the abbreviated version used in the Munsell Student Charts). This can be matched in lightness value by a very pale pink (5R 8/4), a pale green (5G 8/6), pale blue (5B 8/4), and pale purple (5P 8/4).

## Colours in combination

If the total range exceeds four or five classes, it is preferable to employ colour combinations, and this is the basis of the solution. Large numbers of perceptibly different colours can be made by combining two or three colours in an area symbol, using only combinations of solids and tints, and at the same time it is possible to control lightness value within fine limits. The reason for this is that the addition of tints of the darker hues modifies the lightness values of tints of the lighter hues. Thus the intermediate and combined hues provide a much greater range of colours at the same lightness value, without having to introduce shades as well. Very often the possibilities are extended by omitting any pure yellow. The yellow combinations, such as yellow-red and yellow-green are slightly lower in the lightness scale.

Choosing colours of apparently equal values from a printed colour chart is difficult, as the visual judgement of such values is strongly affected by the surrounding colours. One approach is to make a first selection by reference to the Munsell system, and then approximate this by reference to a colour chart. There are several colour charts available which show a range of combinations of solids and tints in the standard process colours (cyan, magenta and yellow), and others which use specified primary hues (red, blue and yellow). The *ITC Colour Chart* (1973) is the one referred to here. This uses 10%, 20%, 40%, 70% and 100% for each hue (magenta, yellow and cyan), as in common with most other cartographic colour charts, it only includes a range of percentage differences which are sufficiently distinct to be useful in composing colour variations. Although tint densities in 10% steps are perfectly

possible, the adjacent pairs are of little use in practice. In addition to the standard process colours of magenta, yellow and cyan it also includes black, to complete the four process colours.

It is very difficult to match hues exactly, as the hues shown in the Munsell book are far purer than printing inks. In fact, exact match of hue is not critical; it is the lightness value that counts. Depending on the overall characteristics of the map, and therefore the level at which the area colours are placed, a particular lightness value can be selected on the Munsell scale, and then variations in hue and saturation used, keeping this lightness value. A selection of these would be approximated on the colour chart shown in the table, all colours having a lightness value of 8 on the Munsell scale.

| Munsell colour | Percentage of standard process inks | | | |
| --- | --- | --- | --- | --- |
| | Yellow | Magenta | Cyan | Total % |
| 5R 8/4 ......... | 20 | 20 | 0 | 40 |
| 5YR 8/8 ...... | 40 | 40 | 0 | 80 |
| 5Y 8/12........ | 100 | 0 | 0 | 100 |
| 5GY 8/10 ..... | 70 | 0 | 20 | 90 |
| 5G 8/6 ......... | 40 | 0 | 20 | 60 |
| 5BG 8/4....... | 20 | 0 | 20 | 40 |
| 5B 8/4 ........... | 0 | 0 | 40 | 40 |
| 5PB 8/4 ....... | 0 | 10 | 20 | 30 |
| 5P 8/4 ......... | 0 | 10 | 10 | 20 |
| 5RP 8/6 ....... | 0 | 20 | 0 | 20 |

It is interesting to note that if the percentages for each colour combination are summed, they provide totals which approximate the relative lightness values of the different hues. This in turn can act as a rough check on the correct progression in the choice of tint combinations.

In composing such a design, the relative areas of the different classes should be taken into account. Classes that occur frequently as small areas require a slightly higher contrast than classes that cover larger areas on the map. This can be achieved by using slightly more saturated colours for small areas. In practice it may well be that one class occurs over both large and small areas, and in this case the specification has to be a compromise.

## Contrast by colour and pattern

With a larger number of classes, it becomes desirable to introduce pattern variations as well as contrast in colour, in order to avoid using colours that are too similar in appearance for clear identification. The ideal arrangement is one where the pattern effect is noticeable, without intruding into the overall graphic balance. For this purpose, fine patterns are essential. Multi-colour patterns provide a great range of possibilities, but they can easily add disturbing effects, especially if there are marked changes in orientation. If the contrast between the pattern elements and the background colour is to be maintained at a low level, then line patterns are more effective than dot patterns. Dot patterns are useful if contrast levels are higher.

The possible approaches include: (1) the addition of a line pattern in one colour to a continuous background colour; (2) the combination of patterns in two different colours; and (3) forming patterns by subtraction. It is also possible to combine (2) and (3).

In order to avoid changing the lightness value of the overall area colour to an appreciable extent, the addition of a line pattern should avoid a high contrast between the pattern and the background colour. A red line pattern over yellow will have a high contrast. A fine red line pattern over a light brown or orange will produce a noticeable texture, without changing the value to the same degree. The same basic principle applies to the second procedure, using two patterns. Fine vertical blue lines and fine horizontal green lines over a yellow-green base will retain a colour unity. Crossed lines of red and blue will not.

From a technical point of view, forming patterns by subtraction (by positive masking) is quite straightforward, and graphically it has the great advantage that it can be used to maintain the overall lightness of the coloured areas. The subtractive effect is also a function of contrast control. A line pattern subtracted from a yellow background will need to use relatively thick lines, in order to allow the white to contrast with the yellow. With darker hues, finer white or light patterns will be perceptible.

Where colours are used in combination, the combined effect of addition and subtraction can be used to advantage. For example, if a green is made by combining a solid yellow with a percentage of blue or cyan, then a line pattern subtracted from the blue will produce a combination of green with yellow lines. The pattern will be noticeable, but the overall lightness value will be relatively little affected. Similarly, if an orange is made by combining red and yellow in suitable proportions, subtracting a pattern from the yellow will leave a combination of orange with red lines.

As in all other cases, the use of patterns is conditioned by the areas of the symbols on the map. Very small areas not only require a higher colour contrast, but coarse or open patterns will be ineffective, because the perception of a pattern depends on having a sufficient area for the repetitive effect of texture to be noticeable.

## Achromatic colour schemes

If variation in hue is not possible, as in a 'black and white' map, then it can be difficult to represent different classes by area symbols which maintain the same lightness level, and thus have the appearance of equal importance. This objective can be approximately achieved over a short range by using patterns. The problem, once again, is to maintain similar lightness values yet provide sufficient contrast. It is compounded if the patterns have to be produced manually, but it can be solved if commercially available pre-printed patterns can be used.

In theory the solution is to employ a range of patterns in which the total proportion of any unit area covered by black ink is approximately the same. It is usually necessary to introduce changes in orientation. For example, a horizontal black line pattern can be contrasted with an identical line pattern placed vertically, and they will be equal in value. Using the intermediate diagonal orientations, in both directions, gives two more possibilities (Fig. 190). Unfortunately, the appearance is often unattractive, especially if relatively coarse patterns are used.

**190.**  Line pattern orientation

Instead of relying solely on orientation of the same pattern, it is possible to choose a limited range of different patterns that have the same percentage value at different rulings (Fig. 191). A series of such line patterns is produced by Mecanorma (1984). For example, at the value of 50%, five usable line patterns with different line widths and spacings are available, and of these certainly the first four have sufficient contrast. The higher density line rulings

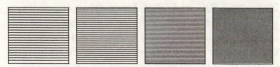

**191.**  Equal pattern density at different screen rulings

approach tint screens in resolution, and as they give the impression of a continuous grey, rather than a pattern, they are of no use for this particular purpose.

A greater range of densities in dot patterns is available, but for dot patterns contrast by different orientation is not possible. These dot patterns are also classified by line rulings, like tint screens, and include 27, 32, 42, 50, 55, 60, 65, 75, 85 and 100 lines. All the adjacent pairs have too little contrast to be effective, but it would be possible to make use of a range such as 32, 50 and 65 at 40%.

# CONTINUOUS VARIATION IN QUANTITY

In this type of representation, the continuous 'surface' may be either relatively static or constantly changing. Graphically it makes no difference to the methods of representation, although it obviously affects the nature of the information. Equally, the 'surface' may be visible and permanent, or intangible. From the point of view of cartographic representation, rapidly changing phenomena, like temperature, have to be treated either by representing conditions as they appear at a given time, or by aggregating time periods or calculating mean values. However the data are processed, the graphic problem remains constant.

Basically, such a surface can be represented by isolines, from which the variation in quantity can be interpreted. Data records relating to observation points, such as climatic record stations, can also be treated individually as point locations, in which case they can be represented like any other point symbols to which values or quantities are related. If the variation in quantity can be directly measured or detected, like the topographic surface, then it is at least theoretically possible to measure it at every point, thus providing complete and consistent information. Either the contour values can be specified, and then these lines individually plotted or measured (as in a photogrammetric contour plot), or the surface can be measured as a series of point values, and then contours derived as required. In practice, of course, these points will not be adjacent (and indeed such detail would be highly uneconomic) but they can be sufficiently close to provide a fine network. Such an assembly of quantitative data can be described as a terrain or elevation model. If such data are stored and

processed digitally, it is usually referred to as a digital terrain model, or digital elevation model.

A slightly lower density of information can be provided if a succession of profiles are obtained, along which the variation in quantity is measured. These data are typical of the hydrographic survey. At its best, the lines of depth measurements will be close together, so approximating the complete data model. On other occasions the lines of soundings will be further apart, but will still provide consistent information over an area.

At the other end of the spectrum, records from climatic stations or boreholes are likely to be taken at relatively few locations, and may be widely dispersed, giving an uneven distribution of basic information.

## Contouring

The interpolation of isolines is closely related to the nature and distribution of the data, and the assumptions made about gradients between points (Peucker 1980). With a complete data model, threading contours or other isolines is a relatively straightforward task. Given the amount of data involved in such a representation, the construction of the isolines is invariably carried out by a computer program, and many 'contouring' programs exist. The interpolated isolines will be close to measured contours in quality.

The task of contouring is most difficult when only an irregular array of dispersed point values is available. In such cases, the normal practice is to use these actual values to construct a grid of points, with either a rectangular or a triangular method. Much depends on the scale of the representation in relation to the distribution of the measured points.

With a Cartesian grid method, the grid points are calculated by reference to a number of adjacent known values that are weighted according to their distance from the grid point (Fig. 192). The known

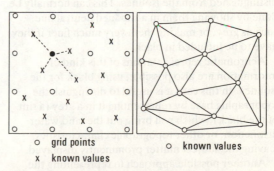

o  grid points
x  known values

o  known values

**192.** Cartesian grid and triangulation

points to be used are selected either as a fixed number, or by including all those within a given radius. In this way the closest known values have the strongest influence on the calculated values. Such a procedure is virtually impossible without computer processing, and again there are numerous programs that will carry out this routine, and subsequently 'thread' a series of isolines or contours (Yoëli 1977).

Clearly the quality of the isolines is essentially a function of the density of the original information in relation to the scale of the map. If there are considerable distances between the measured points, then the isolines will be of a very low order. Unfortunately, this is rarely revealed on maps of this type, although the best maps always show clearly the actual measured points to make this clear to the map user.

With the triangulation method (Elfick 1979; McCullagh and Ross 1980) all the known point values are connected into a series of triangles, and then the isarithms are derived from these. There are several ways in which this can be done. The main argument in favour of this method is that the isolines are calculated directly from the known values, whereas with the gridding method the isolines are calculated from a different set of derived values.

If the object of the representation is to provide information from which variation in the surface can be interpreted, then the policy of a constant vertical interval should be followed, as with contours on topographic maps. For some reason, the method of assisting the user by adding index contours rarely seems to be adopted for special-subject maps, possibly because they usually extend over a much smaller range of values than a topographic map series. However, the graphic representation is comparatively simple, as the continuous lines provide the most important element in the map information, and therefore any relatively dark hue can be used against a light background. If the topographic reference base includes rivers, shorelines, etc., then these should be fully distinguished from the isolines. This can normally be done by showing them in a subdued neutral hue, such as grey, or making them very much finer if they have to be included in black.

Achromatic representations of this kind of information are also possible, using black for the isolines. In this case it is useful to distinguish the topographic base by representing it in a grey (a tint of the black), which will bring out the land/water distinction, or other topographic characteristics, leaving the subject matter prominently displayed.

Another possible approach to representing the three-dimensional form is to use tonal shading, in the same way as the representation of slope in topographic maps. Given a 'contoured' base, this could be done 'subjectively'. Lavin (1986) describes a method of generating this effect directly by using a computer program to plot different dot densities in proportion to the desired tonal values.

## Semi-continuous phenomena

Although strictly speaking precipitation is not continuous in either space or time, the data collected to describe it are of the same kind as those for many continuous distributions. Consequently it is treated graphically in the same manner.

## Graded colour series

The changes in quantity over a surface can also be represented by graded colour series, in which the areas classified by the isolines are further emphasised by the addition of colour. Cartographically this is essentially the same process as using hypsometric layer colouring on topographic maps of the Earth's surface, except that the principle of 'the greater the darker' is invariably followed. In this case the objective is to provide a smooth progression of area colour, and this again is controlled by varying the lightness and saturation of hues.

Such a colour gradation can be a single scale, showing a range above a datum or zero, or a double scale, showing values above and below the datum or zero.

## Single hue methods

The number of possible differences available within a single hue is a function of the lightness value of the chosen hue and the relative areas on the map. Large areas can be shown with a smaller difference in value than small areas which require a higher level of contrast. If it is not necessary to devise the colour scheme so that reference base information must also be perceptible, then the full range of colour contrast can be used, making use of the low lightness value of dark hues such as purple.

If the *ITC Colour Chart* is referred to again, then it is clear that for either magenta or cyan the adjacent tints are too close to be easily distinguished, even though the range is limited to 10%, 20%, 40%, 70%, and 100%. In particular, the 70% and the solid (100%) are visually almost identical. Consequently, a single colour of this lightness value could only be used for a short scale, probably not more than three colours. The range can be increased if a darker hue is used, such as a dark blue or a dark red.

## Two-colour combinations

Instead of depending entirely on contrast in lightness value, contrast in hue can also be employed if two colours are used in combination. A common practice is to keep one colour constant, and to vary the lightness of the other by making a range of percentage tints. Using the same colour chart, such a series combining yellow and magenta would be

Y20 Y20M10 Y20M20 Y20M40 Y20M70 Y20M100 %

The contrast at the lighter end is good, but tends to weaken with the higher densities as the magenta becomes more dominant. Compared with the single colour scale, colour contrast is improved, but gradation in lightness is slightly reduced. The procedure is technically simple, as one hue is held constant, and therefore would be constructed with a single mask.

## Three-colour combinations

Combinations of three hues give a greater range of possibilities, especially in the use of darker colours. Again the method of keeping two of the hues constant and varying the third is commonly adopted, as technically this simplifies the production and reduces the number of tint masks that need to be made. Overall, the series has a better gradation and contrast than that obtained with two hues. A possible series is as follows:

Y10C10 Y10M10C10 Y10M20C10 Y10M40C10 Y10M70C10 %

If it is possible to use the full range of saturation and lightness, then a higher level of contrast can be obtained by choosing colours along a diagonal from upper left to lower right on the colour chart. Such a series would read

C10  M10C20  M20C40  M40C70  M70C100 %

## Double progression scales

For this arrangement, what is wanted is a progression above and below a common datum, the lightest values being adjacent to the zero or datum. If a colour chart is consulted, which shows the lightness variations in magenta and cyan tints, and

then their combinations, on a constant 20% yellow, a scale of at least five values in each direction can be obtained. The difficulty once again is that the dark colours tend to be too close for adequate contrast. The scale can be lengthened as follows.

(Y20) M10 M20 M40 M70 M100C10
C10
C20
C40
C70
M10C100

The yellow 20% on its own would not be used. Further values could be produced by using variations towards the purple; a blue-purple to lengthen the cyan scale, and the red-purple to lengthen the magenta scale. M40C100 and M100C40% would be suitable.

## Colour choice

Theoretically (as with hypsometric colouring), either the 'higher the darker' or the 'higher the lighter' may be used, but colour association also plays its part in the choice of colours. With bathymetric layer colour schemes, blues and purples are invariably used because of the common colour association. Similar considerations apply to many kinds of natural phenomena. For maps of temperature, like elevation, 'cold' colours are associated with low temperatures and cold climates, and 'warm' colours with high temperatures and hot climates. Consequently, such maps normally employ colour scales that proceed from yellow through orange to red for the higher temperatures, and from pale green through blue to purple for the lower temperatures. Similarly, precipitation maps tend to make use of blues, greens and purples, rather than oranges and reds, which suggest 'dry' conditions.

Where there are no natural colour associations then the choice of colour can be independent of the nature of the phenomenon. Even so, large quantities are often associated with red and red-purple, and smaller quantities with light green and yellow. In this, the relative lightness value of the hue seems to be connected with its apparent relationship with quantity or degree, the low value hues such as red and purple, or even dark brown, suggesting the larger quantities.

## Achromatic series

In achromatic maps, which cannot employ contrast

in hue, the range of lightness variations is limited to what can be achieved within black and its intermediate greys. This is similar to the other single colour schemes that use the solid ink and a series of tints. In the case of black, the number of different tints that can be used is greater, although the solid black is rarely of any practical value. Generally, it can only be introduced if the areas of maximum quantity are very small in extent. A range of tints such as 10%, 20%, 40%, and 70% should be adequate, but if any attempt is made to increase the number of gradations by using the intermediate percentage values, there is a risk of insufficient contrast.

If tint screens are not available, and if the maximum tint area is relatively small, then it is also possible to employ pre-printed patterns and tints, in either dots or lines. Mecanorma (1984) provides a series of dot percentage 'tints' at a range of line rulings. Therefore, relatively coarse or open patterns or tints can be chosen, depending on the characteristics of the map. Tints are given in 10% steps, from 10% to 80%. The adjacent pairs of tints lack contrast, especially in the fine rulings, but a series such as 20%, 40% and 70% would be satisfactory. Contrast would be improved by selecting percentages from different screen rulings, which would maintain the gradation but add another visual clue in differentiation. A possible selection would be 85–20%, 65–40%, 55–60% and 42–70%.

If such materials are not available, and the patterns have to be produced by drawing, then two approaches are possible. Either a standard spacing is used, and a series of different line widths constructed to form the gradation, or the same line gauge can be used, and the lines spaced at gradually increasing intervals. For a short range, the former is more likely to achieve a satisfactory visual balance, and is easier to control.

If only black is available, and it is not possible to use produce greys with tint screens or very fine patterns, any reference base information is best shown in fine black lines, and the maximum density of the area symbols restricted to 70%.

## Equal-appearing intervals

In the case of graded series, it is often suggested that if the vertical bands defined by the isolines represent a fixed amount or quantity, then the variations in colour used to represent these bands should appear as equal intervals. Although this basic concept is certainly applied in the control of lightness value to produce a graded series, either achromatically or with different hues, it is made difficult by the unequal response of the visual system to variations in density. Generally speaking, discrimination is easier at the lighter end of the scale compared with the darker: that is, a difference between 20% and 40% is apparently greater than the difference between 70% and 90% or 100%. Some percentage tints do try to accommodate this visual response, being produced at equal-appearing intervals rather than at fixed percentage intervals. These are certainly appropriate with achromatic maps, where the distinction is entirely a matter of relative lightness. With multi-colour schemes this relationship is less critical. As long as the series of colours follows the general order of lightness value, without any sharp discontinuities, it is unlikely that map use will be seriously affected. The greatest danger is in using two colours or percentage tint values that are too close in either hue or lightness to be readily discriminated. This will cause more confusion on the part of the map user than minor variations in the smoothness of the colour progression between light and dark.

# 23 REPRESENTATION OF DISCONTINUOUS PHENOMENA

There are a great number of phenomena located at specific places or within small areas, such as industrial plants. Others, such as human population, exist as discrete individuals. Although variation by type is sometimes involved, the main emphasis is normally on quantity; either the actual quantity at a point, or a quantity in relation to an area. Maps of this kind are based on statistical data. Although large quantities of such data are available, much information tends to be of limited value for map making, as data are essentially collected for administrative, management and research purposes, not specifically for making maps.

Many variations in symbol form and dimension have been devised for representing this kind of information, and the use of the diagram is common.

## Distribution of a single type

In some cases, the map only shows the location of a particular phenomenon or subject, or the locations of series of phenomena. Depending on the nature of the subject, either the actual areas are represented, or simply the point locations at which it occurs. For example a map of peat, or a certain type of woodland, may show the areas against a topographic base. On the other hand, a map of watermills would show the location of the mills by point symbols together with the relevant waterways.

Both kinds of map are cartographically simple, as long as the scale is appropriate to the density of information. Although colour may be an advantage in drawing attention to the subject, and distinguishing it from the base, such maps are also easy to deal with in an achromatic representation. If the areas are relatively small, then solid black can be used to effect, and similarly solid black point symbols for point locations. In some cases a number of different distributions are presented simultaneously to facilitate comparison of location, and for this different patterns may be used for areas, and variations on point symbol form for point locations. A range of such symbols is not difficult to devise, using variations in form, addition and extension.

## QUANTITIES LOCATED AT POINTS OR SMALL AREAS

Although in many cases the actual quantity refers to an area, such as the output of an industrial manufacturing region containing many industrial plants, such areas are often treated as point locations on small-scale maps. The usual method of representing such quantities is by proportional symbols. The term implies that the sizes (magnitudes) of the map symbols will be in proportion to the numerical values being represented.

Cartographically, this raises intractable problems, both in representation and interpretation. On the one hand the two-dimensional map is a poor device for representing differences in quantity, and indeed cannot graphically demonstrate numerical differences unless they are restricted to a very small range. On the other hand, the ability of the map user to discriminate between small symbols which may be very similar in size, and to match two or more symbols in different locations in the visual field, is both limited and individually variable.

Many attempts have been made to determine experimentally the visual response to a range of symbols of different magnitudes (see Baird 1970). Most of them appear to show that comparative estimation of the areas of two-dimensional figures is poor, and in general underestimation takes place. There is considerable variation between different people, and the way in which the experimental test is arranged also influences the outcome. As the visual system cannot actually measure the figures (map symbols) size estimation is a matter of judgement. Discrimination between symbols of similar sizes – the ability to state that two symbols are the same or different – operates at a relatively high level in visual perception, provided that the symbols themselves are not very small, and provided that they are not widely separated in either space or time. Therefore, it is at its maximum if both symbols can be perceived simultaneously in central vision, and at its lowest if one symbol magnitude has to be remembered and then compared with other symbols placed at a considerable distance from it.

Generally speaking, attempts to 'solve' this problem have tried to improve the interpretation performance by modifying the representation. But these can only succeed to the extent that they take into account the fundamental limitations of the graphic representation of quantities.

Unless the numerical range is severely limited, the proportional symbols cannot be made to match the range of numerical values exactly. If the statistics only included quantities or values covered by the numbers 1 to 10, then a series of symbols could be constructed proportional to the digits, and it should be possible to perceive the differences between them. Even so, the difference between 8.8 and 9.1 would be indistinguishable. If a large numerical range is involved, and with statistical maps this is very often the case, then the symbol sizes needed to represent large quantities in correct proportion would soon become excessively large, and indeed could be larger than the map itself. For example, if 5 is represented by a circle with a radius of 1.26 mm, then 10 would require a circle of radius of 1.78 mm (Fig. 193). It would be possible to represent 5000 and 4900, but the difference between them would be imperceptible. A range of 1000 times (5 to 5000) is quite small numerically, yet the circle representing 5000 is large graphically. On the same basis it would be impossible to represent 5 million on a normal map.

Theoretically, the user should be able to perceive that the largest circle in Fig. 193 is 1000 times larger than the area of the smallest. Of course, the user can name no such judgement, and minor variations to accommodate either underestimation or overestimation (both of which can happen) would still not make this possible.

The representational problem can only be partially solved by severely limiting the number of

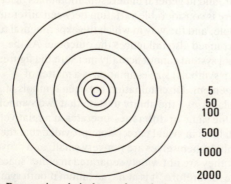

**193.** Proportional circles and symbol size

symbols used and their maximum and minimum sizes. It must be possible to discriminate between all the symbols, and the map user must be able to identify each symbol by reference to the legend. Two different approaches are possible. Either the symbols used on the map are given sizes relative to actual quantities, and then a set of symbols at particular values are shown in the legend; or the total range of data is divided into sub-classes or groups, each represented by a common symbol (Fig. 194). In the first, a selection of symbols will be shown at appropriate intervals, such as 10 000, 25 000, 50 000, 100 000, etc. In the second, the legend will contain a series of symbols in rank order, a given symbol representing a range such as 100 000 to 250 000.

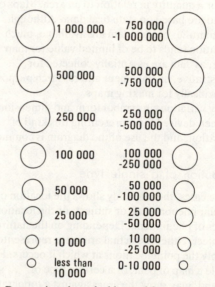

**194.** Proportional symbol legend for selected values and for classified ranges

The first method may come closer to representing the true quantities, but the map user is left with the task of trying to judge the value of a particular symbol on the map in relation to those shown in the legend. The second should make symbol identification easier, but often has to group a large numerical range under one symbol. In both cases, the actual symbol dimensions may be devised on an arbitrary 'scale', ensuring that they are sufficiently different, but not attempting to base them on their numerically correct 'sizes'.

Dividing the total range into a set of sub-classes, each represented by a symbol of distinguishable size, also raises the problem of how these divisions of the total data range should be made. Numerical symmetry suggests that sub-classes should be equal, or constant, rather like a vertical interval for contours. Although such an arrangement is easiest to understand, and to produce, it often fails to accommodate the characteristics of the phenomenon. Where there is a great variation in size, a better practice is to attempt to sub-divide the total range into sub-classes according to the actual grouping of the data. A common method is to make a frequency distribution graph, in order to locate any groupings and breaks in continuity, so that values that are ranged round a common level are represented by the same symbol, and not divided arbitrarily into different groups. In most cases it is possible to arrive at a solution that provides sensible groupings without involving very uneven sub-classes numerically.

The fact that certain round figures, in decimals, such as thousands, tens of thousands and hundreds of thousands, seem to be obvious choices, should not disguise the fact that this may bear no significant relationship to the phenomenon. For example, in dealing with the populations of large cities, figures such as 1 million, or 250 000 have little to do with the actual population sizes which characterise many large cities.

None of these approaches actually solves the problem of visual interpretation. Although the map user can distinguish between two similar symbols if they are seen in close proximity, he cannot measure by how much they differ in actual quantity. The judgements are comparative, not absolute. If the difference in size between two symbols is great, then one is perceived as 'much larger' than the other, but deciding whether it is actually twenty-two or twenty-five times as large is quite beyond the capacity of human visual perception.

This in turn raises the question of the primary function of the map. Obviously it would be highly desirable if the map could show location and represent all the quantities correctly. But maps are used primarily to demonstrate spatial arrangement and distribution. If the actual quantities are significant, the only way to accommodate this is to show them in figures, either on the face of the map adjacent to the symbols, or in a separate table.

## 'Three-dimensional' symbols

Because of the difficulty of attempting to represent a large numerical range by proportional symbols, and because the finite map space means that the dimensions of the largest possible symbol are limited, various expedients have been tried to accommodate very large numerical ranges. One is to use 'three-dimensional' symbols, such as spheres or cubes. Of course, the symbols are not actually three dimensional: the illusion is contrived either by shading a sphere or using perspective geometry to suggest a cube. The relationship between a quantity and its representation by such a symbol is still an act of interpretation. As users generally find the relative comparison of even two-dimensional symbols difficult, not surprisingly the use of apparently three-dimensional symbols is even more ambitious. The shaded sphere, in particular, can be graphically attractive, but it does nothing to solve the problem of interpretation of actual size.

## Multiple variables

Several different types of data may be related to individual point symbol locations. It is possible to distinguish between: (a) sub-divisions of a total, shown by sub-dividing a single symbol; (b) several different totals grouped together at the 'point'; and variations in either time or direction at the point.

In the first, the proportional symbol may be sub-divided into sections, each representing a certain proportion or quantity by type (Fig. 195).

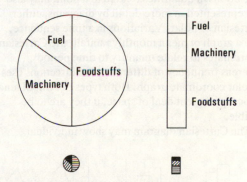

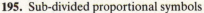

**195.** Sub-divided proportional symbols

For example, the exports from a port may be described as a total volume divided into categories, such as foodstuffs, machinery, fuel, etc. A proportional circle can be divided into sectors, and a square or column may also be sub-divided into sections, each distinguished by colour or pattern. In the second, a group of columns, squares or rectangles may be placed together, each representing a quantity for one particular type of sub-group (Fig. 196). Although these representations take on the appearance of diagrams related to points, they differ from other diagrams in that only one dimension is used.

With all symbols of this type, problems are encountered with very small symbols, which tend to

**196.** Multiple proportional symbols

become illegible if attempts are made to sub-divide them into small sections, whether the symbols are circles or squares. The single symbol emphasises the total quantity better, but can make the identification of small proportions difficult. The multiple symbols represent the individual component sizes better, and facilitate direct comparison, but tend to make the comparison of total sizes more difficult. Where large symbols of this kind are placed in large administrative divisions, the map is little more than a locational outline with diagrams. For such maps the reference base itself should be highly simplified, to avoid giving any impression that the map is dealing adequately with the locational characteristics of the subject.

## Diagrams at points

Variations in quantities related to a point may also be represented in more detail by diagrams, either Cartesian or polar. Variations in a time sequence, like a graph of mean monthly rainfall, use Cartesian coordinates to relate quantity to time periods, whereas frequency of different wind directions uses a polar coordinate graph. Both types of symbol tend to occupy a great deal of space, if they are to be legible.

The Cartesian diagram may show individual

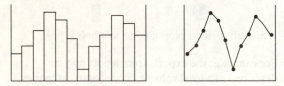

**197.** Cartesian diagrams

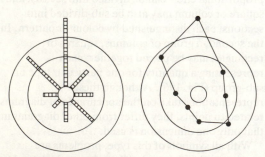

**198.** Polar diagrams

values for each stage, leaving the interpretation of the general change or trend to the user, or, if the phenomenon is continuous, may use a curve through the points (Fig. 197). Similarly, the polar diagram can show the individual quantities at particular vectors, or emphasise the overall pattern with an isoline through the points, giving a continuous 'shape' (Fig. 198).

## Overlapping symbols

Because the proportional symbols take up space, and because the larger ones occupy comparatively large areas of the map, actual geographical location may cause problems, as it frequently happens that one symbol intrudes into the space occupied by another, or a large symbol conceals a smaller one. The general rule graphically is that the larger symbol is interrupted to show the smaller one completely. With solid symbols, such as solid circles, small symbols contained wholly within the area of a larger symbol are shown by outlining.

Where the quantities are shown by columns, the problem is more difficult, as any marked interruption in a narrow column would reduce its legibility considerably. In this case the location of the symbol may be slightly displaced to ensure that it is unbroken, and to avoid interruption (Fig. 199).

**199.** Overlapping proportional symbols

Partly to avoid this problem, a series of quantities related to a point may be arranged at angles about the point. Although these are similar in appearance to vector diagrams, the angles of placing do not have any specific function, apart from differentiating the various quantities. Of course, the same angle must be maintained for each component.

## QUANTITIES RECORDED FOR AREAS

In those cases where the data are only available for areas, usually administrative divisions, a limited number of possibilities exists. The difficulty is that the true location of the phenomenon remains unknown, although it may be deduced from other informational sources. For example, agricultural returns may only be published in terms of administrative divisions, or agricultural regions. This does not reveal the actual location of the production areas, which may be a very small part of the total administrative division.

Three approaches are possible. First, the total quantities can be represented by proportional symbols, and these 'located' simply by placing them within, or over, the administrative division. In this method the data are treated graphically as though actually located at a point. Second, the total quantity may be expressed in relation to the total area, providing a mean value or density. And third, either of these two may serve as a basis for interpolating isolines, which can then be used to represent a 'surface', as described previously.

Maps that show discontinuous distributions in terms of totals applied to arbitrary administrative units are themselves quite arbitrary, and lose the essential informational basis of any map, which is to show location. Such maps are better described as 'diagram' maps, as they combine some of the properties of the map with the properties of a diagram. Consequently, they are crude devices, and it is to be hoped that map users understand the nature of this misrepresentation. Graphically they are constructed on the same basis as those described above.

## Choropleth maps

Where the total quantity is distributed over the entire area of the administrative division, giving an average or mean value, there are similar consequences for interpretation. Unfortunately, large amounts of data are available in this form, and the production of the choropleth map is comparatively simple and straightforward. What the map states graphically is that the phenomenon concerned is spread evenly and continuously over the area, and that any change in the density occurs only at the bounding lines of the administrative divisions. The phenomenon is represented as though it were continuous. This is not intended to be taken literally, and indeed the map user is not even expected to believe what the map states. Lack of better information and ease of construction make this kind of representation all too popular.

Cartographically these maps pose the same problem as any other graded series, where a limited number of area symbols are used to indicate differences in value, normally on the 'darker the greater' principle. In an achromatic map, these value differences can be expressed either by a range of tints of the black, or by a set of patterns that follow the lightness scale. Care must be taken to ensure that adjacent percentage tints or graded patterns are not too similar in density to be effective for this purpose. If tint screens are not available, the series of values can be produced by using pre-printed patterns, the best of which come close to screen tints in resolution. Although such materials are normally available in 10% steps, these also are too close visually to have sufficient contrast, and it is difficult to obtain more than four steps (such as 20%, 40%, 60% and 80%) on this basis. If the patterns are available in various rulings, then the effect of texture can be used to reinforce discrimination (Fig. 200). Coarser rulings can be used for the ligher tints, and the tint density difference accompanied by a ruling difference as well, thus reinforcing the contrast.

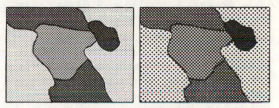

**200.** Graded patterns (same ruling) and graded patterns with different rulings

Considerable attention has been paid to the problem of obtaining visually equal intervals (see Williamson 1982), but the informational basis of such maps is so low, and the representation of the actual phenomenon so approximate, that this concern with the details of representation, in contrast with the deficiencies of the information, is quite unwarranted. The fact is that the map is largely a creation of the boundaries of the administrative divisions, and several different 'maps' of the same phenomenon can be made, depending entirely upon how the data are treated. This has been shown by many researchers, and the point is clearly made and illustrated by Forbes (1984).

If the subject has a clear relationship with topography and/or land use, then sometimes it is possible to improve the representation by first determining the likely areas of occurrence, and then re-calculating the mean value or density in relation to these areas. This variation is often referred to as a dasymmetric map. Agricultural statistics, for example, giving figures for total output or crop yield for administrative districts, can be dealt with in this manner, if it is possible to approximate the areas of crop production. The same approach can be applied to small-scale maps of population density.

## Data based on regular units

In some cases, the data may be collected on a regular grid square basis. This has the great advantage that the locational basis is consistent, and if a suitable scale for the map is chosen, then a comparatively fine resolution of the overall variations can be presented. If the unit square occupies a small area (for example a 2 mm square) at map scale, then the real variation in density becomes apparent. Graphically, such maps still follow the same principle, in that differences in quantity are expressed by variations in lightness of the area symbols. The sub-division of the total numerical range will be a function of the number of perceptibly different area symbols that can be constructed. If colour can be used, then the problem is fundamentally the same as for sub-divisions of a continuous phenomenon, but a close array of coloured small symbols can be difficult to produce technically. For an achromatic map, a series of four or five tints is usually the maximum, depending on whether the solid black can be employed for the highest value. If the unit areas are small at map scale, which they should be for a good representation, then it is virtually impossible to use patterns because of the small symbol dimensions.

If the interest lies in a general impression of the total variation, then a small degree of difference between the values can be employed, even though this may make the identification of an individual square difficult. But if it is desired to make each square clearly identifiable in relation to the legend, then additional contrast will be needed, to compensate for the decreased size of the visual targets.

Data compiled on this basis can also be used for the interpolation of isolines, treating each grid unit as a point (Fig. 201), and this may be easier to deal with graphically than the construction of different densities in a fine raster.

## Isolines interpolated from area data

A map of density based on average or mean values inevitably contains sharp discontinuities if produced as a choropleth map, which is dominated by the outlines of the administrative or other divisions. But if the units for which the information is available are relatively small in area at map scale, and do not differ greatly in size, it is possible to use these data to construct a representation as a continuous 'surface' by isolines (Fig. 201). The method is to place the mean value centrally in each area, and then treat it as though it were a point value. The normal procedure for interpolating isolines is then followed. Clearly, much depends on the actual distribution and the density of the calculated values. Once the isolines have been constructed, the map can either show these as a 'contoured' map of the assumed surface, or reinforce the visual impression by adding a graded series of colours or tints between the isolines. Graphically, it is the same as any other map using isolines to describe a surface.

This method needs to be treated with extreme caution. Provided that the areas are relatively similar, and that the subject itself is widely distributed, the construction of a continuous surface representation may be justified. If the areas for which data are available differ greatly in size and shape, and if the actual phenomenon is limited to only a small part of the total area, such a map becomes grossly misleading.

# MAPS OF DISCRETE INDIVIDUALS

Population of either human beings or animals is recorded as aggregates of individuals. Although they may occur in dense groups, they remain individuals, and as such differ from other totals in which any numerical value is possible. Population figures can only be obtained by recording the population in terms of a fixed location at a given time, which is the normal basis of any census. Although it is theoretically possible to map individual people in terms of their place of residence, in practice the totals are recorded in relation to areas, using small areal units where large concentrations exist, and larger areas where the

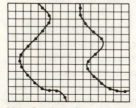

**201.** Isolines from area data

population is more dispersed. The resolution of locational detail is therefore a function of the initial collection stage. There are occasions when a regular grid square basis is used, especially in urban areas, and this approach has at least been demonstrated experimentally in a population census (Forbes and Robertson 1967). Populations are not evenly dispersed over any area, and they change location over time. To represent them as though they were continuously and evenly distributed is geographically incorrect, even though it may be statistically 'true'.

Three possible methods of representation are the same as those used for other data collected for areas. In the first, the total enumerated for a given area can be represented by a proportional symbol, located within the area concerned. Such a map makes no attempt to show actual location, and is a diagrammatic map. Cartographically the procedure is the same as for any other map using proportional symbols. Second, the mean value or density for an area can be shown, using the choropleth method. As the population is not continuously distributed, it suffers the same deficiencies as other choropleth maps of this type. Third, the construction of isolines, treating the mean or average for any area as a point value, is also possible. But in addition the dot distribution method can be used, which does attempt to portray the actual distribution in an approximate fashion.

## Dot distributions

This method involves three principal factors: the choice of dot value (the number that each dot represents); the choice of dot size, form and colour; and the distribution of dots to suggest the actual distribution of population.

Immediately the problem of total range of data becomes apparent. It is very difficult to deal with both heavy concentrations and wide dispersions of human population by a single method, so that in many cases two complementary methods are used. The populations of urban areas are usually treated as totals based on point locations, and the dot method is limited to the population in semi-urban and rural areas. Other data of this type, such as that applying to livestock, does not usually extend over such a range in density, and the dot method alone may suffice.

The graphic problem is to decide on a dot value and size which will be sufficient to represent small concentrations of population, yet will be legible where the population is widely dispersed, with some

dots being isolated. At the same time, the maximum concentration should not exceed the point where numbers of dots are adjacent, showing the areas of greatest density. Small dots with a low value make possible a detailed treatment of distribution, but risk illegibility where the dots are separated. A larger dot with a higher value has a greater graphic effect, but makes the problem of distribution more difficult (Fig. 202). It is a classic example of conflicting requirements in design and representation, and inevitably the decision must be a compromise. It is essential that the individual dot should be legible, and therefore the best dot size is the minimum that can be readily detected. For this reason, maximum contrast is needed, and generally dots are shown in a solid black, or any hue that is dark in the lightness scale, such as dark red or purple.

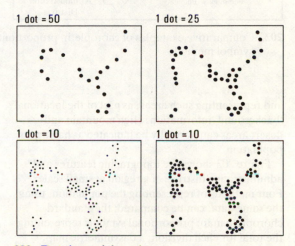

**202.** Dot size and dot value

The question of distribution is largely a matter of the scale of the map and the relative sizes of areas for which data are available. If the scale is such that a topographic map, which represents all buildings and built-up areas, can be used as a reference base, then the dots can be distributed in accordance with residences, as the census data for human population are collected on this basis. A primary difficulty is that the boundaries of the collection areas rarely have any straightforward connection with the actual variation in density. The edges of rural areas close to urban areas may have relatively high concentrations, but these may not be immediately evident from the data themselves. Even so, given a sufficiently small unit value, it is possible to make a reasonable approximation of actual distribution.

In sparsely populated areas it is clear that much of the surface will have no resident population at all. The base map can often be improved by identifying

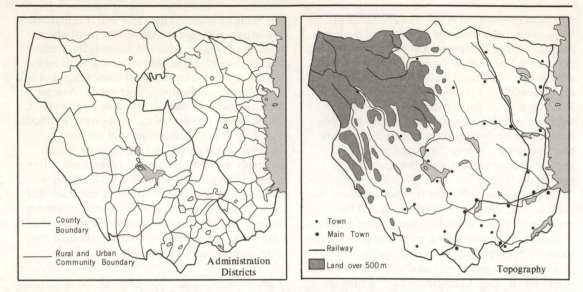

**203.** Comparative examples of choropleth, proportional symbol, dasymmetric and dot plus proportional symbol methods

and representing such areas, as part of the locational 'background' information. High mountain and desert areas can normally be indicated as having no population.

Figure 203 shows the topographic features and administrative districts of a region at small scale. Four methods of representing the population, using the same data, can be compared: the standard choropleth map; proportional symbols representing the total for each division; a dasymmetric map, taking into account relief and settlement; and a combined dot/proportional symbol map representing urban and rural districts respectively. The latter gives much the best representation of the actual distribution, and the difference between concentrated and dispersed populations.

Very small-scale maps based on population data for large areas, which themselves may contain a great variation in population density, are more difficult to deal with. Ideally, such maps should be constructed as generalisations of more detailed, larger scale maps, but often this is not possible. For example, a population distribution map of an entire continent not only suffers from great disparities in data collection methods, census dates, etc., but only comparatively large administrative divisions can be plotted and used at compilation scale. In such cases, the cartographer must be guided by what is geographically sensible. Where necessary, the obvious areas of nil or low population density (water areas, high mountains, extensive jungle, etc.)

should be marked out and excluded, and so far as possible larger scale maps consulted to discover the presence of towns and villages. The quality of such a map is therefore dependent partly on available information sources, and partly on the amount of time and effort that can be devoted to finding and examining other geographical evidence.

## Dot construction

For relatively simple maps with a small total number of dots, it may be possible to draw them individually, using a standard technical pen of a suitable gauge. For very fine dots, such pens are unlikely to produce identical round dots, but this will not be critically important. If larger numbers of dots are needed, and especially if they are large enough in diameter for their true shapes and sizes to be identified, then dry transfer is much to be preferred. The dots will gave a consistent size and shape, and can be constructed much more rapidly.

## Unit symbols based on areas

For some subjects, such as agriculture, the statistics may describe the total area under the crop in standard units, such as hectares. The data are often only available for administrative divisions or agricultural districts, so the same problem arises of attempting to represent location within each area. It

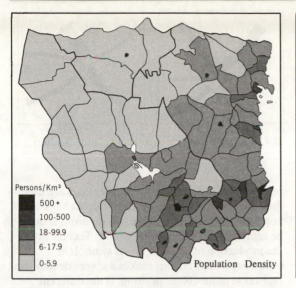

Persons/Km²
500 +
100-500
18-99.9
6-17.9
0-5.9

Population Density

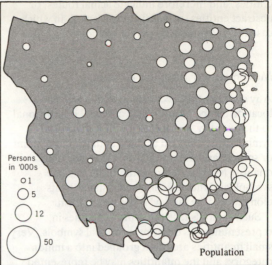

Persons
in '000s
○ 1
○ 5
○ 12
○ 50

Population

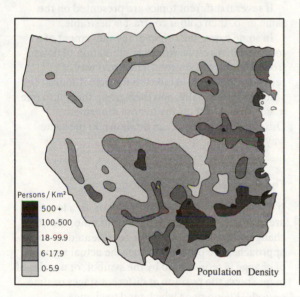

Persons / Km²
500 +
100-500
18-99.9
6-17.9
0-5.9

Population Density

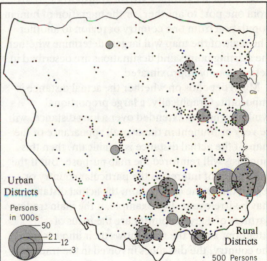

Urban
Districts

Persons
in '000s
50
21
12
3

Rural
Districts

• 500 Persons

is possible to use the dot method for this, each dot representing a certain fixed number of areal units on the ground, depending on the scale of the map. The square symbol is common, its size being made equivalent to actual dimensions at map scale, so that if one square equals 1000 hectares, the size of each square corresponds to the area of 1000 hectares at map scale. For distribution, reference is made to topographic and other evidence in order to approximate the true location of the crop areas.

## CHANGES IN LOCATION AND CHANGES IN QUANTITY

The representation of change may involve either a change in total quantity over a period of time at the points or areas of location, or the movement of goods, people or services from one location to another. This movement may involve physical entities, and therefore indicate actual routes, or abstract entities such as financial capital, in which case the 'movement' is purely symbolic.

## Movement to another location

Any quantity that occurs at or within a given location can be described by a symbol proportional to the quantity. In the case of movement such symbols are extended to show both the initial point or source, and the destination. Therefore, the quantity is represented by the width of a line symbol, and the direction and distance by its elongation. The 'arrow' is the typical form.

Such symbols face the same difficulties in representation as other proportional symbols. Very small quantities are often grouped into a minor category, and the quantities may be represented individually, or may be divided into numerical classes. The largest symbol possible will be affected by the actual dimensions of the source areas at map scale. Typical examples are the transport of goods from one port to another, or the migration of human population from one country or region to another. The scale of the map will largely determine whether the starting points and destinations are described in detail, or only approximated.

Much depends on whether the actual distance is important. Graphically, a large proportional symbol, which is extended over a long distance, will be very prominent in the overall appearance of the map. If the actual distance is significant, then this emphasis will reinforce the map purpose. But if the chief point of interest is the particular country or region that is the destination, the actual distance may be irrelevant. For example, if the main trading partners of a given country are the focus of attention, and if the map shows their imports into the country, the distances involved in the trade may be of relatively small significance. There is no graphic solution to this problem, as the design cannot use size to show relative importance in one sense, but then avoid the inference that it is also important in relation to distance.

If only one topic is included, then achromatic maps can normally represent this kind of information satisfactorily, the proportional symbols being shown in black against a basic outline (Fig. 204). Emphasis can be improved if the 'background' information can be shown at a lower visual level, using grey lines and/or tints. The volumes can be sub-divided into subsidiary types of quantity, as with other sub-divided proportional symbols. Again, problems will occur if small proportions are

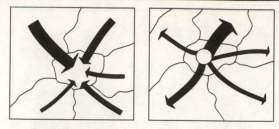

**204.** Proportional symbols showing immigration and emigration

represented by narrow symbols. If colours are used, then the relative lightness of the different hues needs to be considered, as the sub-divisions should be represented as equally important. Technically, the production of adjacent linear symbols in different printing colours makes a severe demand on register in multi-colour printing. If the different colours can be made by colour combination, this difficulty may be reduced.

If several different topics are presented on the same map, then colour contrast is desirable.

In some cases, the origin of the movement of a total quantity results from a combination of flows from a number of sources. The usual way of representing this is to indicate the actual source volumes individually, and then group them into one main symbol to show the overall movement. Different destinations can be shown in the same way.

## The representation of change

If quantities related to either point or area locations are available for different periods of time, then the change over the period can be represented. Two approaches are possible. Either the actual amount of change is represented by the symbol, or a series of maps shows the position at different stages or time periods. Phenomena which tend to change frequently or continuously over time are more commonly shown by the latter method.

The representation of change at a point or within an area makes use of the same methods as those used for representing quantities. Proportional symbols related to either point locations or areas, area symbols representing density or mean values, and even the dot method, can all be used. The basis of representation may be either the actual degree or amount of change at each point or area, or the degree or amount of change as a proportion of the initial quantity.

For example, if an administrative division had a

population of 1 million in 1970, and in the decade to 1980 the population had decreased by 100 000, either the quantity of 100 000 could be represented as the amount of decrease, or the total of 1 million could be shown, and the proportion of one-tenth indicated by a sub-division of the symbol. However, if the population density of a region was described as 50 per hectare at some previous date, and this had decreased to 45 per hectare over a period, then the choropleth map of change over this period could only show the actual amount by a symbol that represented a decrease of 10%. Graphically, such maps are no different to those that represent other quantities, and the same principles for representation apply.

A series of maps over a time period is often used for climatic phenomena, such as temperature and precipitation. The variation over the year may be shown by a series of maps, each describing values for each month. More specific information is provided by such a map series, but the detection and interpretation of change depend on the map user. Direct representation of the amount of change is easier for the map user, but is restricted to the specific period for which the amount of change is shown.

# 24 ATLASES

Maps in atlases are not confined to any particular type, either of subject or purpose. Although the term 'atlas' is frequently associated with the concepts of 'world' and 'small scale', there are atlases with large-scale plans (city street guides), special-purpose atlases (such as road atlases for motorists), and special-subject atlases (such as an atlas of agriculture). Even the familiar world atlas, composed mainly of small-scale topographic reference maps, is likely to have specialised maps as well, and there is an overlap between the reference atlas for general public use and the educational atlas, which often contains many of the same maps. Regional and national atlases are primarily collections of special-subject maps, and are therefore specialised in coverage as well as subject matter.

In its most simple form, an atlas may be no more than a map of an area sub-divided into a series of atlas pages, placed in consecutive order. A modern national atlas is likely to consist of a collection of separate folios, containing both maps and texts, issued in a publication programme covering many years. What they have in common is that all atlases have the characteristics of books: a uniform format and a consecutive page order, with any 'opening' consisting of a left-hand and a right-hand page (apart from the first and last pages). It is this specific form that gives the atlas its particular character, and that provides common cartographic problems regardless of the subject matter, purpose or area of the particular volume.

## THE ATLAS AS A BOOK

A book has a fixed format. For a printed text, the division into a series of pages is no disadvantage. There is a linear sequence, and the text is continuous. For a two-dimensional image, such as a map, the fixed format is of great consequence, for it imposes itself on the two variables of scale and area.

Like any other book, the pages in an atlas are produced by printing a set of them on a single large sheet, usually on both sides, and then folding this sheet to the desired format. This printed section therefore has a fixed number and sequence of pages. Most atlases are created by taking a set of printed sheets and folding them into four-, eight-, or sixteen-page sections. Sections of more than sixteen pages are uncommon, because of the difficulties this introduces for folding and binding. So the total number of pages in the atlas is likely to be thirty-two, forty-eight, sixty-four, eighty, or any multiple of eight.

The further consequence is that the size of the printed page depends on the size of the printed sheet and the number of folds. All layout and general design stem from this factor. Most atlases have a large format by book standards, and very large atlases are among the largest books published. Indeed there is a division between large books, which are meant to be portable (normally up to about the A4 paper size), and even larger books, which are regarded as essentially 'library' reference works, and not normally carried about. The cartographic problem is that the smaller the page, the more the total information is broken up into separate units, and the smaller the scale at which any given geographical area can be shown as an entity. Many map users, and certainly librarians and bookshop proprietors, object to very large books, which are difficult to store, handle and display. They do not fit in to the 'normal' range of book formats and shelf divisions. This is in contrast to the natural desire of a cartographer to achieve the largest possible scale for an area, and to avoid a multiplicity of sheet edges and overlaps.

### Imposition

When the individual pages are laid out on the printed sheet (imposed), they have to be arranged so that they will appear in correct page sequence after the sheet has been folded. The imposition

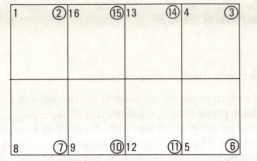

**205.** Page imposition

therefore depends on the number of folds, and their direction. In a common arrangement for an atlas, eight pages will appear on each side of the sheet, making a sixteen-page section (Fig. 205). This requires three consecutive folds. The first produces a half size, the second a quarter, and the third an eighth. The imposition also depends on whether these folds are 'up' or 'down'. Whatever method is used, it is clear that only in the centre fold will two consecutive pages actually be adjacent on the printed sheet. For all other 'openings' the two facing pages will be printed in different places. It also follows that both the first and last pages of the book will be single pages.

## Page use and page sequence

Although in a printed text, the type is always placed horizontally (apart from very unusual exceptions) and each individual page is treated as a separate unit, in an atlas the pages may be used in several different ways. Apart from the first and last pages of the entire volume, every opening provides two facing pages. These may be used individually as single pages for separate maps, or one map may be extended over the double page. The single page may present the map in either the upright (portrait) or downright (landscape) orientation, and this may be done with the double page 'spread' as well. The left-hand page (verso) has an even number, and the right hand page (recto) has an odd number. In the sequence, any double-page map must be arranged so that it consists of a verso and recto page.

If a double-page map is included, then only in the centre of the section will this actually extend continuously on the printed sheet. In all other cases, the two halves will be in different places in the imposition, or even in different sections. In order to avoid losing part of the map in the binding (as it can be difficult to open the bound volume so that it lies completely flat), and to allow for slight imperfections in the placing of the folds, the normal practice is to separate the two halves of the map by a

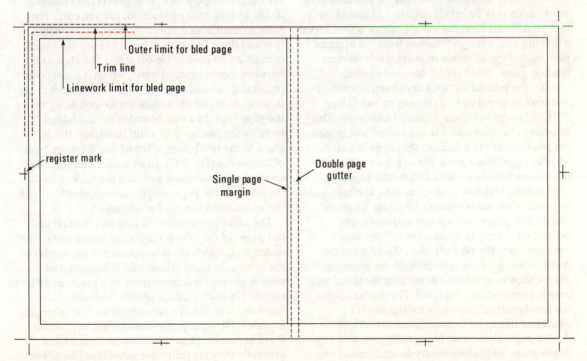

**206.** Page layouts

Outer limit for bled page
Trim line
Linework limit for bled page
register mark
Single page margin
Double page gutter

narrow blank band, three to five millimetres wide, called a 'gutter' (Fig. 206).

## Use of printed area

The extent to which the available page area is used for the map is also a question of the general 'style' of the atlas. The normal typographic style for books isolates the text area within margins, and indeed the relative size of these margins is regarded as important in book design. With maps, the use of margins further reduces the area of the page available for the maps, and consequently may affect scale and area. With small atlases in particular, it is a common practice to use the 'bled off' system in which the printed map extends to the edge of the page. The map is made slightly larger than the trimmed page size, and cut through along the edges when the complete book is guillotined to final size. Many medium-sized atlases also employ this style, which increases the available printed space for the maps.

A full 'bleed' for every page is however only one of the possibilities, and some atlases use a combination of margin and bleed. For example, a marginal band at either the head, foot or outside edge of the page may be reserved, usually for the page number and map title, and the other two outer edges bled off. The bled style runs the risk that some information may be cut through or lost during the final trim. Therefore, the usual practice is to observe a limiting line, a few millimetres from the trimmed page edge, beyond which important information (such as names and figures) does not extend.

After the printed sections have been assembled and sewn or glued on to a supporting band (Fig. 207), the entire volume is trimmed uniformly. This also has to be allowed for in the layout, and indeed the positions of the trim lines should be included. The full page layout guide for an atlas must contain guide lines for single pages and double pages, in both upright and downright positions. If margins are included, then the positions of margins for single and double pages, and upright and downright orientations, must be marked out. If the page is wholly or partially bled off, then the edges of the maps allowing for the trim through the bleed must also be shown, as this will determine the actual map area in construction (Fig. 206). For double pages, the usual practice is to allow for the gutter by deducting this space from the overall width of the double page, as the map is normally produced as a single image and subsequently divided into its two halves. This page layout is a necessary operation at the initial phase of specifying the page arrangements and style of the atlas.

## Sections and map production

If sixteen pages are imposed on the two sides of a single printed sheet, it follows that all these pages must be ready at the same time for plate making and printing. There are several consequences to this. A double page may consist of the last page of one section and the first page of the next. They will be printed at different times and on different sheets. Therefore, the control of printed impression must be to within fine limits, as otherwise discrepancies may show up, especially with any area colour. It may also mean that a small scale of an area has to be produced before a larger scale, which may occur at some other place in the atlas. Theoretically, it would be desirable to produce all smaller scales of areas after any larger scale coverage, in order to observe the principle of consistent generalisation, but in practice this may not be possible.

## Binding

Although binding as such does not involve any cartographic operations, it does exert an influence on the overall atlas production, use and cost. The conventional atlas is normally given a reasonably firm and durable cover, in order to keep the pages complete and protect the contents. The choices lie between a conventional 'case', which is a stiff cover projecting beyond the page edges, to a soft or flexible cover, which is often guillotined along with the page trim. In a case bound atlas, the folded sections are first sewn to a stiff band, and this is glued to the two pieces of board that form the basis of the covers (Fig. 207). In a cheaper binding, the sections are guillotined and then the backs glued to the band, or the pages and cover are assembled and the whole book trimmed as one unit.

The junction between the case and the first and last pages of the atlas is reinforced by the addition of endpapers, which are sheets pasted to the insides of the covers. In many atlases, the colour-printed section of maps is accompanied by a preliminary text section (including title, half-title, contents, foreword, etc.), and a subsequent section, normally an index. Obviously, it is uneconomic to produce these as part of the multi-colour section, so that generally they are printed separately in black. For reasons of economy, some cheaper atlases avoid

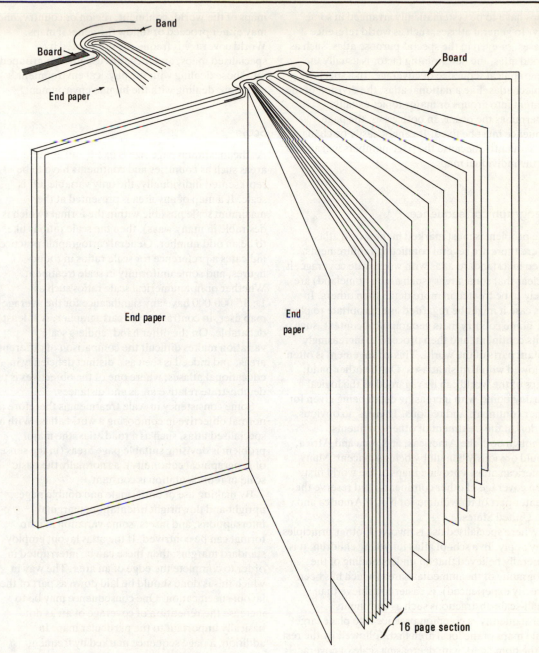

**207.** The structure of a case-bound book

such additional preliminary pages, and start with the map sequence. In this case, the endpaper may be used for title, contents, etc.

For a commercially published atlas, the cost of a full case binding is considerable, and affects the published price of the volume.

## GENERAL PLANNING

### Order of contents

In any atlas, the fixed page sequence means that the

maps have to be systematically arranged in some way. In general atlases, such as world reference atlases, or even in the special-purpose atlas, such as a road atlas, the determining factor is usually the geographical sequence of coverage. In a special-subject atlas, like a national atlas, the division of the content into groups or major topics usually determines the order. In both cases, the page sequence must be fixed. It cannot easily be changed subsequently, because the page numbers will appear on the individual maps.

## Geographical sequence

The problem is most marked in the world atlas, where there are several conflicting requirements. Even in a standard atlas with worldwide coverage, it is clear that some areas (such as the homeland) are likely to be treated in more detail than others. In this case it might be regarded as appropriate to place the home country in its geographical context, such as its continent, and then proceed to increasingly distant parts of the world. This arrangement is often followed with British atlases. On the other hand, maps of the world can be regarded as the logical starting point, with increasing detail being given for other continents and regions. There is no obvious 'order' in the treatment of other continents, although both the Americas, and Asia and Africa, would obviously benefit by being adjacent. Many American atlases provide maps of the world first, then coverage of other continents, and reserve the greater part of the volume for North America and the United States.

Where specialised use is involved, other principles may apply. In a school atlas for young children, it is generally believed that an understanding of the geography of the immediate area (which has been directly experienced), is easier to achieve than small-scale abstractions such as continents. Consequently, a common practice is to place large-scale maps of the local area first, followed by the rest of the homeland, with decreasing scale of coverage for other parts of the world.

The geographical sequence is also affected by the consideration of major 'divisions' on the basis of geographical regions or political units. Most world atlases involve a combination of both, but the irregular shapes and sizes of continents and different countries pose considerable problems in fitting them to a fixed rectangular format.

Many world atlases contain both general topographic maps and specialised maps. The specialised maps may be grouped with the general maps of the world, continent, region or country, and may either proceed or follow the general maps. World coverage is frequently dominated by specialised maps, and these are usually sub-grouped into those dealing with the physical environment, and those dealing with the human environment.

## Scale

As the maximum page size is fixed, and as finite areas such as countries and continents have to be represented individually, the only variable left is scale. If a map of any area is presented at the maximum scale possible within the format (which is desirable in many ways), then the scale ratio is likely to be an odd number. General cartographic practice indicates a preference for scale ratios in round figures, and some uniformity in scale treatment. Whether or not numerical scale ratios such as 1 : 15 000 000 have any significance for the average map user, in contrast to the cartographer, is at least debatable. On the other hand, endless scale variation makes difficult the comparison of different areas, and indeed is seen as a distinct deficiency in educational atlases, where one of the objectives is to demonstrate relative areas and distances.

Some consistency in scale treatment is therefore a normal objective in composing a world atlas. With a specialised atlas, such as a road atlas, the major problem is devising suitable page areas (in the sense of geographical continuity), as normally the basic scale of representation is constant.

By making use of both single and double pages, upright and downright orientation, marginal interruptions, and insets, some variation in map format can be contrived. If the atlas layout employs standard margins, then these can be interrupted in order to complete the edge of an area. The way in which this is done should be laid down as part of the layout specification. One consequence may be to increase the repetition of coverage of areas not basically important to the particular map. In addition, a page sequence marked by frequent changes in orientation and numerous insets is unattractive in appearance and often tedious to the user. Insets, in particular, need to be treated with care.

## Scale use

In a world atlas, too great an attachment to regular scales can lead to excessive duplication of coverage, and extensive overlapping. As the total space for maps is limited, it is preferable to try to use constant scales for the major areas, such as continents, and then to treat individual regions at the largest scale possible. But in any world atlas areas at continental edges are likely to appear on many different pages. For example, the eastern Mediterranean will be included on maps of Europe, Africa and Asia, at several different scales.

## Overlaps

Because of the fixed format, and the repetition of the same areas at different scales, any world atlas is likely to have numerous overlaps between different maps, in terms of both scale and area. The cartographic consequences of this are significant. As the land masses are the main centre of attention (although there may be special maps of oceanic areas as such), the total amount of space occupied by 'water' is in many ways wasteful. On the other hand, reaching agreement in generalisation and representation for several different maps of the same area is difficult, especially if these are compiled from different source material, on different projections, and by different people. In this respect it is good practice to compile an outline of all page layouts on a map of the world, to serve as an index of the overlap and coverage of different pages. Some atlases include a finished version of this graphic index on one of the endpapers, as a guide to the content and coverage of the atlas pages.

It is particularly important to obtain consistent treatment of simplification and selective omission, the classification of towns, the representation of international boundaries and the selection of names.

## Coverage and sequence

Once the general principles of content and coverage have been agreed, the first task is to attempt a 'mock-up' or preliminary layout of all the pages. This makes clear the sequence and use of single and double pages, their orientation, and any need for marginal interruptions and insets. At the same time, any remaining difficulties in fitting the desired content to the available space should be revealed. Once this provisional arrangement has been agreed, it is usually necessary to explore this more thoroughly, compiling outlines of the maps at scale, to ensure that they can be fitted into the pages as intended.

## Indexing system

Virtually all world atlases, and many others, include a reference system for indexing place names. Search is assisted by using the divisions of any page formed by the geographical graticule (or grid at larger scales), and coding these in two dimensions. In this way the columns may be lettered across the head and foot of the page, and the rows may be identified by figures at each side. This alphanumeric indexing system is easy to use and understand, but it must be allowed for in the page layout and design, particularly for maps which are bled off in style.

Although universal indexing systems, involving full geographical coordinates, may also be used, these tend to require relatively long and complex figures, and location to even one degree is not necessarily straightforward if the map projection only includes full lines of latitude and longitude at five or ten degree intervals. In the alphabetical index the names are followed by the page number and reference, such as 19 A4. This confines the user's search to a relatively small area.

As in many cases a named place or feature will occur on several different maps and scales, the normal practice is to index it at the largest scale at which it appears. Because of page interruptions, large physical features, such as major mountain ranges, may only occur completely on the smaller scale maps. Although the actual alphabetising of the index may be done rapidly with a computer program, identifying the names and pages to be included still needs some selective judgement. For this a complete graphic index of the different maps and scales, and their overlaps, provides the best starting point.

## Basic map styles

In most general world atlases, and many regional or special ones, there is a limited number of basic styles, related to the principal map types. In many cases, these styles are deliberately designed to give a particular character or identity to the volume.

In a world reference atlas, most of the maps will be small-scale topographic maps, or small-scale plans, and will have the same basic specification, depending on scale and scale groups. The sections on the home country, continents, and the world may also have a number of special-subject maps, and although these differ in content, they are likely to be unified by a particular design approach. Sometimes, general physical maps of the continents add a third variation.

This aspect of style is based on the consistent use of certain graphic variables and characteristics. For example, the general maps may have a neutral colour background, or elevation colouring, or political colouring, or hill shading, as a common element. The treatment of water areas may range from pale to a dark blue. The size, typographic style and weight of lettering will strongly influence the general appearance. In this respect, the specification for each map is considered not just individually, but in relation to its group. This uniformity of style helps to give the volume harmony.

Style is also affected by the approach to generalisation and the overall level of detail. The contrast between atlases from different countries makes this clear. Some have a very detailed topographic base, numerous settlements and names, and a high density of detail. Others are more open, more selective, and usually employ larger and bolder type for fewer names.

## Standard page styles

Apart from general design factors, each of the basic map styles will also be expressed by a standard page style. This includes the treatment of the reference information: map title, scale and scale title, page number, and any explanation of symbols. The page number in particular should be placed uniformly throughout the volume, as the user subconsciously expects to find this in the same place. For ease of reference, either the upper or lower outside margin is the preferred position.

Where a marginal band is used, this helps to provide a fixed position for the main title, scale and page number, regardless of map type. With the bled-off style, a major problem is the need to use map space to locate the title, scale, page number and any legend. This can only be done by interrupting the map, which assumes that some part of the map can be dispensed with. Although oceans and/or areas of overlap will often provide the opportunity for this, it inevitably means that standard information will appear at different places on different pages. The design and location of the indexing codes also has to be included as part of the basic reference information.

## Production and design

An important factor in design is the influence of the number and choice of printing colours. If this is restricted to four, then it is essential to obtain the maximum from a limited palette. Although the four process colours can be employed, magenta has a particularly strong effect on the visual appearance, and many users find it overpowering. In addition,

the cyan tends to be on the light side for a blue line image, although it is satisfactory for blue tints. In many cases, the absence of a single neutral colour, such as grey or brown, is a severe constraint. For example, if it is desired to use hill shading as a background, the choice lies between a half-tone of the black or a complex combination of either cyan and black or magenta and black. As the black impression cannot be varied, any half-tone has to be placed exactly on the characteristic curve, or it will be either too light or too dark. In addition, the combination of a black half-tone with lines and lettering makes revision correction difficult. Many general atlases incorporate a grey, which is useful for background images, both in relation to topographic maps and specialised maps. Although this can be made from the black, a continuous fine grey line, and a grey half-tone, have a smoother effect than black, and the grey itself can be modified towards a blue, yellow or green. It is unfortunate that what is in many ways the most useful colour selection – black, grey, blue, yellow and red – contains five colours, which does not correspond with standard machine printing arrangements of two, four or six colour units.

# COMPILATION

The compilation of the individual pages of any atlas is a major cartographic task, and in many cases apparently simple maps of some parts of the Earth's surface, or subjects, may require a great deal of time in compilation.

The first consideration is the choice of compilation scale. Same scale compilation is more economic, and speeds the subsequent production, but it demands accuracy. A larger compilation and working scale may be chosen, especially if more than one map is to be derived from the same originals. If this is done, then the effect of scale exaggeration on the entire specification must be taken into account, as well as the allowance for generalisation.

All compilations must fit the page layout correctly, and preferably any additional sections, for example insets, should be included in position.

## Projection and framework

The starting point for any small-scale topographic

map is the reference framework. For most atlas maps this will be a projection, but a grid may be more useful for some larger scale maps, such as those in a road atlas. In theory, each projection should be chosen in relation to area, scale, map purpose and the format of the page. In practice, projections are rarely constructed individually, for a variety of reasons.

Most of the source material for regional maps and coverage of individual countries will have been based on small-scale maps published for those countries. For these, it can be assumed that suitable projections will have been chosen, especially if the source maps are products of national mapping agencies. Some international series are also frequently used, and again the compromise projections of these will be suitable for general-purpose map coverage. For maps of continents and large areas, a number of common projections are widely used, and copies of maps based on them are available. Therefore, there is little point in constructing them afresh. So far as projections (more correctly 'conventional arrangements') of the whole Earth are concerned, there is little that has not been explored at one time or another. Despite claims to 'special' projections with superior properties, these all turn out to be particular cases of well-known projection types.

The particular problems of world atlases are more a question of fitting to the format and map type than originality of choice. The rectangular format of the normal printed page, whether a single- or double-page map, presents problems for maps of the whole Earth. For many projections, the format is too deep, so that a world map that fits in longitude is 'short' of the page size in latitude. A better use of page area and a larger scale can be achieved by interrupting the projection. If oceanic areas are not important in the subject matter, then interruptions through the oceans can be used effectively. For double-page maps, a projection which is symmetrical about the centre line (and therefore the gutter) is visually preferable to one in which the meridians cross the centre obliquely. This is another reason for the frequent use of conic projections for mid-latitudes.

Other practical problems arise in the relationship between the coverage of the main source map, and the page area of the atlas. In most cases, one particular map is chosen as the main compilation base, even though additional material will be added from other sources, or extensive revisions made. It may happen that the sheet lines of this source map do not coincide with the new format. A common problem therefore is the need to slightly extend a projection. If the source map is a small-scale map,

then only a limited number of meridians and parallels will be shown, and there may be no sub-division of these in the map border. In addition, the projection may not be fully described. Normally scale-correct lines can be identified by measurement and comparison. If the projection is a conic, and if the meridians are equally spaced on a given parallel, then it is possible to add additional meridians by measurement. Most of the standard projections are so well known that they can be identified, and if necessary their construction formulae consulted as a guide to any extension.

In some cases the selection of full meridians and parallels on the source map may be inappropriate for the derived map. The construction of additional graticule lines can pose problems. If the scale along certain lines is constant, then they can be sub-divided by measurement. But if the scale changes in one direction, then the positioning of new lines must be calculated.

## Source material

For a world atlas, a great deal of source material could be consulted, and indeed ought to be consulted, if correct, consistent and up-to-date maps are to be produced. In practice there are many limitations. Much of the world's medium-scale topographic mapping is unavailable, as it is restricted. Small-scale maps of countries and regions will differ in projection, content, classification and level of generalisation. All published maps will be out of date to some degree, and the most useful source map, from the point of view of scale, projection, coverage and content, may be quite old.

Small-scale derived maps of individual countries with good national mapping agencies pose no great problem. Their official small-scale maps or series will be of a high standard. But a map of any continent may have to bring together material from many different sources, and of varying qualities. The usual practice is to select a suitable map as a base, and then add to and improve it, so far as time and resources permit. In this there are priorities. Changes to international boundaries, country names, and names of towns and cities have high priority. Improvements in details of the physical environment tend to have a lower priority. This reflects the chief usage of the maps. If hypsometric colouring is included as a general 'background', with a standard specification for the continental maps, then this in turn will depend on compiling the same limiting contours for the elevation bands. Various source maps of suitable scale are likely to have different vertical intervals, and indeed none of them may show all the contours desired for the new map. This kind of interpolation can consume a great deal of time and effort, and in the long run depends entirely on the conscience of the cartographer, for the quality of the compilation is unlikely to be examined in any detail. Those items that are most likely to change, such as place names, need to be examined carefully, and many different sources of information may need to be consulted.

## Compilation from several sources

Although the source maps may be photographically reduced to approximate the scale of compilation, at least achieving a corresponding fit of local detail, this leaves unresolved the differences in projection and generalisation. Traditionally, this has involved scale change, projection change and generalisation simultaneously. It is a good example of high-level visual processing, and one that demands concentration. It is inevitable that its performance is uneven, and one great weakness of small-scale maps derived from disparate material is the inconsistency in representation that occurs. Although much of this might be reduced by transforming the source material to a digital data base, on which metrical transformations could be more easily performed, there is still a problem of which version would eventually be stored as the digital data for subsequent use, and its internal quality. Poor generalisation has a progressive effect, because its defects may be incorporated in subsequent derived maps. Such basic cartographic information could be greatly improved if small-scale versions of continental areas were to be prepared carefully from much larger scale material: that is, if it was possible to invest a great deal of time in basic preparation.

## Copyright

Although copyright applies to all published material, there are considerable variations in its interpretation with regard to derived maps. The general principle is that the specific form is copyright, but the information itself is not. Direct reproduction of another publisher's map is an infringement of copyright, but this does not prohibit the use of the material to create a new product. There are exceptions to this principle. In Britain there is no specific law on copyright as applied to maps. Any questions are considered on the general basis of copyright law, and cases. The Ordnance Survey maintains copyright protection on the

grounds that its maps are a product of original survey, and that there is a precedent because other publishers have accepted this over a long period, thereby recognising the claim. Therefore, copyright permission and royalties are normally required for the use of the material. In several other countries, such as the United States, exactly the opposite policy is adopted. All publications funded through federal government are unrestricted, on the grounds that the public has already financed them. In many other countries (now in the majority), large- and medium-scale topographic maps are protected under military security, so that the question of access to the information does not arise.

# PRODUCTION

Atlases make use of normal production procedures, but as many of them are intended to have a relatively long life in publication, there are some special requirements. Commercially published atlases, both general and educational, are also constrained by a number of related technical and financial factors (see Thompson 1977).

The large amount of material produced for a world atlas frequently means that it will be used for more than one publication, and may serve as a basis for a variety of other maps. This puts a premium on proper organisation, labelling, storage and maintenance of all the original materials. Compilations should not be trimmed close to page edges, as a subsequent demand may involve slight extension of the original map. All production materials should be properly labelled and dated. It can be an advantage if source materials are noted on compilations for subsequent reference.

## Checking

The checking requirements for a world atlas involve more than the normal checking of individual maps. Although the check list should be employed for each map, it is the agreement between overlaps and scales which demands special attention. For example, the Aleutian Islands may well appear on at least one map of North America, a map of the Soviet Union, and a map of the Pacific. There will be extreme differences in shape because of projection, as well as differences in scale. Ensuring that the version shown by the atlas is consistent requires collaboration and thorough checking at all stages. Starting technical production before all the maps of one area have been compiled normally engenders subsequent problems, even though it may be inevitable in terms of the production schedule.

## Revision

The maintenance of a world atlas means that the up-dating of information is an important task, and one that needs considerable organisation. With worldwide coverage, there is a constant flow of new information. Revision can be regarded as a separate operation for each product or publication, or a policy of continuous revision can be employed for at least one basic atlas. In either system the normal procedure is to take the latest printed copy of each map or page, and mark this up with deletions and additions, usually distinguished by colour coding. In many cases, the location of the correction is indexed, and reference made to a separate document that will contain the details. Any correction should be applied immediately to the revision documents for all other maps on which it occurs. Most of the larger organisations publishing world atlases necessarily maintain a map collection, and other sources of information, both for compilation of new maps and revision purposes.

## NATIONAL AND REGIONAL ATLASES

National and regional atlases, which usually aim to give a complete cartographic account of a great range of subjects, are among the most expensive and sophisticated of all cartographic products. The resources required for their production and publication are usually so great that they need to be subsidised, at least in part, by governmental or other institutions. Because they need to give the largest possible scale of treatment to the whole subject area, they often require a very large format by book standards, and there are many variations in the way in which the total material is assembled or 'packaged'. They normally involve the cooperation of a large number of specialists, and as such their organisation is a complex task, extending over many years of effort. Many of the best examples of cartographic representation can be found in national atlases, and they are an important source of

information about methods of representation for any student of cartography.

## Publication form and policy

A single case-bound volume at a large format is often a rather inconvenient product, normally regarded as a library reference work only. The case-bound volume has the advantage that the whole work appears simultaneously, and the contents are complete and protected. The disadvantages are that the length of time taken for production means that some maps will be 'out of date' before the volume is published; it can be difficult to refer to or work on any single map within the whole volume; and changes to the content cannot be made subsequently. Consequently, the method of serial publication is widely used, in which specific sections are issued independently, and eventually collected in a box or other form of container. This means that individual maps or sections can be withdrawn for consultation or display, and that subscribers can be issued with sections as they appear. It also means that sections may be removed, replaced in the wrong order, or not replaced at all. Clearly, there is no obvious single solution.

Separate serial issue also means that it is easier to combine maps with text and other illustrative material, and that the number and choice of printing colours can be varied with different topics. For example separately printed text material in black and white can be added to colour-printed maps. Some national atlases practise this extensively, giving detailed descriptions and analyses of topics, as well as the map coverage. Others concentrate entirely on maps.

## Scale and layout

At a very large format, the printed sheet may consist of only one folded unit, giving a single page front and back, and a double page spread in the centre. Depending on format and printing press, two such sheet units may be printed simultaneously, and then guillotined before folding. This means that the largest scale at which the whole country or region can appear is controlled by the maximum double page size and the shape of the country. This may be convenient, or highly inconvenient, in terms of a fixed rectangle. Long narrow countries may need to use two maps side by side in order to make full use of the page area. The normal procedure, therefore, is to produce a series of different scale treatments and page arrangements, to a standard format, so that the specific area will appear as a complete double page map, single page map, or any multiple of small-scale versions arranged on the double or single page. For example, a monthly series of climatic phenomena may be shown by repeating the map unit twelve times on the page.

Although the majority of the maps are likely to be devoted to the whole country or region, there may also be a need for more detailed treatment of particular areas or regions. Larger scales can be obtained by dividing the territory into regions on different pages. On the other hand, the choice of scale, and therefore detail of treatment, is also a function of the range of information available for particular topics, which may differ considerably for different subjects.

As most of the maps are likely to be of special subjects, space is also needed for any legend and explanatory information. For subjects with many classified symbols, this itself will require a considerable amount of space, and has to be allowed for in the page layouts.

Cartographically, an initial task is to construct a series of such scale and page arrangements, to serve as a basis for compilation and production. The arrangements may be formal, with neat lines and map borders, or they may be left free or 'floating'. The latter can achieve a greater degree of harmony, but requires careful planning and arrangement.

## Basic styles

As most of the maps will be of special subjects, various base maps will be needed on which to present them. These have to be chosen both to provide the correct reference information, and also to be complementary in terms of design. Therefore, a number of variations on the reference base need to be considered. These may range from a simple administrative 'outline' to the use of a hill-shaded version. Because in many cases the data shown on the map are based on administrative units, and because the naming and identification of these units may intrude into the map information, some national atlases use a separate transparent overlay, or series of overlays, which can be placed over the maps for locational reference. They can also include names of places or features, which helps to increase the total reference information provided.

The design and construction of these bases, and their cartographic style, will have a strong influence on both the appearance and usefulness of the atlas. A neutral grey or brown is often regarded as essential for these basic outlines, and consequently few atlases of this type are produced solely with the four process colours.

## Compilation and editing

Although the editorial committee will begin with a list of desired contents, what is eventually included will depend on the level of information available, and the extent to which it can be properly presented in map form.

Normally each map will be researched and devised by a specialist in the subject field. There are three different methods of organisation for this research and compilation stage. If the atlas organisation has a full cartographic staff, then the task of compilation may be carried out by staff cartographers, directed by the specialist. If frequent consultation is possible, this approach can be very effective, and eases the task of production. On the other hand, the specialist may compile a 'rough' presentation of the desired material, and then leave it to the staff cartographer to refine this and present it correctly on the proper reference base. These initial compilations by map authors may be at a larger scale, or consist of several pieces of material. This approach inevitably leads to misunderstandings, confusions, and repetition of work, and should be avoided if possible. Finally, the complete compilation can be produced by the specialist on base maps provided by the cartographic organisation. This is highly effective, but depends on the cartographic ability of the specialist.

Compilation bases will often need to be at a larger scale, and considerably more detailed than the final reference base for each map. It is clearly an advantage if these are in agreement to begin with, rather than being produced by individual authors at the initial stage.

Although the planned production and publication of a national atlas has to be considered in terms of a production schedule and expected date of publication, the organisation and planning of such a work is extremely difficult, as so many contributors are involved. The maintenance of standards of cartography and technical production throughout the work means that a strong central direction is essential, and allowance must be made for the inevitable delays and changes to the original plan.

The second half of the twentieth century has been one of considerable change and development both in map making as a whole and in cartography in particular. There are many reasons for this. Although it is customary to describe such developments in relation to specific cartographic issues, it is essential to realise that many of them are a consequence of changes in the larger society in which map making exists, which influence both the attitudes and interests of map makers and map users. Indeed, many of the problems that have been raised and examined as important in cartography have their close parallels in other applied sciences and applied arts.

## Historical background

The end of the Second World War saw the release of a great deal of energy for the rebuilding and reorganisation of both societies and landscapes. A huge backlog of incomplete mapping programmes had accumulated, and many fundamental map series had to be entirely reconstructed. The initial phase was therefore a drive towards achieving new or replacement topographic map coverage, and this was a prime objective in the national mapping of many countries. In turn, this material resulted in the production of new derived and smaller scale maps and atlases as well. Because of the emphasis on production, a great deal of attention was given to technical methods and their improvement, and to better working practices.

The investment of resources in production was also accompanied by developments in education and research in cartography. This was particularly noticeable in the English-speaking world, which in general had little of the tradition of publication on cartographic matters which was more characteristic of continental Europe. The foundation of cartographic societies, of the International Cartographic Association, and the introduction of increasingly numerous conferences at both national and international level reflected these developments. As part of the same movement, the establishment of new journals devoted to cartography in the English language was highly significant.

Lacking any tradition which associated cartography with the other fundamental areas of map making – geodesy, surveying and photogrammetry – much of the increased publication on cartography and maps in the English-speaking world was initially located most often within university departments of geography. This stress on the connection between the making and the use of maps – both of which could be of interest to geographers – emphasises the map as a collection of geographical information, in comparison with the European engineering tradition, which sees the map (and therefore cartography) as essentially a matter of technical graphic production (Keates 1985). The divergence between the two, although rarely commented on, accounts for the many differences in fundamental attitude between the two 'cartographic' groups.

Initially, both the conferences and publications concentrated mainly on the description and analysis of cartographic methods and technical processes, and especially on new technical developments. This was natural in view of the general concentration on improved production methods characteristic of this first post-war phase.

## Technical development after 1950

In the immediate post-war period, and particularly between 1950 and 1960, there were considerable and rapid improvements in what may be called analogue graphic arts technology. From a cartographic point of view, the most important were the introduction of plastic sheet materials as bases (first polyvinyl and subsequently polyester); the development of light-sensitive coatings in many reprographic fields relevant to cartographic production needs; the widespread adoption of scribing as a means of line construction; the use of photolettering output on film in place of type impressions on paper; the production of better and larger tint screens; and at the end of this period the development of strippable or peelable coloured surface films on polyester supports. Through these the current practices of scribed lines, photoset lettering, and the construction of area colours by mask and tint screen became standard practices. Stable translucent base materials, direct production of line and mask negatives, and the high quality of photoset lettering, also meant that contact copying replaced process camera photography as the principal application of reprographic processing. Coloured dye proofs became the norm, and at the end of this period pre-sensitised printing plates began to replace zinc plates for the lithographic printing surface.

These developments were essentially a consequence of general progress in graphic arts technology, resulting from research into printing, light-sensitive materials and their applications.

The consequence of this technical progress was considerable, in terms of speed and quality of output. Allied to the significant changes in the technologies of data collection, and especially the increasing application of photogrammetry, the types of map and mapping programmes that had been the objective of map making for a long period came nearer to realisation. Broadly speaking, it can be said that in many areas of map and atlas production high quality of output became the norm, and what may be described as the phase of analogue cartographic production reached a very high level. Many of the sheet map series, national atlases, and specialised maps produced at this time made full use of the improved technology to reach impressive standards of cartographic representation and expression. So much so that it is the elaborate and technically sophisticated images generated by these methods that have set the visual standard for other technical changes that occurred subsequently. Yet even as this period of technical and graphic sophistication reached its climax, events were taking place that would eventually challenge its foundations.

## Map making under pressure

Despite this technical progress, changes taking place in the map-using community were to profoundly influence the next two decades of map making. These can be described in relation to two major areas of influence, which to some extent existed separately. On the one hand the needs of the users of official map and chart series, and of many important basic specialised maps, would lead to reconsideration of goals and objectives. On the

other hand the interests of geographers and other Earth and social scientists were to have a significant effect on both the objectives and concepts of cartographic activity.

The separation of these two areas of influence is necessary. Compared with the practice of cartography, the study of geography as an academic subject in universities is a relatively recent development, in most cases dating from the earlier part of this century. In the English-speaking world in particular, this was dependent in the first place on the provision of maps at both large and small scales, and indeed many of the early academic geographers regarded the use of maps as central to geographical study. In addition, the expression of geographical ideas through maps was also important. Most of these maps were produced as illustrations in published papers and books, and normally consisted of relatively simple black-and-white representations. Often derived from available information, such statistical maps became an important component in geographical description and analysis. They were usually produced by the cooperation of the geographical or other expert (the map author) and a 'technical cartographer' who actually carried out the production, essentially as a manual drawing operation. These circumstances still exist today in many cases.

In the wider field of official map production, and in the production of series of maps and charts for internal or external clients, with its dependence on long-term planning and public investment, major changes in the speed and range of data collection had a serious effect on the role of cartography. The central problem was that the sophisticated cartographic production of multi-colour maps was time consuming, and could not match the increase in speed of production and output which became characteristic of many measurement and data collection methods. Photogrammetry, the availability of aerial photographs that could be studied and interpreted directly by specialists, and the development of electronic measuring devices, all emphasised that the cartographic stage of map production to a standard specification was relatively slow in response. Gradually a range of alternatives was tried, and sometimes adopted. Photomaps and orthophotomaps; provisional editions of standard map sheets, sometimes without field checking or colour completion; black-and-white versions of some sheets in a multi-colour series; planimetric versions without contours; all these were introduced essentially as a means of providing a product for prospective users more quickly than the standard cartographic production schedule would allow.

Similar conditions applied to nautical charts, as these began to be influenced by a greater rate of production of basic hydrographic data.

## Social and economic changes

Throughout the period of modern map making (which dates essentially from the late eighteenth century in Europe), map production has been both stimulated and affected by the needs of society. Public investment in maps, and individual use of maps, eventually determines what maps are made and under what conditions.

From the end of the nineteen-fifties in particular, the whole concept of social and economic planning underwent a rapid transformation in many developed countries. Important changes in population density and distribution; agrarian reforms; rapid urbanisation and industrialisation; all these had consequences for the planned management of the environment. Although basic concepts of planned location and development had been established at the end of the nineteenth century, this concern with the management of the environment – both physical and social – expanded to the point where regional and local planning became the established practice in many countries. Planning at this level demanded the investigation of environmental factors, which depended on information. This concern with the environment was also the traditional focus of geographical study, but it began to move from a dependence on observation and description to analysis and the use of scientific models and theories. In a later phase, concern with the environment expanded beyond the initial horizons of the control of industrial and residential location. Concern with both pollution and conservation began to have a greater influence on public attitudes, which in turn reinforced the need for more and better information about the environment.

In addition, the development of psychology and social studies was rapid. As in so many other areas of academic activity, there was an increasing attachment to the notion of 'scientific' methods – objective analysis and systematic description. The analysis of human behaviour, both individual and social, moved from its initial point of theoretical enquiry to the introduction of psychological testing methods, deliberately based on a more rigorous scientific approach. This concern with human behaviour had great economic and political repercussions, as well as social ones. Eventually the concept of the analysis and prediction of human

responses to controlled stimuli – and especially visual stimuli – became fundamental to both advertising of products and services and political persuasion, represented increasingly by the output of the mass media, and television in particular.

## The quantitative 'revolution'

Although the first phase of this major change in geographical description and analysis is often referred to as a 'revolution', in fact the adoption of statistical analysis and prediction was only the starting point for a much more fundamental shift in focus. The emergence of this is thoroughly and entertainingly described by Harvey (1969). As in many other fields, and not only in geography, the adoption of new methods in turn led to a major re-thinking of basic concepts, and in this case to the scrutiny of scientific models and methodology as a whole. Because maps themselves were an important component in 'spatial data', it was perhaps inevitable that the geographer's view of maps should also be affected by the new attitudes to scientific analysis. This new influence was initially expressed by Board (1967) in his account of 'Maps as models'. It inaugurated a period of widespread discussion and the publication of many new 'theories' about cartography. Because it seeks to examine the whole basis of cartography and the use of maps, the emergence and development of these new ideas needs to be examined in some detail. But before this, another important strand also has to be taken into consideration.

## Electronics and digital computers

Despite their importance for the future, the early computers were large, cumbersome devices, and only a limited number of 'experts' could manage them. But developments in electronics brought about the miniaturisation of fundamental components – especially the micro-chip – which in turn led to the rapid proliferation of small, high-speed and much more sophisticated devices, which themselves could be applied to data processing and the management of a vast range of operations.

Initially, the conversion of detailed two-dimensional images to digital data – the process of digitising – was very difficult, and research and investment in this area was predominant in the early stages of development. But the combination of high-speed digital processing and control with electronic scanning systems began to overcome this difficulty, at least in some circumstances. More recently, the development of computer-driven graphic displays of increasing sophistication has made possible the operation of 'mapping' systems of great potential.

In this field too the division between geographical illustration and large-format multi-colour map production led to different lines of development. This is reflected in the terms that evolved to describe map making based on digital data. 'Computer mapping', essentially the processing of statistical data followed by relatively simple achromatic map presentation in display or 'hard copy', focused on quite different problems to 'automated cartography', which sought to achieve the same type of elaborate image as that produced by the traditional cartographic and reprographic methods.

On the one hand these developments can be seen simply in technical terms. A major driving force was the desire to output maps much more quickly, and therefore help to satisfy the needs of map users. If the slow procedures of manual and reprographic construction could be replaced by computer-controlled operations, then cartographic production could reach the point where it would match the advances in information collection. Specialist map authors could use computer programs not only to analyse and process large quantities of statistical data, but also to generate maps at high speed, and indeed to produce many different versions with facility. At the same time, their dependence on a 'cartographer' to contribute to 'design' problems and to prepare the 'artwork' would be reduced or eliminated. But such major changes in the basis of map production inevitably had far greater effects than those initially conceived.

This electronic–digital revolution on the cartographic side of map making could be regarded as the equivalent to the major developments in instrumentation in data collection, and especially in measurement, which in turn had radically changed land surveying, photogrammetry and remote sensing.

## Science, computers and technology

Remarkable developments in scientific enquiry and discovery in many fields, and the influence of technology on both production and consumption in social and economic terms, has brought profound changes in societies as a whole. In order to see the present evolution of cartography in context, it is essential to take this into account.

Many historians have characterised the 'Western' world since medieval times as being increasingly dominated by science and scientific developments. Bringing with it huge increases in material gain, it also has the consequence of both influencing and being influenced by basic concepts and attitudes in any period. In the modern world, the desirability of greater technical control of the environment, increased output of sophisticated goods and services and the achievement of material wealth is rarely challenged, although occasionally a voice is raised in protest (Mumford 1952). Therefore, science is associated with 'progress', and indeed the word itself has connotations of superiority that are constantly exploited in advertising.

This reliance on scientific methods and procedures is an essential development in the 'pure' sciences, but increasingly it has become the dominant theme in applied 'sciences', which themselves have become largely dependent on advanced technology. Consequently, improvements in technology are themselves regarded as a desirable goal.

The increased level of investment in consumable products and services also raised problems in appraising the markets for them. This in turn influenced the production and publication of maps, both by official organisations and private publishers. Many national surveys began to devote more resources to investigating the needs of consumers, using the research methods that had been adopted in general for advertising and marketing (see McGrath 1984).

The notion that the purpose of production was simply to satisfy a known or potential customer demand was only the starting point. It soon became obvious that demand for goods could also be stimulated artificially by extensive and high-level promotion. Regarding maps as important sources of information in their own right began to be replaced by the idea that maps should be seen as marketable commodities, and should therefore closely reflect the interests of the consumers, assuming that these could be properly analysed and determined. These changes reinforced the belief that it was also important to understand the problems map users had in the use and interpretation of maps, in order that the product could be manufactured in the most suitable manner. This again led back to ideas of psychological analysis of map-using activities.

Although many of these developments had quite diverse origins, it is possible to bring together the several strands described so far. These include the widespread adoption of quantitative methods of description and analysis; the great improvements in this field made possible by computers; the reliance on scientific principles as a basis for theoretical development; and the belief that understanding how human beings behave should also be approached on the basis of scientific enquiry. As it was postulated that maps are made so that prospective map users can obtain useful information from them, it seems logical that the provision of this information should be seen as an activity that carries out the necessary operations according to properly verified principles in order to satisfy properly verified processes of map use. In so doing, map making could adopt the approved philosophy of scientific method, making full use of advanced technology, and thus satisfy the fully researched 'market' that is taken to be properly representative of the users' needs and interests.

## Map use and the 'scientific' approach

After about 1960 developments began to occur along several different lines. The first was the introduction of psychological testing and experiment to the analysis of visual response to map symbols, a procedure which at that time seemed vitally important to gaining a better understanding of how people actually used maps, and therefore how effective map design really was. But this revealed a peculiar difficulty. In order to conform with the agreed principles of scientific testing, an experiment could only deal with one variable, all others being held constant. Many experiments pursued this line of objective scientific testing (for a good example see Groop and Cole 1978). With real map-using tasks, individual differences in geographical knowledge and understanding affect responses, so the tests have to be arranged so that they prevent, or at least minimise, such influences. This can only be achieved by conducting the tests so that the visual displays or targets are deliberately controlled. But in any real map-using activity people combine their existing knowledge with the interpreted map information, and this is strongly affected by their understanding of map structure. Once these differences are removed, the conclusions reached by such tests had a very limited validity, a point which many of the researchers clearly pointed out. Consequently, many people who were involved in the broader questions of map composition and design increasingly questioned the value of such experiments.

The movement towards experimental tests using realistic tasks gradually gained momentum, and indeed many such experiments have been conducted. This was directly influenced by changes in psychological methodology itself, which in the same period was reacting against the ideas of

behaviourism, and developing what is now generally described as cognitive psychology. The application of cognitive concepts to cartographic research is described by Petchenik (1977) and Olson (1979). As Petchenik points out, it is still difficult to reach any general or overall conclusions from the examination of the responses of a group of individuals to the use of a particular map, bearing in mind that any design is only one out of an unknown number of possibilities, and that many maps may be used for quite different purposes by different people. Even trying to decide the nature and effect of the factors that influence map interpretation and thereby map design is difficult, as Castner (1979) explains.

## Computers and cartographic theory

The introduction of computer processing involved three related but separate problems. All the input data has to be digital in form, and this can be achieved either by providing the initial information in digital form, or by digitising existing graphics. The processes by which this can be done also affect any subsequent use of the data.

Second, what may be called the compilation and construction of the map as a scaled graphic image still has to meet the requirements of map structure. This means that cartographic processes of construction and compilation have to be described in specific terms in order to be programmed. For those tasks that can be defined mathematically, computer processing is a distinct advantage; for example, in the construction of projections, and the interpolation of contours through the generation of Cartesian or triangular grids. But for those that depend on complex assessments of relative importance and the relationships of shapes and spaces – such as generalisation – the many variables involved raise very difficult issues. In addition, although some tasks, such as the arrangement of names on a particular map, can be undertaken with sophisticated programming, it is so difficult to do properly that little is to be gained by carrying it out.

Although it is hard to assess the degree to which computer programs of cartographic processes have been successful, there is no doubt that the need to construct such programs has led to a renewed discussion of many basic cartographic ideas. The large number of research publications on generalisation are not only of interest in the sense that they solve particular problems, but also because of the contribution they make to a better understanding of the fundamental character of cartographic procedures. In this respect the series of contributions to digital generalisation published in the foreign-language series of *Nachrichten aus dem Karten- und Vermessungswesen* (for example Christ 1976; Schittenhelm 1976; Berger 1976; Lichtner 1978; Christ 1978) illustrate the value of detailed studies of generalisation processes.

If the digital data can be properly organised and stored, then at least in theory the map in its visible graphic form can be output by digitally controlled plotting machines. Here there are two related requirements. Large-format images can be exactly constructed by high-level plotting machines working to a very fine tolerance, although the technology is sophisticated. But the initial display of many types of map, or parts of maps, is essentially confined to a viewing screen. Constructing the image by transmission is fundamentally different to constructing a coloured image by reflection. The design rules and conditions that apply to one are not identical with those that apply to the other. And although very high-quality colour graphics displays have become available, it is not easy to realise the visual correspondence between these and printed maps.

In the case of simple small-format choropleth and other types of statistical map, such limitations are of little consequence. Low resolution and a very limited graphic output are no hindrance, bearing in mind the low quality of input information itself. But achieving the visual level typified by the large-format multi-colour printed map is still a quite different technical problem

## Maps and digital data

Although the initial starting point for the application of computers to cartographic processes was to replace graphical methods of map construction, in some respects the most important developments have resulted from the assembly of digital data itself. Assuming that it is produced on the same coordinate system, then it is possible to combine map data with other spatial data sets. The long-term implications of this may be more important than the direct contribution of computing to map making. This is particularly obvious in the development of digital data bases for planning and administration, especially in urban areas, but in a few countries at national level (Rystedt 1977; Eklund 1977). There are many examples from cities and metropolitan areas in Europe and North America in which the administrative and planning departments, and public utilities, have combined both their mapping operations and the storage and analysis of both

physical and human data on a unified topographic base, frequently that of the land parcel. Those countries that have – or can introduce – a complete cadastral property register as a basic element in this data, may realise immense advantages in the planning and management of the environment.

Because of the rate of change in the environment, and the need for up-to-date information, it is in urban areas in particular that the published printed map, to a standard specification, is least useful, and most difficult to keep up to date. The ability to produce different versions or selections of basic information in map or other form demonstrates the real advantages of digital data bases over graphic ones. In this sense, it is also likely that for such purposes standard maps will no longer be a requirement, and other types of graphic display and output will be more relevant to particular tasks. Even so, devising such multi-purpose digital data bases also raises interesting questions of data structure.

Because of the advantages which digital bases have in relation to map production in this type of urban management, it has also been suggested that the printed map will become obsolete. It is more likely that the portable, flexible, semi-permanent map in its traditional form will continue to be used where it is appropriate, but will not be used, as was often the case previously, for tasks to which it is unsuited.

## Cartography and scientific theory

The reconsideration of geographical thinking associated with the quantitative 'revolution', and reflected in the many treatises on geographical concepts and methodology (for example Haggett, Cliff and Frey 1977), inevitably led to the projection of these ideas into cartography, or at least into theoretical considerations that seemed important to geographers. This in turn was affected by the widespread adoption of the term 'communication', in social sciences, psychology and in the transmission of information by broadcasts. Like so many other terms of this nature (such as information and perception) it is used in so many ways that it is no longer susceptible to any firm definition. The importance of 'communication' was stressed, partly in the sense of human interaction and the exchange of knowledge, and partly in the technical sense of information transmission. Initially the two were frequently confused. As at other times, different concepts of 'cartography', assumed but rarely overtly stated, strongly influenced different views about the nature and relevance of such theories.

# CARTOGRAPHIC COMMUNICATION

The central idea of 'cartographic communication' is that the provision of information in map form, and its subsequent interpretation by map users, should be regarded as a single 'system', in which the graphic form of the map representation strongly affects the efficiency of map use. Fundamental to this is the assumption that satisfying the needs of the map user is the only requirement of importance, and that therefore 'cartographers' should attempt to discover and then practise methods to facilitate map use. The description of these interacting components or stages naturally led to the construction of diagrammatic 'models' that would represent this synthetic 'system'.

Of the many contributors to these ideas, the works of Board (1967), Koláčný (1969) and Ratajski (1973) are usually accorded a central position. The development of these theories and the responses they aroused is summarised by Board (1983), and a more detailed treatment of the different elements involved in the evolution of communication theories is contained in his 1981 paper. The starting points of many contributors were diverse. For Koláčný it was a general dissatisfaction with existing maps; for Board it was an attempt to model cartography on concepts of geographical analysis; and for Ratajski it was a desire to find a systematic structure for a diverse and complex subject.

Alongside this central idea, other contributions to a more theoretical analysis of cartography resulted from different approaches. Linguists in particular have been concerned for a long period with the analysis of signs, and as maps consist of symbols, it seemed proper to develop similar analyses for maps, or map 'language'. This approach is usually associated with the works of Bertin (1967), although in fact he concerned himself with a very limited range of map types. The importance of the analysis of graphic signs was dealt with by Keates (1982), and Freitag (1971, 1980) attempted to combine communication and sign theories into a single structure.

## Information theory and cartography

A quite different notion was introduced with the proposition that information theory, using the analogy borrowed from telecommunication engineering, could be relevant to the analysis of cartographic processes. Sukhov (1970) suggested that it could be used to examine generalisation, and since then various attempts have been made to incorporate it in general communication theories. Its influence is shown in the inclusion of references to 'messages', 'signals' and 'noise', all of which have quite specific technical meanings in signal transmission theories, but which many cartographers strongly criticised when applied to interpreted information and human understanding (Robinson and Petchenik 1975).

## Cartography and geography

The influence of ideas current in geographical circles is clear. This has even reached the point where it is claimed that cartography has an 'emerging identity', it being assumed by some that 'cartography' is 'part of' or dependent on geography. The very existence of such a proposition underlines the importance of fundamentally different concepts of cartography. Theorising about scientific models was a natural accompaniment to theorising about geography. As the identity of geography was frequently under question in this period, similar uncertainties were extended into cartography, an area often treated in geographical circles as little more than a collection of 'techniques'. Therefore, it is not surprising that many of the publications on cartographic communication concentrate primarily on theoretical questions of map design and methods of representation, and do not raise any serious discussion of the relationship between cartography and map making.

The danger of this is clearly explained by Bos (1982). In most cases 'cartography' and map making are regarded as synonymous. If the satisfaction of the map user is the primary objective, then the quality and availability of information of the required type is just as important as its cartographic representation. A comprehensive theory would have to take this into account as part of the 'system', and indeed any communication 'system' would need to encompass map making as a whole. If 'information' about the real world is the basis of the input into the communication system, then presumably what is being described or modelled is 'communication by maps'. As the primary information depends fundamentally on the applied sciences of geodesy, surveying, photogrammetry and possibly remote sensing, it is difficult to see why these should be excluded from communication theories.

However, it is evident that the concepts that treat cartography by itself as a 'communication system' correspond closely to the developments described so far. It is modelled on 'scientific' principles; it regards the satisfaction of the user as the prime motive; it requires the detailed 'scientific' analysis of map use in order to improve methods of cartographic representation; and if all this could be achieved it would certainly be possible to organise map production by computer processing, as it would consist of inputting data into a closed and fully-defined 'system' to reach specified objectives. As Board (1983) points out, although these ideas have gained a wide degree of acceptance, some cartographers continue to express serious reservations, not only about the details, but also about the fundamental concepts.

## THEORETICAL AND APPLIED CARTOGRAPHY

The widespread circulation of the ideas associated with communication did not go unchallenged. Although several of the new developments in cartographic teaching and research in European universities originated in departments of geography, the question of the existence of cartography as a subject in its own right could only be resolved in most cases by developing cartographic education and research apart from the limitations generally imposed on it within academic geography. On the one hand cartography needed time and space in a university system; on the other hand it had to reflect the whole field of map and chart production, and not just the special interests of geographers.

As Kretschmer (1978) observed, it is necessary for cartographers to develop more clearly and consistently the structure of cartography, in order to provide a proper basis for research at university level. In her view this can only be done by treating cartography as a specific subject field, with its own methodology of representation and construction, which can operate on the subject matter offered by other sciences and studies. The parallel she draws is with mathematics. Her view of cartography as a 'formalistic discipline', developed from the work of Arnberger (1970), accepts the importance of developing cartographic methods based on rigorous

scientific investigation, and the desirability of devising appropriate methods of testing the efficiency of maps, but stresses the close connection between theoretical schemes and the production of real maps. In her words, 'The tasks of cartography should be the development of forms, methods and rules to transform spatial data into cartographic representations, to reproduce them and to evaluate them.'

From a rather different point of view, Petchenik also deals with questions that arise in considering the ways in which cartography has been developing, and the problems that remain unsolved. But her main focus is the real complexity of the map design process, and the difficulty of devising research of the 'scientific' type, which will truly illuminate the relationship between the map design as a solution to a set of problems, and the possible responses of users. Like Kretschmer, her observations are rooted in the practice of cartography, and differ radically from many of the very superficial references to 'design' characteristic of communication theories. In a major contribution to this discussion (Petchenik 1983) she includes a detailed analysis of the problems of map design in the broadest sense. It is also significant because it is in sharp contrast to the main thrust of cartographic communication, which is to identify map 'design' as the least effective or successful aspect of cartography, and therefore the area that is most to blame for any deficiencies in map use. But it is by no means evident that the normal methods of scientific research can deal with this area. Petchenik (1983) poses the central question:

'The chief problem, however, lies in determining and stating unambiguously, in advance of choosing the graphic characteristics, what impression the design is intended to create. What *is* a 'correct' impression? The process of scientific research requires that one begins with a hypothesis. The process by which hypotheses in cartographic research may be formulated has been largely unexamined in the cartographic literature. In order to obtain information about the subjective (internal) impressions of a map user, the researcher must ask the user to do something that can be observed externally. What it is that the viewer is asked to do with the map will determine what one learns about his impressions.'

In the same paper she also deals in detail with the difficult question of whether design can be analysed scientifically, and makes an important point: 'Two assumptions are implied: one, that it is possible to state one or more sharply defined functions for every map, and two, that a clarification of function would lead unambiguously to particular design characteristics' (Petchenik 1983).

This deeper analysis of what is really involved in design, and its function in relation to things made by Man, is in sharp contrast with the current adherence to the promotion of scientific research as the main vehicle of human progress, which is generally tacitly assumed, even if not stated. Consequently, it is not surprising that cartography and map making are also frequently seen in this way, especially as the rapid developments in computing and graphic technology provide exciting and interesting challenges. Yet historically the importance of the other strand in cartography – its creative artistic element – finds its place. Imhof (1982) states this clearly:

. . . the following facts are clearly established: we demand, of a good map, the highest measure of legibility and clarity; we demand of it a balanced expression which emphasizes the significant and subdues the insignificant; and we demand a well-balanced, harmonious interplay of all the elements contained. It is in accordance with practical experience, however, which the author has personally observed over many decades, that in cartographical affairs, as in all graphic work, the greatest clarity, the greatest power of expression, balance and simplicity are concurrent with beauty. Beauty is, to a large extent, irrational. Artistic talent, aesthetic sensitivity, sense of proportion, of harmony, of form and color, and of graphical interplay are indispensable to the creation of a beautiful map, and thus to a clear, expressive map.

Despite the limitations of the translation into English, the appeal of this statement is clear. It raises the important question of whether cartography – and especially cartographic representation and design – can be fully described by a scientific 'model' which by its nature cannot accept individual creativity as a prime component. If Imhof is correct, then the full rationalisation of cartography is neither possible nor desirable. In contrast to being modelled on a scientific system, the design of a map in its full sense can be seen as depending on the reaction of an individual cartographer to certain conditions of information, subject matter, scale and purpose, within which he can construct an original composition. If this is so, then it goes beyond mere 'communication' because it depends on original creative expression (Keates 1984). It also means that the desire of a cartographer to construct an object which he regards as aesthetically satisfying is not an unscientific aberration, but a proper and necessary cartographic objective.

As has been explained in Chapter 3, the nature of creative design is difficult to fit into any activity described in scientific terms, in which all elements must be fully defined. In originating a design solution to a proposed map, a cartographer can only operate by solving problems in representation, both for single elements in the informational content, and then for the map as a whole. It remains to be seen whether the study of 'communication' will improve this process, and if so, how.

## Map design and artificial intelligence

Until quite recently, even the most sophisticated computer was used essentially as a high-speed processor directed by a humanly conceived program. But the power of the computer to both store and process information rapidly has also led to consideration of the degree to which a computer can solve complex problems, if these are presented and structured in the right manner. The use of computers and computer graphics in industrial design has steadily increased. Not surprisingly, bearing in mind the emphasis on computing as a means of solving map-making problems, the idea that a computer could be programmed to carry out map design, or at least to assist with it, is highly intriguing.

Given the existence of a data base containing the necessary information, the theory is that the computer, guided by an 'expert system', could provide solutions to problems of map composition, representation and graphic design. Initially this might consist of little more than preventing would-be map authors from committing gross mistakes – such as choosing a quite unsuitable scale for the desired map output. This would require in the first instance a detailed and logical description of the relevant factors, and the sequence of decisions that a cartographer would take under given circumstances. It would also have to be flexible enough to propose different possible solutions that would be compatible with a complete map design. In this sense it has to be modelled initially upon the practice and experience of an 'expert', hence its common title.

The concept is of interest, because as soon the emphasis moves from general theoretical notions to actual decisions, it is clear that deciding which problems are particular to the representation of certain information for a given area, subject and scale are the most important considerations. In this respect a proper map composition must precede any specific design solution, for the quality of the design will depend on a correct solution to the map composition in the first place. The actual specification of design parameters – once the basis and form of the representation has been settled – is likely to be less difficult than might be supposed, provided that the objective is stated in realistic terms. Establishing equal colour values for area colours, and compensating for small areas by increasing contrast; ensuring that all lettering is at least at a legible size; making sure that the operator is warned if fine lines appear over dark backgrounds: these are all typical processes that can be described within reasonable limits. Despite what is often implied by theories of cartographic communication, there is a great body of empirical knowledge about 'everyday' cartographic design

problems, and a considerable understanding of the principles by which they are resolved.

Such a system might prove invaluable in the generation of usable maps from digital data directly by specialists or map authors, for it is in the field of graphic composition and design that they experience the greatest difficulties. It should also help to reveal the complexity of many apparently simple maps. Even so, to design a map which is sufficiently legible for an interested user to interpret is one thing; creating an original and aesthetically attractive composition is another. One can be undertaken with understanding and skill; the other requires individual talent.

# REFERENCES

ACIC (1964) (Aeronautical Chart and Information Center) Technical Paper 13 *Production Control in Cartography*, USAF, St Louis

**Arnberger, E.** (1970) Die Kartographie als Wissenschaft und ihre Beziehungen zur Geographie und Geodäsie, in *Grundstazfragen der Kartographie*, ed. Arnberger, Osterreichischen Geographischen Gesellschaft, Wien

**Baird, J. C.** (1970) *The Psychophysical Analysis of Visual Space*, Pergamon Press, Oxford

**Bantel, W.** (1973) Der reproduktionsweg vom einfarbigen relieforiginal zur mehrfarbigen relief-karte, *International Yearbook of Cartography*, Vol. XIII

**Berger, A.** (1976) Computer-assisted generalisation and its possibilities of numerically-determined manipulation by parameters of design and generalisation, *Nachrichten aus dem Karten- und Vermessungswesen*, Series II: **33**

**Bertin, J.** (1967) *Sémiologie Graphique*, Gauthiervillars/Mouton, Paris

**Blok, D. P.** (1986) Terms used in the standardisation of geographical names, *World Cartography*, **18**

**Board, C.** (1967) Maps as models, in *Models in Geography*, eds. Chorley and Haggett. Methuen, London

**Board, C.** (1983) The development of concepts of cartographic communication with special reference to the role of Professor Ratajski, *International Yearbook of Cartography*, Vol. XXIII

**Board, C.** (1984) Higher-order map-using tasks: geographical lessons in danger of being forgotten, Monograph 31, *Cartographica*, **21**(1)

**Bond, B. A.** (1973) Cartographic source material and its evaluation, *Cartographic Journal*, **10**(1)

**Bos, E. S.** (1982) Another approach to the identity of cartography, *ITC Journal*, (2)

**Breu, J.** (1982) The standardisation of geographical names within the framework of the United Nations, *International Yearbook of Cartography*, Vol. XXII

**Breu, J.** (1986) Exonyms, *World Cartography*, **18**

**Brown, A.** (1980) The use of densitometry in determining equal-interval screened value scales for maps and charts. *I.T.C. Journal*, (1)

**Brown, A.** (1982) A new colour chart on the Ostwald system, *ITC Journal*, (2)

**Buttimer, A.** (ed.) (1983) *Creativity and Context*. Lund Studies in Geography. Series B, no. 50. Royal University of Lund, GWK Gleerup

**Castner, H. W.** (1979) Viewing time and experience as factors in map design research, *Canadian Cartographer*, **16**(2)

**Castner, H. W. and Eastman, J. R.** (1984/1985) Eye movement parameters and perceived map complexity, *American Cartographer* Part 1, **11**(2), Part 2, **12**(1)

**Christ, F.** (1966) Reduction in number of printing inks for the preparation of multi-colour maps, *Nachrichten aus dem Karten- und Vermessungswesen*, Series II: **20**

**Christ, F.** (1976) Fully-automated and semi-automated interactive generalisation, symbolisation and light drawing of a small-scale topographic map, *Nachrichten aus dem Karten- und Vermessungswesen*, Series II: **33**

**Christ, F.** (1978) A program for the fully-automated displacement of point and line features in cartographic generalisation, *Nachrichten aus dem Karten- und Vermessungswesen*, Series II: **35**

**Cuenin, R.** (1972) Cartographie Generale: Tome 1, Notions generales et principes d'elaboration Éditions Eyrolles, Paris

**Davis, J. C.** (1986) *Statistics and Data Analysis in Geology* (2nd. edn.), Wiley

**Eklund, O.** (1977) GISA – a geographic based information system, Monograph 20, *Cartographica*

**Elfick, M. H.** (1979) Contouring by use of a triangular mesh, *Cartographic Journal*, **16**(1)

**Forbes, J.** (1984) Problems of cartographic representation of patterns of population change, *Cartographic Journal*, **21**(2)

**Forbes, J. & Robertson, I. M. L.** (1967) Population enumeration on a grid square basis: the census of Scotland, a test case, *Cartographic Journal* **4**(1)

**Forrest, D. & Castner, H. W.** (1985) The design and perception of point symbols for tourist maps, *Cartographic Journal*, **22**(1)

**Freitag, U.** (1971) Semiotik und Kartographie: über die Anwendung Kybernetsiche Disziplinen in der theoretischen Kartographie, *Kartographische Nachrichten* 21

**Freitag, U.** (1980) Can communication theory form the basis of a general theory of cartography? *Nachrichten aus dem Karten- und Vermessungswesen*, Series II: **38**

**Geodætisk Institut, København** (1986) *Produktionsdiagrammer*, Topografisk Avdeling, København

**Groop, R. E. & Cole, D.** (1978) Overlapping graduated circles/magnitude estimation and method of portrayal, *Canadian Cartographer* **15**(2)

**Hagen, M. A.** (1986) *Varieties of realism*, Cambridge University Press

**Haggett, P., Cliff, A. D., & Frey, A.** (1977) *Locational Models,* Edward Arnold

**Harvey, D.** (1969) *Explanation in Geography*, Edward Arnold

**Hettner, A.** (1962) Die Eigenschaften und Methoden der kartographischen Darstellung, *Geographische Zeitschrift* 1910 (Leipzig), and *International Yearbook of Cartography*, Vol. II

**Hodgkiss, A. G. & Tatham, A. F.** (1986) *Keyguide to information sources in cartography*, Mansell

**Hydrographic Department, Taunton** (1976) Symbols and abbreviations as used on Admiralty charts

**Imhof, E.** (1977) Tasks and methods of theoretical cartography, Monograph 19 *Cartographica*

**Imhof, E.** (1982) *Cartographic Relief Representation* (ed. transl. H. Steward) Walter der Gruyter & Co. Berlin

**International Training Centre (ITC)** (1973) *Colour Chart*, Dept. of Cartography, ITC

**Jones, J. C.** (1981) *Design Methods*, Wiley

**Karssen, A. J.** (1975) The production of a cartographic colour chart, *ITC Journal*, (1)

**Keates, J. S.** (1978) Screenless lithography and orthophotomaps, *Cartographic Journal*, **15**(2)

**Keates, J. S.** (1982) *Understanding Maps*, Longman

**Keates J. S.** (1984) The cartographic art, in *New Insights in Cartographic Communication*, ed. Board, *Cartographica*, **21**(1)

**Keates, J. S.** (1985) Cartographic education in the mapping science field, in *Progress in Contemporary Cartography*, Vol. 3, Wiley

**Kers, A. J.** (1978) Experiments in screen printing for map production, *ITC Journal*, (2)

**Kers, A. J.** (1980) Review of registration systems, *ITC Journal*, (1)

**Knöpfli, R.** (1970) *The Representation of Rocks in the Official Plans and Maps of Switzerland*, Topographic Service of Switzerland

**Koláčný, A.** (1969) Cartographic information – a fundamental concept and term in modern cartography, *Cartographic Journal*, **6**(1)

**Kretschmer, I.** (1980) Theoretical cartography: position and tasks, *International Yearbook of Cartography,* Vol. XX

**Lavin, S.** (1986) Mapping continuous distributions using dot-density shading, *American Cartographer*, 13(2)

**Lawson, B.** (1980) *How Designers Think*, Architectural Press

**Lichtner, W.** (1978) Locational characteristics and the sequence of computer-assisted processes of cartographic generalisation, *Nachrichten aus dem Karten- und Vermessungswesen*, Series II: **35**

**Lundqvist, G.** (1958) Generalisering – några synpunkter i en betydelsefull fråga, *Globen*, 3

**McCullagh, M. J. and Ross, C. G.** (1980) Delaunay triangulation of a random data set for isarithmic mapping, *Cartographic Journal*, **17**(2)

**McGrath, G.** (1984) Using official topographic and special-purpose maps to support outdoor recreation, and user views on map revision, *Technical Papers*, Vol. 2, Twelfth ICA Conference

**McGrath, G.** (1980) The dialogue between the Ordnance Survey and its map users, *Proceedings Applied Geography Conference*, Vol. 3

**McGrath, G. (ed.)** (1984) Symposium on the marketing of cartographic information, Queen's University, Kingston, Ontario

**McMaster, R. B.** (1986) A statistical analysis of mathematical measures for linear simplification, *American Cartographer*, 13(2)

**Makowsky, A.** (1967) Aesthetic and utilitarian aspects of colour in cartography, *International Yearbook of Cartography*, Vol. VII

**Maling, D. H.** (1966) The principles of selection (translation, with commentary of F. Töpfer and W. Pillewizer Das Auswahlgesetz, ein Mittel zur kartographische Generalisierung (*Kartographische Nachrichten*, **14**(4), 1964), *Cartographic Journal*, **3**(1)

**Mecanorma Industries** (1984) *Mecanorma Graphic Book*

**Morrison, A.** (1980) Existing and improved road classifications for British road maps related to the speed of travel, *Cartographic Journal*, **17**(2)

**Morrison, A.** (1981) Using the Department of Transport's road network databank to produce route planning maps, *Cartographic Journal*, **18**(2)

**Muller, J. C.** (1987) Fractal and automated line generalisation, *Cartographic Journal*, **24**(1).

**Mumford, L.** (1952) *Art and Technics*, Oxford University Press

**Munsell Book of Color** (Neighboring hues edition) (1976) Macbeth Division, Kollmorgen Corp., Baltimore

**Neumaier, K.** (1966) The interpretation of tests 'Bedford' and 'Waterbury' – common report established by all participating centres of Commission E of OEEPE, Publ. 2, European Organisation for Experimental Photogrammetric Research

**Newson, D. W.** (1983) International standardisation of nautical charts, *Proceedings (E4) Conference of Commonwealth Surveyors*

**Olson, J. M.** (1979) Cognitive cartographic experimentation, *Canadian Cartographer*, **16**(1)

**Ormeling, F. J. (jr).** (1978) Procedures for standardisation in cartographic representation. *I.T.C Journal*, (2)

**Ormeling, F. J. (snr.)** (1980) Exonyms: an obstacle to international communication. *I.T.C Journal*

**Palm, C.** (1972) Maps for orienteering, *International Yearbook of Cartography*, Vol. XII

**Pannekoek, A. J.** (1962) Generalisation of coastlines and contours, *International Yearbook of Cartography*, Vol. II

**Permanent Committee on Geographical Names** for British Official use, *Principles of Geographical Nomenclature*

**Petchenik, B. B.** (1977) Cognition in cartography, Monograph 19, *Cartographica*,; *Proceedings International Symposium on Computer-Assisted Cartography*, ACSM, 1975

**Petchenik, B. B.** (1983) A map-maker's perspective on map design research, 1950–1980, in *Progress in Contemporary Cartography*, Vol. 2, ed. Taylor, Wiley

**Petrie, G.** (1977) Orienteering maps, *Cartographic Journal*, **14**(1)

**Peucker, T. K.** (1980) The impact of different mathematical approaches to contouring. *Cartographica* **17** : 2

**Piket, J. J. C.** (1972) Five European topographic maps, *Geografisch Tijdschrift*, **VI**(3)

**Ratajski, L.** (1973) The research structure of theoretical cartography, *International Yearbook of Cartography*, Vol. XIII

**Robinson, A. H.** (1952) *The Look of Maps*, Wisconsin University Press, Madison

**Robinson, A. H.** (1967) Psychological aspects of colour in cartography, *International Yearbook of Cartography*, Vol. VII

**Robinson, A. H. and Petchenik, B. B.** (1975) The map as a communication system, *Cartographic Journal*, **12**

**Russom, D. and Halliwell, H. R. W.** (1978) Some basic principles in the compilation of nautical charts, *International Hydrographic Review*, **LV**(2)

**Rystedt, B.** (1977) The Swedish land data bank – a multipurpose land information system, Monograph 20, *Cartographica*

**Schittenhelm, R.** (1976) The problem of displacement in cartographic generalisation: attempting a computer-assisted solution, *Nachrichten aus dem Karten- und Vermessungswesen*, Series II: **33**

**Shearer, J. W.** (1982) Cartographic production diagrams: a proposal for a standard notation, *Cartographic Journal*, **19**(1)

**Shearer, J. W. and Weinreich, H.** (1973) An examination of some photomechanical methods of producing colour proofs, *ITC Journal*, (3)

**Spiess, E.** (1978) *Some graphic means to establish visual levels in map design*, Institut für Kartographie, Eidgenössische Technische Hochschule, Zürich

**Stenhouse, H.** (1979) Selection of towns on derived maps, *Cartographic Journal*, **16**(1)

**Sukhov, V. I.** (1970) The application of Information Theory in generalisation of map contents. *International Yearbook of Cartography*, **X**

**Thompson, J. D.** (1977) Economic factors in the production of school atlases. Paper presented to British Cartographic Society Technical Symposium

**Tobler, W. R.** (1964) An experiment in the computer generalisation of maps, Technical Report 1, Department of Geography, University of Michigan

**Unwin, D.** (1981) *Introductory Spatial Analysis*, Methuen

**White, E. R.** (1985) Assessment of line generalisation algorithms using characteristic points, *American Cartographer*, **12**(1)

**Williamson, G. R.** (1982) The equal contrast gray scale, *American Cartographer*, **9**(2)

**Yoëli, P.** (1977) Computer executed interpolation of contours into arrays of randomly-distributed height points, *Cartographic Journal*, **14**(2)

**Zuylen, L van** (1969) Production of photomaps, *Cartographic Journal*, **6**(2)

**Zuylen, L van** (1974) Some remarks about the development of the application and reproduction of photomaps during the last four years, *International Yearbook of Cartography*, Vol. XIV

# INDEX

Finkelston

*Single impressions from individual printing plates, and progressive impressions of two, three and four printed colours.*

*Based on a four-colour topographic map of the Ythan Estuary, scale 1:7,500, Glasgow University, 1972.*